7th International Conference of Materials and Manufacturing Engineering (ICMMEN 2020)

MATEC Web of Conferences Volume 318 (2020)

Thessaloniki, Greece
2 - 3 July 2020

Editors:

Gabriel Mansour
Dimitrios Tzetzis
Panagiotis Kyratsis

ISBN: 978-1-7138-1594-5

Printed from e-media with permission by:

Curran Associates, Inc.
57 Morehouse Lane
Red Hook, NY 12571

Some format issues inherent in the e-media version may also appear in this print version.

For additional information, please contact EDP Sciences – Web of Conferences
at the address below.

EDP Sciences – Web of Conferences
17, Avenue du Hoggar
Parc d'Activité de Courtabœuf
BP 112
F-91944 Les Ulis Cedex A
France

Phone: +33 (0) 1 69 18 75 75
Fax: +33 (0) 1 69 28 84 91

contact-edps@webofconferences.org

Additional copies of this publication are available from:

Curran Associates, Inc.
57 Morehouse Lane
Red Hook, NY 12571 USA
Phone: 845-758-0400
Fax: 845-758-2633
Email: curran@proceedings.com
Web: www.proceedings.com

TABLE OF CONTENTS

Editorial

Gabriel Mansour[1], *Dimitrios* Tzetzis[2], and *Panagiotis* Kyratsis[3]

[1] Department of Mechanical Engineering, Laboratory of Machine Tools and Manufacturing Engineering, Aristotle University of Thessaloniki, Greece
[2] International Hellenic University, School of Science and Technology, Greece=
[3] University of Western Macedonia, Dept. of Product and Systems Design Engineering, Kila Kozani, GR50100, Greece

The 7th International Conference of Materials and Manufacturing Engineering (ICMMEN2020) was a great opportunity to change information of mutual scientific interests and to develop new collaborations among participants. The aim was to strengthen synergies between people involved in scientific research and those having to meet the rapidly increasing demands of industrial applications. The conference was hosted by the Department of Mechanical Engineering, Laboratory of Machine Tools and Manufacturing Engineering, of Aristotle University of Thessaloniki, in Thessaloniki, Greece from 2nd to 3rd July 2020.

The proceedings of this year's conference include 53 papers that have been carefully selected through a rigorous review process. This process entailed detailed reading of the papers, reporting of comments to authors, modification of papers by authors, and re-evaluation of the revised papers. Manuscripts were checked for plagiarism, proofread, and professionally edited to ensure quality of the contents. The role of the Scientific Committee was to ensure that the final accepted papers incorporated the reviewers' comments. Of the 81 abstracts originally received, 53 papers were finally accepted at the conference, representing an acceptance rate of 65.5%.

These papers address theories and applications in a variety of areas i.e. Advanced Materials Engineering, Advances in Machining and Manufacturing Technologies, 3D Scanning and Additive Manufacturing, Non-Conventional Technologies in Manufacturing, Coatings, Surface Engineering and Tribology, CAD/CAM/CAE Technologies and Geometrical Representations, Automation and Robotics, Learning and education in Robotics, Mechatronics and New Technologies, Human Mechatronic Systems Interaction, Product Design, Sustainable Design and Circular Economy, Computational Design and Digital Fabrication, Energy Consumption Aspects in Manufacturing Processes.

Sincere appreciation is due to all authors, members of the international scientific committee, keynote speakers and the organizing committee for their great effort. We would also like to express our deep gratitude to the editorial team Prof. Gabriel Mansour, Prof. Dimitrios Tzetzis and Prof. Panagiotis Kyratsis and Mrs Maria Zoumaki the conference secretary. It is the contribution of all involved that made this conference an important event that strengthens the interdisciplinary links between manufacturing engineering and material science.

Organizers

Conference Editors

Prof. G. Mansour, Department of Mechanical Engineering, Laboratory of Machine Tools and Manufacturing Engineering, Aristotle University of Thessaloniki, Greece
email address: mansour@eng.auth.gr

Prof. D. Tzetzis, School of Science and Technology, International Hellenic University, Greece
email address: d.tzetzis@ihu.edu.gr

Prof. P. Kyratsis, Dept. of Product and Systems Design Engineering, University of Western Macedonia, Greece
email address: pkyratsis@uowm.gr

Conference Chairmen

Prof. G. Mansour, Greece

Prof. P. Kyratsis, Greece

Prof. M. Placzek, Poland

Prof. K. Salonitis, Greece

Prof. M. Banica, Romania

Prof. A. Antoniadis, Greece

Prof. A. Wróbel, Poland

Prof. M. Blatnický, Slovakia

Prof. O. Dodun, Romania

Prof. M. Botsaris, Greece

Prof. D. Karalekas, Greece

Prof. E. Budak, Turkey

Prof. P. Bartolo, England

Honorary Chairs

Prof. Emeritus K.-D. Bouzakis, Greece
Prof. Emeritus S. Mitsi, Greece

Presidents of Organizing Committee

Prof. G. Mansour, Greece
Prof. P. Kyratsis, Greece
Prof. D. Tzetzis, Greece
Prof. A. Korlos, Greece
Prof. A. Tsagaris, Greece

Organising Committee

Department of Mechanical Engineering, Aristotle University of Thessaloniki, Greece
(https://www.auth.gr/en/meng)
School of Science and Technology, International Hellenic University, Greece
(https://www.tech.ihu.edu.gr/index.php/en/)

Department of Product and Systems Design Engineering, University of Western Macedonia (https://ide.uowm.gr/)

International Scientific Committee

Keynote Speakers

K.D. Bouzakis

Advanced methods facilitating the prediction of coated tools' cutting performance in milling considering entry impact loads' duration

G.T. Virgil

Complementary methods of profiling the tools that generate by enwrapping

P.J. Bartolo

New trends on additive manufacturing

Statement of Peer review

In submitting conference proceedings to Web of Conferences, the editors of the proceedings certify to the Publisher that

1. They adhere to its Policy on Publishing Integrity in order to safeguard good scientific practice in publishing.
2. All articles have been subjected to peer review administered by the proceedings editors.
3. Reviews have been conducted by expert referees, who have been requested to provide unbiased and constructive comments aimed, whenever possible, at improving the work.
4. Proceedings editors have taken all reasonable steps to ensure the quality of the materials they publish and their decision to accept or reject a paper for publication has been based only on the merits of the work and the relevance to the journal.

7th International Conference of Materials and Manufacturing Engineering (ICMMEN2020)
2-3 July 2020, Thessaloniki, Greece

Proceedings editor(s):
Gabriel Mansour, Dimitrios Tzetzis, Panagiotis Kyratsis

Date and editor's signature

3 July 2020

MATEC Web of Conferences **318**, 01001 (2020) https://doi.org/10.1051/matecconf/202031801001
ICMMEN 2020

Virtual pole method applied at the profiling of the gear shaped cutter tool for generating a K-type bore

Georgiana Alexandra Costin[1], and *Virgil Gabriel* Teodor[1,*]

[1]"Dunărea de Jos" University of Galati, Faculty of Engineering, Department of Manufacturing Engineering, Domnească street, no. 111, Romania

Abstract. The "virtual pole" method allows profiling of the cutting tools that generate by rolling. This mode of generation is advantageous in terms of the accuracy of the geometric shape obtained. The mentioned method combines the analytical accuracy of profiling with a simplified calculation mode, compared to the established algorithms. The paper presents an application made on the basis of the "virtual pole" method, for the profiling of the gear shaped tool, designed to generate the bore of a bush with a K-profile. Based on this methodology, in house software was developed for the type of profile mentioned above. The program allows the numerical and graphical representation of the profile to be generated, the contact curve and the profile of the generating tool.

1 Introduction

The processing of the bores profiled by generating, by the rolling method, has the advantage of a good precision of the shape of the surface to be obtained. For this type of generating, the primary peripheral surface of the generating tool and the generated surface are reciprocally enveloping surfaces. This determines that the surface to be obtained is generated by a large number of cuts and, moreover, between the contour of this surface and the profile of the generating tool, to be a relative sliding.

For these two reasons, the real surface obtained is less likely to be affected by the shape errors of the generating surface. One of the classic methods of obtaining the profiled bores is the processing with gear shaped cutter tools. In the case of these tools, a problem of major importance is the profiling of the cutting edge of the tool, so that it will obtain the desired shape for the piece profile [1].

The paper presents an application of the „virtual pole" method [2] made for the profiling of the gear shaped cutter tool, designed to generate the profiled bore of a bush with K-profile. Also known as polygonal sections, the K-profiles of the shafts and bores have the advantage of being self-centering, having no areas with tension concentrators and ensuring an easy assembly and disassembly [3].

* Corresponding author: virgil.teodor@ugal.ro

Huang et. al [4] states that this type of assembly ensures a lower processing cost with 40÷50% compared to the splined shaft assembly and also a lower vibration and noise level compared to the key assembly.

Polygonal profiles can have a wide variety of shapes, given their number of lobes, but those with three lobes (P3G) and four lobes (P4G) are standardized [3, 5].

2 The applying algorithm for virtual pole method

The “virtual pole” method, widely presented in the papers [2, 6], is a variant of the Willis method, according to which: “the profiles that transmit rotational movement between two parallel axes admit, at the contact point, a common normal, which passes through the gearing pole”. The “virtual pole” method is based on the observation that, in the case of tools that generate by rolling, the enwrapping condition can be defined by identifying a „virtual pole” of gearing. This virtual pole is actually the intersection point between the normal taken to the generated profile, in the current point and the centrode associated with the piece.

Figure 1 shows the reference systems: xy is the fixed reference system; XY - the mobile system, joined with the profile to be generated, Σ; C_l - the centrode associated with the piece.

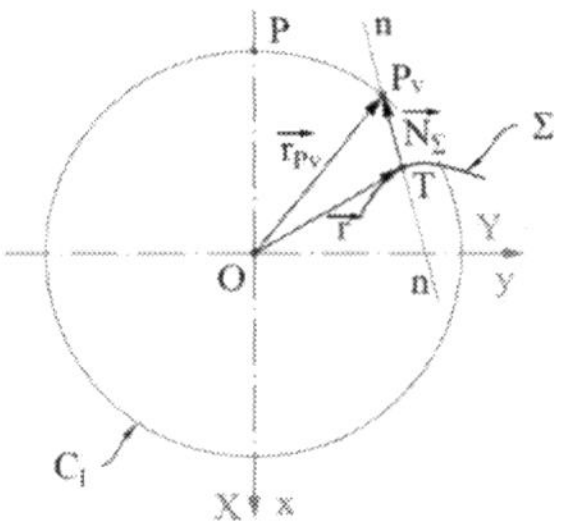

Fig. 1. Identifying the virtual pole

In the current point T, belonging to the profile to be generated, Σ, the normal n-n has been taken. It intersects the centrode, C_l in the point P_v, representing the virtual pole.

If it is admitted that the profile to be generated, Σ, has the parametric equations:

$$\Sigma:\begin{vmatrix} X = X(u); \\ Y = Y(u), \end{vmatrix} \tag{1}$$

with u, independent variable parameter, then the position vector of the current point, T has the form:

$$\vec{\mathbf{r}} = X(u)\cdot\vec{\mathbf{i}} + Y(u)\cdot\vec{\mathbf{j}}\,. \tag{2}$$

In this point, the normal vector will have the form:

$$\vec{\mathbf{N}}_\Sigma = \lambda\cdot\left(\dot{Y}_u\cdot\vec{\mathbf{i}} - \dot{X}_u\cdot\vec{\mathbf{j}}\right), \tag{3}$$

λ - being a scalar dimension which represents the distance between the current point, T and the virtual pole, P_v.

The position vector of the point P_v is obtained by summing the vectors $\vec{\mathbf{r}}$ and $\vec{\mathbf{N}}_\Sigma$ [2], obtaining the form:

$$\vec{\mathbf{r}}_{P_v} = \vec{\mathbf{r}} + \vec{\mathbf{N}}_\Sigma = \left[X(u) + \lambda\cdot\dot{Y}_u\right]\cdot\vec{\mathbf{i}} + \left[Y(u) + \lambda\cdot\dot{X}_u\right]\cdot\vec{\mathbf{j}}\,. \tag{4}$$

The size of the distance λ is determined from the belonging condition of the point P_v to the centrode C_l.

The centrode equations are:

$$C_1:\begin{vmatrix} X = -R_{rp} \cdot \cos\varphi_1; \\ Y = R_{rp} \cdot \sin\varphi_1, \end{vmatrix} \tag{5}$$

and thus the belonging condition becomes:

$$\begin{cases} -R_{rp} \cdot \cos\varphi_1 = X(u) + \lambda \cdot \dot{Y}_u; \\ R_{rp} \cdot \sin\varphi_1 = Y(u) + \lambda \cdot \dot{X}_u. \end{cases} \tag{6}$$

The parameter λ can be eliminated from the system of equations (6):

$$\left[-R_{rp} \cdot \cos\varphi_1 - X(u)\right] \cdot \dot{X}_u - \left[R_{rp} \cdot \sin\varphi_1 - Y(u)\right] \cdot \dot{Y}_u = 0, \tag{7}$$

representing the enveloping condition, meaning the connection between the independent parameters u and φ_1.

Once known the specific value $\varphi_1 = \varphi_{1u}$, for which the point P_v coincides with the gearing pole, the position of the T point, in its absolute movement, can be determined:

$$\mathbf{x} = \boldsymbol{\omega}_3^T(\varphi_{1u}) \cdot \mathbf{X}(u), \tag{8}$$

meaning the position which it occupies on the contact curve, in the rolling process:

$$T_F\begin{vmatrix} x(u,\varphi_{1u}) = X(u) \cdot \cos\varphi_{1u} - Y(u) \cdot \sin\varphi_{1u}; \\ y(u,\varphi_1) = X(u) \cdot \sin\varphi_{1u} + Y(u) \cdot \cos\varphi_{1u}. \end{vmatrix} \tag{9}$$

Considering the rolling condition, namely the condition that the two centrodes roll without sliding:

$$R_{rp} \cdot \varphi_1 = R_{rs} \cdot \varphi_2 \Rightarrow \varphi_2 = \left({R_{rp}}/{R_{rs}}\right) \cdot \varphi_1 = i \cdot \varphi_1, \tag{10}$$

it can be determine the position of the tool when the piece rotates with the angle φ_{1u}, see also figure 2.

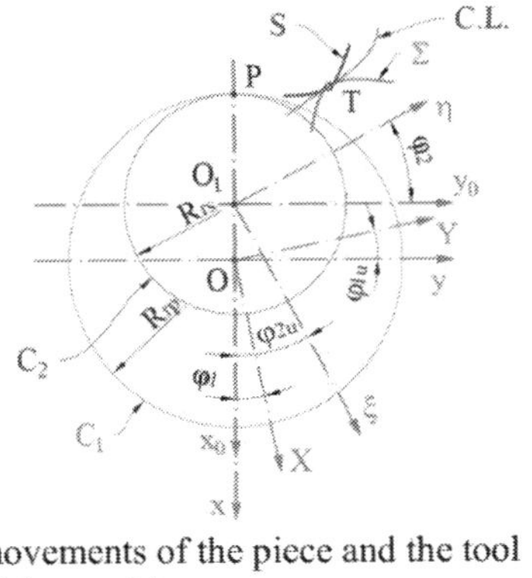

Fig. 2. Correlation between the movements of the piece and the tool

The absolute movement of the tool is:

$$\boldsymbol{\xi} = \boldsymbol{\omega}_3^T(\varphi_2) \cdot \mathbf{x}_o = \boldsymbol{\omega}_3^T(i \cdot \varphi_1) \cdot \mathbf{x}_o, \tag{11}$$

and the correlation between x_o and x systems is given by:

$$\mathbf{x}_o = \mathbf{x} - \mathbf{A}; \qquad \mathbf{A} = \begin{pmatrix} A_{12} & 0 \end{pmatrix}^T; \qquad A_{12} = R_{rp} - R_{rs}. \tag{12}$$

From the equations (9), in the fixed reference system, the T_F-point coordinates are obtained - be these x_{T_F} and y_{T_F}.

If the coordinate transformation (12) is taken into consideration, the T_F-point coordinates can be determined in the fixed system x_0y_0, namely $x_{T_{F0}}$ and $y_{T_{F0}}$.

Applying to this point the absolute movement of the tool, given by (11), its coordinates are obtained in the reference system $\xi\eta$, associated with the tool:

$$T_S \begin{vmatrix} \xi_{T_S} = x_{T_{F0}} \cdot cos(i \cdot \varphi_{1u}) - y_{T_{F0}} \cdot sin(i \cdot \varphi_{1u}); \\ \eta_{T_S} = x_{T_{F0}} \cdot sin(i \cdot \varphi_{1u}) + y_{T_{F0}} \cdot cos(i \cdot \varphi_{1u}). \end{vmatrix} \quad (13)$$

When the current point T go through the whole profile Σ, the T_S-point completely crosses the profile S, reciprocally enveloping to profile Σ.

2.1 The profiling of the gear shaped cutter tool designed to process a K-type bore

The K-type profile equations are given by:

$$\Sigma: \begin{vmatrix} X(u) = [R - e \cdot cos(n \cdot u)] \cdot cos\, u - n \cdot e \cdot sin(n \cdot u) \cdot sin\, u; \\ Y(u) = [R - e \cdot cos(n \cdot u)] \cdot sin\, u + n \cdot e \cdot sin(n \cdot u) \cdot cos\, u, \end{vmatrix} \quad (14)$$

where n is the number of lobes, and e is the eccentricity.

In figure 3 are represented the reference systems, the two centrodes, C_1 - associated with the piece and C_2 - associated with the tool, the profile Σ, on which the current point T was considered, as well as the normal to the profile, n-n.

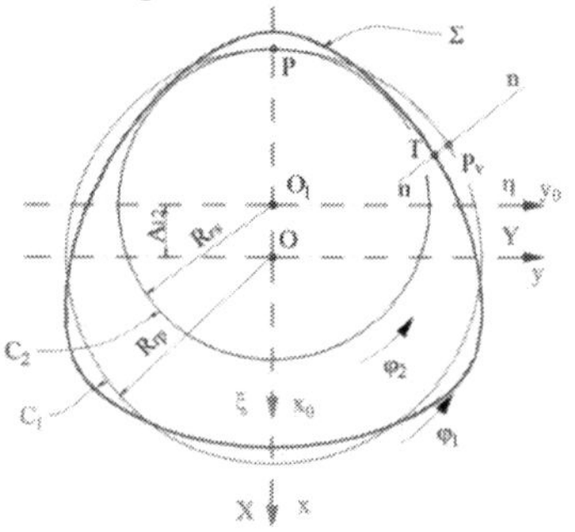

Fig. 3. Reference systems, piece profile and conjugated centrodes

The reference systems used are: xy is a fixed reference system, having the origin in the centre of the centrode C_1 of radius R_{rp}; x_oy_o - fixed reference system, having the origin in the centre of centrode C_2, of radius R_{rs}; XY - mobile reference system, joined with the Σ profile and which initially has the axes superimposed on the xy system; the absolute movement parameter of this system is φ_1; $\xi\eta$ - mobile reference system, joined with the S profile (of the tool), initially having the axes superimposed on the x_oy_o system. The movement parameter of this system is the angle φ_2, related to the angle φ_1, is given by the rolling condition (10).

Taking into account the partial derivatives of the Σ profile equations in relation with the variable u:

$$\dot{X}_u = \left[e\left(1-n^2\right) \cdot cos(n \cdot u) - R\right] \cdot sin\, u,\ \dot{Y}_u = -\left[e\left(1-n^2\right) \cdot cos(n \cdot u) - R\right] \cdot cos\, u \quad (15)$$

According to equation (3), in T-point, the normal at the profile Σ is:

$$\vec{\mathbf{N}}_\Sigma = \lambda\left\{-\left[e\left(1-n^2\right)cos(n \cdot u) - R\right]cos\, u \cdot \vec{\mathbf{i}} - \left[e\left(1-n^2\right)cos(n \cdot u) - R\right]sin\, u \cdot \vec{\mathbf{j}}\right\}. \quad (16)$$

The position vector of the T-point is given by:

$$\begin{aligned} \vec{\mathbf{r}} = &\left\{\left[R - e \cdot cos(n \cdot u)\right] \cdot cos\, u - n \cdot e \cdot sin(n \cdot u) \cdot sin\, u\right\} \cdot \vec{\mathbf{i}} + \\ &+\left\{\left[R - e \cdot cos(n \cdot u)\right] \cdot sin\, u + n \cdot e \cdot cos(n \cdot u) \cdot cos\, u\right\} \cdot \vec{\mathbf{j}}. \end{aligned} \quad (17)$$

As result according to (4), the position vector of the virtual pole is:

$$\begin{aligned}\vec{\mathbf{r}}_{P_v} = &\left\{\left[R - e\cos(n\cdot u)\right]\cos u - ne\sin(n\cdot u)\sin u - \lambda\left[e\left(1-n^2\right)\cos(n\cdot u) - R\right]\cos u\right\}\vec{\mathbf{i}} + \\ &+\left\{\left[R - e\cos(n\cdot u)\right]\sin u + ne\sin(n\cdot u)\cos u - \lambda\left[e\left(1-n^2\right)\cos(n\cdot u) - R\right]\sin u\right\}\vec{\mathbf{j}}.\end{aligned} \quad (18)$$

Intersecting this vector with the C_1 centrode:

$$C_1 : \left| X = -R_{rp}\cdot\cos\varphi_1;\ Y = R_{rp}\cdot\sin\varphi_1, \right. \quad (19)$$

is obtained:

$$\begin{aligned}-R_{rp}\cos\varphi_1 &= \left[R - e\cos(nu)\right]\cos u - ne\sin(nu)\sin u - \lambda\left[e\left(1-n^2\right)\cos(nu) - R\right]\cos u; \\ R_{rp}\sin\varphi_1 &= \left[R - e\cos(nu)\right]\sin u + ne\sin(nu)\cos u - \lambda\left[e\left(1-n^2\right)\cos(nu) - R\right]\sin u.\end{aligned} \quad (20)$$

From (20), by removing the λ parameter, is obtained the specific form of the enwrapping condition:

$$\varphi_1 = -u - \arcsin\left[\left(n\cdot e\cdot\sin(n\cdot u)\right)/R_{rp}\right]. \quad (21)$$

The value of the angle φ_{1u} for which the condition (21) is accomplished, represents the value of the angular parameter for which the P_v point coincides with the gearing pole, P.

For a certain value of parameter u and a corresponding value φ_{1u}, determined from the enveloping condition (21), the T-point, belonging to the Σ profile, will have, in the fixed reference system, the coordinates given by the absolute movement of the blank, (8), are:

$$T : \left| x_{T_F} = X_T\cdot\cos\varphi_{1u} - Y_T\cdot\sin\varphi_{1u};\ y_{T_F} = X_T\cdot\sin\varphi_{1u} + Y_T\cdot\cos\varphi_{1u}. \right. \quad (22)$$

Note: In this position, T-point also belongs to the gearing curve, meaning $T \equiv T_F$.

While the blank moved with the angular parameter φ_{1u}, the tool rotates with the parameter φ_{2u}, whose value is given by the rolling condition (10).

Therefore, the position of the tool reference system is given by its absolute movement, (11), by developing obtaining:

$$T_S : \left| \xi = x_0(u)\cdot\cos\varphi_2 + y_0(u)\cdot\sin\varphi_2;\ \eta = -x_0(u)\cdot\sin\varphi_2 + y_0(u)\cdot\cos\varphi_2, \right. \quad (23)$$

For u, taking values from u_{min} to u_{max}, the T_S-point go through the entire profile of the tool, S.

2.2 Numerical example:

For a profile defined by R=60 *mm*; e=5 *mm*; n=3 *lobes* (according to DIN 32711, [7]), the points coordinates on the generating tool profile are presented in table 1, and the profile to be generated, the gearing curve and the generating tool profile are shown in figure 4.

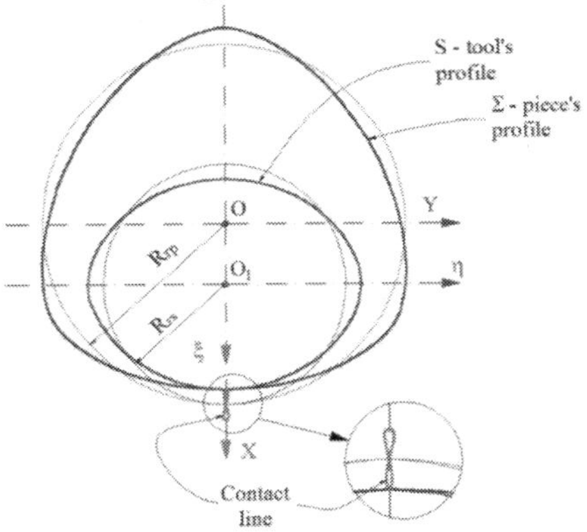

Fig. 4. The profile to be generated, the gearing curve and the generating tool profile

Table 1. Coordinates of the points on the tool profile and on the gearing curve.

Crt.	Tool's profile		Contact line		Crt.	Tool's profile		Contact line	
no.	ξ [mm]	η [mm]	x [mm]	y [mm]	no.	ξ [mm]	η [mm]	x [mm]	y [mm]
1	0.000	-45.000	55.000	0.000	14	34.323	8.277	64.945	0.321
2	1.838	-44.909	55.306	0.304	15	32.363	16.161	64.763	0.600
3	3.899	-44.597	56.171	0.482	16	29.322	23.302	64.410	0.792
4	6.372	-43.957	57.450	0.458	17	25.502	29.440	63.830	0.849
5	9.379	-42.823	58.949	0.233	18	21.267	34.429	62.979	0.741
6	12.945	-40.990	60.467	-0.116	19	16.978	38.249	61.845	0.476
7	16.978	-38.249	61.845	-0.476	20	12.945	40.990	60.467	0.116
8	21.267	-34.429	62.979	-0.741	21	9.379	42.823	58.949	-0.233
9	25.502	-29.440	63.830	-0.849	22	6.372	43.957	57.450	-0.458
10	29.322	-23.302	64.410	-0.792	23	3.899	44.597	56.171	-0.482
11	32.363	-16.161	64.763	-0.600	24	1.838	44.909	55.306	-0.304
12	34.323	-8.2775	64.945	-0.321	25	0.000	45.000	55.000	0.000
13	35.000	0.000	65.000	0.000					

3. Conclusions

The algorithm based on the "*virtual pole*" method represents an improved method for profiling of tools that generate by enwrapping, by the rolling method.

The application presented in the paper demonstrates the accuracy and simplicity of the "*virtual pole*" method and, at the same time, allows verifying the results obtained with this method.

Based on the presented methodology, a calculation program was created, which allows the profiling and graphical representation of the gear shaped cutter tool, designed to process a *K*-type bore.

References

1. V.G. Teodor, N. Baroiu, F. Susac, *The Synthesis of New Algorithms for CAD Profiling of Cutting Tools* (Lambert Academic Publishing, ISBN 978-613-7-08923-1, 2018)
2. G.A. Costin, V.G. Teodor, N. Oancea, *Virtual Pole Method - Alternative Method for Profiling Tools Which Generate by Enwrapping*. Ann. of "Dunarea de Jos" Univeristy, TMB, **5**, 31 (2019)
3. R. Bhatta, *Comparison of Load Carrying Capacity of Three and Four Lobed Polygonal Shaft and Hub Connection for Constant Grinding Diameter*, Masters Theses, (2016)
4. B. Huang, J. Li, Q.J. Sun, L.Q. Wang, J. Zhejiang Univ. Sci. B, **3**, 039, (2010)
5. E. Ghiţă, *Teoria şi tehnologia suprafeţelor poliforme (Theory and technology of poliform surfaces)*. (Bren Publishing House, Bucharest, 2001)
6. V.G. Teodor, G.A. Costin, *Virtual Pole Method - Alternative Method for Profiling Rack Tools*, Ann. of "Dunarea de Jos" Univeristy, TMB, **5**, 5 (2019)
7. *, *DIN 32711-1. Welle-Nabe-Verbindung - Polygonprofil P3G - Teil 1: Allgemeines und Geometrie (Shaft to collar connection - Polygon profile P3G - Part 1: Generalities and geometry)*, (2009)

Effect of the crystallinity of NCD and MLD diamond coatings characterized by same level of residual stresses after annealing on their fatigue strength and wear behavior in milling

Georgios Skordaris[1,*], *Tilemachos* Kotsanis[1], *Apostolos* Boumpakis[1], and *Fani* Stergioudi[2]

[1]Laboratory for Machine Tools and Manufacturing Engineering, Mechanical Engineering Department, Aristotle University of Thessaloniki, Greece
[2]Physical Metallurgical Laboratory, Mechanical Engineering Department, Aristotle University of Thessaloniki, Greece

Abstract. Nano-composite (NCD) and multi-layered (MLD) diamond coatings were deposited on cemented carbide tools using hot filament chemical vapour deposition (HFCVD) techniques. Appropriate annealings were conducted on the examined diamond coatings in order to be characterized by the same level of residual stresses. The crystalline structure of the employed diamond coatings was investigated by conducting Raman spectra. Inclined impact tests at ambient and elevated temperatures were carried out for assessing their temperature-dependent fatigue strength. Moreover, the wear behaviour of diamond coated inserts was investigating in milling aluminum foam. Raman spectra were also conducted on the treated diamond coatings for capturing potential crystalline changes developed due to the exercised thermal and dynamic mechanical loads during cutting. According to the attained results, the co-existence of sp^2– and sp^3-bonded phases in the cases of MLD diamond coatings results in an accelerated wear development, despite their structure capability to decelerate the crack propagation. As a result, nano-crystalline diamond coatings characterized only by sp^3-bonded phase exhibit an improved wear behaviour. The cutting performance of the NCD coated inserts is further improved due to the enhanced tribological properties of the NCD coatings.

1 Introduction

Nano-composite (NCD) and multi-layered (MLD) diamond coatings can be effectively applied in machining of nonferrous materials such as of aluminium alloys, carbon fibre reinforced plastics etc. [1-4]. Due to the superior adhesion characteristics of micro-crystalline diamond layers and to improved tribological properties of nano-crystalline diamond ones, various micro- and nano-crystalline layer coating systems on cemented carbide substrates are manufactured and used effectively in machining procedures [5,6].

* Corresponding author: skordaris@meng.auth.gr

Moreover, recent investigations revealed that MLD coatings withstand more effectively the repetitive impacts during milling than the NCD ones, since the layer's interfaces obstruct the crack propagation. Thus, MLD coatings are characterized by reduced brittleness and improved cutting performance [4].

A significant issue affecting the brittleness of the produced diamond coatings is the level of residual stresses in the film structure [4]. The high level of residual stresses in the diamond film structure is attributed to epitaxial crystal differences (structural stresses) and thermal expansion coefficients mismatch of the diamond coating and its cemented-carbide substrate (thermal stresses) [4,7]. An effective way to reduce the structural residual stresses in the diamond film structure is attained by appropriate annealing. In the frame of this work, NCD and MLD coatings were produced and appropriately annealed in vacuum at a temperature of 400°C for possessing the same level of residual stresses [8]. As a result, by decoupling the effect of residual stresses in the diamond film structure on the films' fatigue strength, the impact of a nano-crystalline structure or a combination of nano- and micro-crystalline diamond layers on the wear behaviour of diamond coated tools was assessed.

2 Experimental details

Cemented-carbide inserts of HW-K05/K20 SPGN120308 ISO specifications were coated with nano-composite (NCD), or multi-layered (MLD) films of approximately 5 μm thickness via the hot filament method using a CC800/9Dia CEMECON coating machine. Cobalt-etching procedures were carried out prior to diamond coating deposition for attaining sufficient adhesion [1]. The MLD coatings consist of ten successive and alternate nano- and micro-structured layers. These nano- and micro-structured layers had approximately 0.34 and 0.66 μm thickness, respectively.

The fatigue strength of the NCD and MLD coated cemented-carbide inserts was investigated by conducting inclined impact tests at temperatures of 25°C and 300°C [4]. The impact tester device was designed and manufactured by the Laboratory for Machine Tools and Manufacturing Engineering of the Aristotle University of Thessaloniki in conjunction with the company Impact-BZ (see figure 1a) (London, UK) [9,10]. The Raman spectroscopy on the untreated as well as treated diamond coatings was carried out using a LabRAM HR spectrometer. The milling investigations were conducted employing a three-axis numerically-controlled milling center using aluminum foam as workpiece material. This workpiece material consists of various hard phases, as related optical microscopy observations using standard metallographic techniques revealed (see figure 1b). Moreover, due to the structure of the workpiece material, intense dynamic loads are developed on the cutting edge of the coated tools during cutting.

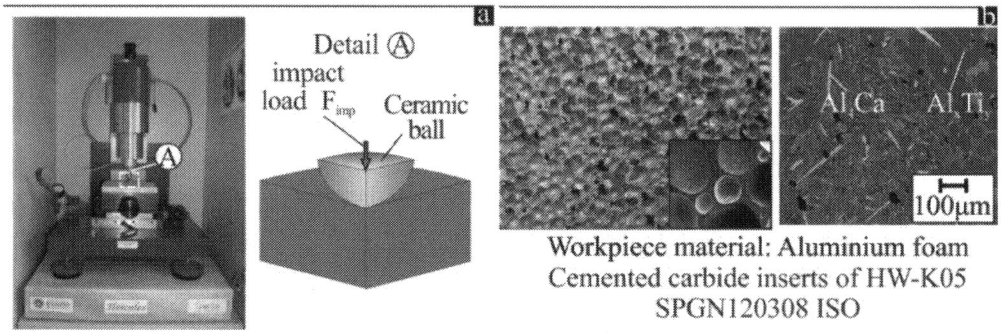

Fig. 1. (a) The employed impact tester for conducting experiments; (b) the workpiece material used in milling experiments.

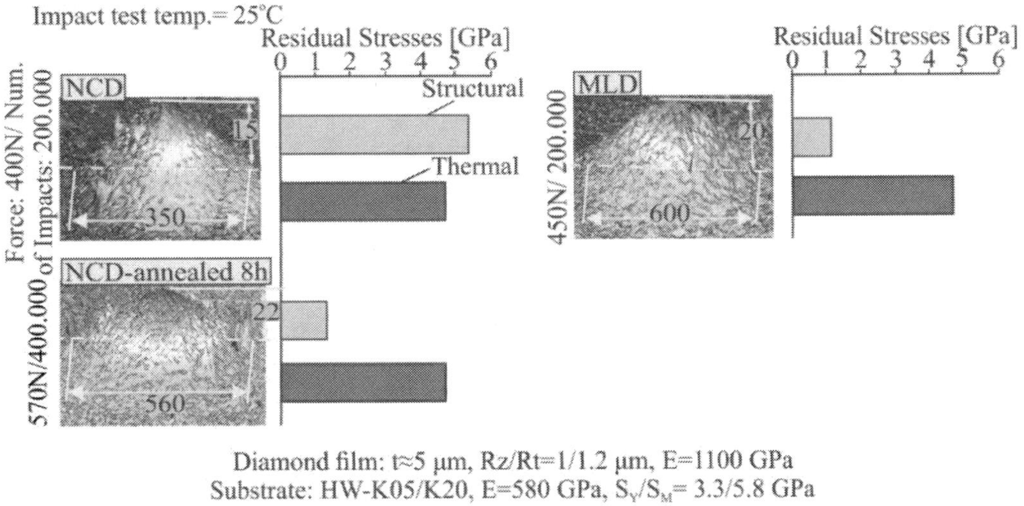

Fig. 2. Determined thermal and structural residual stresses in the investigated diamond film cases after inclined impact tests.

3 Characterization of the investigated diamond coatings

3.1 Residual stresses

For determining the overall residual stresses in the investigated diamond film cases, inclined impact tests at various loads and cycles were conducted at room temperature. A fatigue damage of the diamond coating-substrate interface develops after a certain number of repetitive impacts depending on the impact load. After the interface fatigue failure, the high compressive residual stresses of the diamond structure are released, and the detached coating hikes up at a certain maximum height (bulge formation). Based on a developed FEA model described analytically in [11], which simulates the film lifting after the fatigue failure of its interface and the residual stress release, the overall residual stresses in the film structure were calculated (see figure 2). Moreover, the thermal residual stresses were determined with the aid of an additional FEA model [11]. These stresses are shown in figure 2.

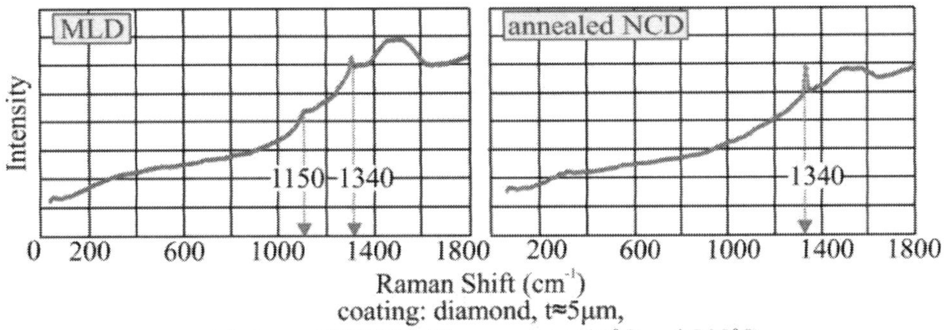

Fig. 3. Raman spectra of the investigated MLD and annealed NCD films.

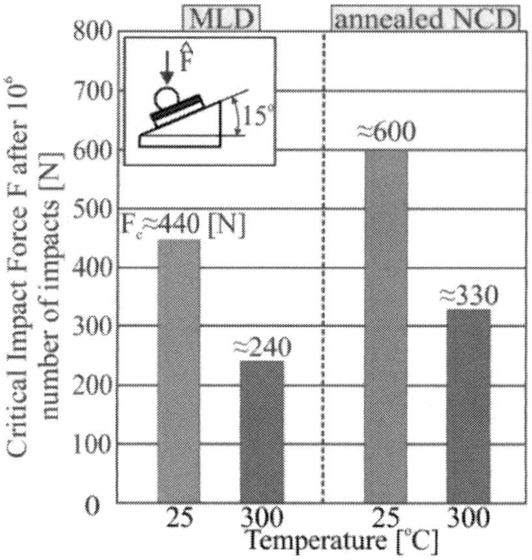

Fig. 4. Determined critical impact force for avoiding coating damage after 10^6 impacts at 25°C and 300°C.

As it can be observed, NCD coatings are characterized by a comparably larger amount of residual stresses in the film structure. In order for both coating types to possess the same amount of residual stresses, NCD coatings were annealed in vacuum at a temperature of 400°C for a duration of 8 h. Due to the applied annealing procedure, a portion of the structural stresses is released, whereas the thermal residual stresses in the diamond film structure remain invariable. As a result, both MLD and NCD coatings are characterized by the same level of residual stresses.

3.2 Crystallinity

The crystallinity of the produced diamond coatings was characterized by conducting Raman spectroscopy. The recorded spectra are presented in figure 3. Both MLD and annealed NCD films exhibit a narrow peak at around 1340 cm^{-1} confirming the high diamond crystalline quality of the coatings and the existence of an sp^3-bonded phase [12,13]. One more peak appears pronounced at roughly 1150 cm^{-1} in the Raman spectra of the MLD coated specimens. This extra feature originates from co-existing sp^2-phases, called as trans polyacetylene. The trans polyacetylene possesses obviously inferior strength properties compared to the sp^3-bonded phases. As a consequence, a decomposition of the sp^2-bonded trans polyacetylene phases takes place when repetitive thermal and mechanical loads both leading to elastic film deformations are simultaneously applied. This could happen during the cutting procedure [13].

3.3 Fatigue

For evaluating the fatigue strength of the MLD and the annealed NCD coatings, inclined impact tests were carried out at ambient temperature and at 300°C. The critical impact force for fatigue failure of the examined diamond coatings at 25°C and 300°C is displayed in figure 4. In both examined coating cases, an increase of the operational temperature results in a deterioration of the fatigue strength. Moreover, NCD coatings possess an improved fatigue strength compared to MLD ones, when both coatings are characterized by the same level of residual stresses and thus brittleness.

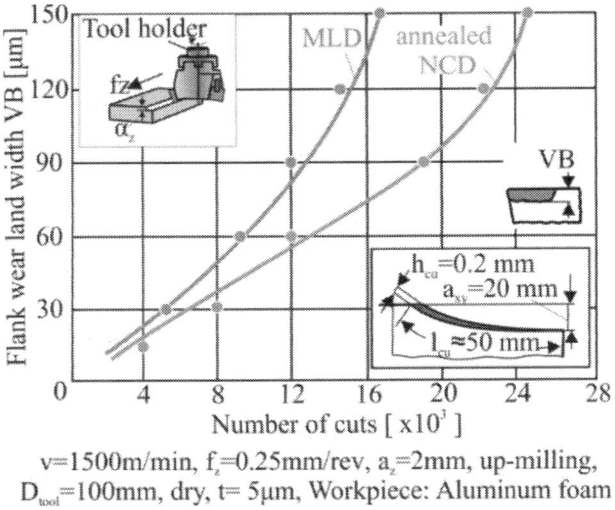

Fig. 5. Flank wear development versus the number of cuts of the investigated diamond coated inserts.

4 Cutting performance of the diamond coated inserts in milling and explanation of the obtained results

The cutting performance of the manufactured diamond coated inserts was investigated in milling without coolant or lubricant for attaining a more intense wear evolution. After a prescribed number of successive cuts, the cutting insert wear was recorded. In the case of the NCD coated inserts, after approximately 24000 cuts, a flank wear width of 0.15 mm developed (see figure 5). Compared with this, the wear behavior of the MLD coated inserts characterized by the same level of residual stresses after annealing worsens significantly. More specifically, the MLD coated inserts managed to cut only approximately 16000 cuts up to the same flank wear width. The obtained larger number of cuts when annealed NCD inserts are used exhibit their superior wear resistance compared to the MLD ones.

In order to explain these results, Raman spectroscopy was conducted on the remaining coatings on the tool rake after milling (see figure 6). In this way, potential crystallinity changes of the diamond coatings due to the exercised thermal and mechanical loads could be captured. In the case of the MLD coatings, the weaker spectral peaks in the diagram of figure 6 indicate a decomposition of the trans polyacetylene sp^2-bonded phases (peaks at

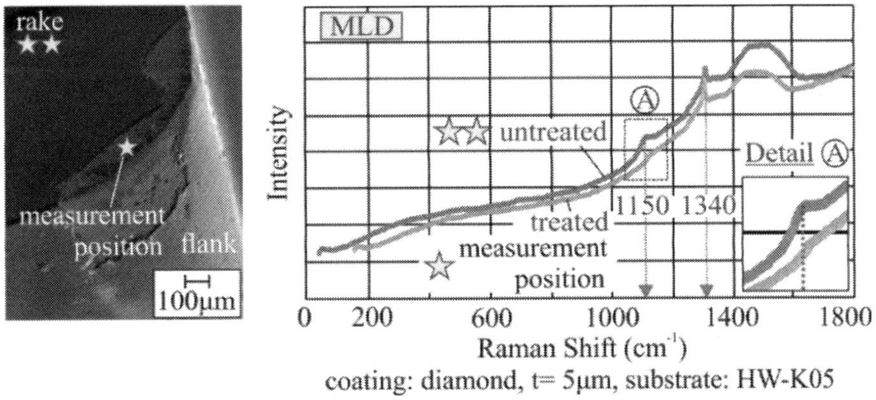

Fig. 6. Raman spectra of the remaining coating for MLD film case after milling.

around 1150 cm^{-1}), when thermal and mechanical loads during milling are simultaneously applied. Consequently, a diamond coating strength weakening arises resulting in a more rapid wear evolution compared to NCD coatings during the cutting processes.

5 Conclusions

In this paper, the effect of the structure of NCD and MLD coatings characterized by the same level of residual stresses after annealing on their fatigue strength and wear behaviour was investigated. In this context, NCD coated inserts were annealed in vacuum at a temperature of 400°C for 8 h for possessing the same level of residual stresses as the MLD coated ones. Thus both coatings are characterized by the same brittleness. Based on the attained impact and milling results, when mechanical and thermal loads are simultaneously exercised, the decomposition of an sp^2-bonded trans polyacetylene phases, existing in the MLD diamond coating structure, leads to more intense wear compared to a trans polyacetylene free NCD film. Moreover, the improved tribological properties of the NCD coatings further enhance their cutting performance.

This research is co-financed by Greece and the European Union (European Social Fund- ESF) through the Operational Programme «Human Resources Development, Education and Lifelong Learning 2014-2020» in the context of the project "Optimization of the cutting performance of nano- and micro- structured diamond coatings via appropriate thermal treatments" (MIS 5047906).

References

1. R. Haubner, W. Kalss, NT J REFRACT MET H, **28**: 475–483, (2010)
2. E-G Ng, D. Szablewski, M. Dumitrescu, M.A. Elbestawi, J.H Sokolowski, CIRP ANN – MANUF TECHN, **53**: 69-72, (2004)
3. R. Polini, Thin Solid Films **515**: 4 – 13, (2006)
4. G. Skordaris, K.-D. Bouzakis, P. Charalampous, T. Kotsanis, E. Bouzakis, O. Lemmer, CIRP ANN – MANUF TECHN, **65**, pp. 101-104, (2016)
5. R. Dumpala, N. Kumar, CR Kumaran, S. Dash, B. Ramamoorthy, Ramachandra Rao MS, Diam. Relat. Mater., **44**:71–77, (2014)
6. R. Dumpala, M. Chandran, N. Kumar, S. Dash, B. Ramamoorthy, Ramachandra Rao MS, Int. J. Refract. Met. H, **37**:127–133, (2013)
7. N. Woehrl, T. Hirte, O. Posth, V. Buck, Diam. Relat. Mater., **18**: 224–228, (2009)
8. M H Staia, M D'Alessandria, D T Quinto, F Roudet, M Marsal Astort, J. Phys.: Condens. Matter, **18**: S1727–S1736, (2006)
9. Available online: www.impact-bz.com
10. G. Skordaris, F. Stergioudi, A. Boumpakis, D. Stergioudi, COATINGS, **9**, 855, (2019)
11. K.-D. Bouzakis, G. Skordaris, E. Bouzakis, S. Makrimallakis, S. Kombogiannis, O. Lemmer, Surf. Coat. Technol., **237**, pp. 135-141, (2013)
12. AC Ferrari, J. Robertson, Phys. Rev. B., (2001)
13. G. Skordaris, K.-D. Bouzakis, T. Kotsanis, A. Boumpakis, F. Stergioudi, D. Christofilos, O. Lemmer, W. Kölker, M. Woda, CIRP ANN – MANUF TECHN, **68**: 65-68, (2019)

Influence of stiffness characteristics of a railway track on output parameters in a multibody model

Ján Dižo[1,*], and *Miroslav* Blatnický[1]

[1]University of Žilina, Faculty of Mechanical Engineering, Department of Transport and Handling Machines, Univerzitná 8215/1, 010 26 Žilina, Slovak Republic

Abstract. The article is aimed at the research of the influence of stiffness characteristics included in a model of a railway track, which is the part of a multibody system. The other part of the multibody model is a railway vehicle. Authors are focused on the investigation of response of some selected output parameters under various values of input of stiffness and damping coefficients. The interaction of a railway vehicle and a railway track is studied. A passenger railway vehicle has been chosen for presented research. Outputs parameters are chosen in the passenger ride comfort point of view and the running safety point of view. The passenger ride comfort can be evaluated either by the direct method, when a real vehicle runs on a track and passengers evaluate a vehicle by means of their feelings during the ride or by means of the indirect method, when accelerometers are used for measuring accelerations in various positions of a tested wagon and subsequently values of accelerations are processed in required way. Then, the ride comfort is calculated and indexed by means of ride comfort indices. In the presented work, the indirect method has been used. In the computer multibody model of the wagon accelerations on a floor have been detected and the mean ride comfort for a person is assessed. The ride safety is most often determined by waveforms of vertical wheel forces, lateral wheel forces and the derailment quotient.

1 Introduction

A rail vehicle is a relatively complicated mechanical system. It is specified by its properties and together with a railway track form basic elements of railway transport system. In terms of mechanics, a railway vehicle is a multibody system, whose behaviour in relation to a railway track and surrounding environment is analysed and evaluated based on given criteria [1-4]. In process of design of railway vehicle, it is necessary to meet as far as possible the requirements and operating conditions, in particular with regard to running safety and level of passengers ride comfort [5, 6]. Simulation calculations of railway vehicles running are focused on investigation of dynamics of a railway vehicle mechanical system, which is most often excited due to track irregularities [7-9].

* Corresponding author: jan.dizo@fstroj.uniza.sk

The objective of the article is to analyse running properties of a railway vehicle by means of simulation computations in commercial simulation software. For these purposes, a passenger wagon model and a railway track model are created. The research examines, how the change of stiffness-damping parameters of the railway track influences dynamic properties of the passenger wagon. These dynamic properties are evaluated mainly from the passengers ride comfort point of view and from the running safety point of view.

2 Description of an analysed multibody system

A virtual model of the passenger car and the railway track has been created in the Simpack software package. In this software, a user can create various virtual multibody models from a simple wheelset to complicated high-speed trainsets, perform extensive dynamic simulations, animate movements of individual elements of an entire model etc. The Postprocessor interface offers extensive options of visualization of reached results by means of 2D or 3D graphs, animations as well as sophisticate mathematical operation with calculated data using implemented algorithms [1, 10-12]. They allow to compute and to evaluate passenger ride comfort, to assess the running safety etc.

2.1 A railway vehicle model

The multibody model of the analysed passenger wagon consists of two subsystems of bogies and one subsystem of the body of wagon (Fig. 1). Every subsystem includes several construction elements. The bogie model is formed by one frame, two wheelsets and four axleboxes. These elements are interconnected by means of force elements, joints and constraints. Coils springs and hydraulic dampers act in vertical direction between axleboxes and bogie frame (primary suspension system) as well as between bogie frame and the wagon body (secondary suspension system). The suspension system comprises also additional hydraulic dampers of yaw and lateral oscillations.

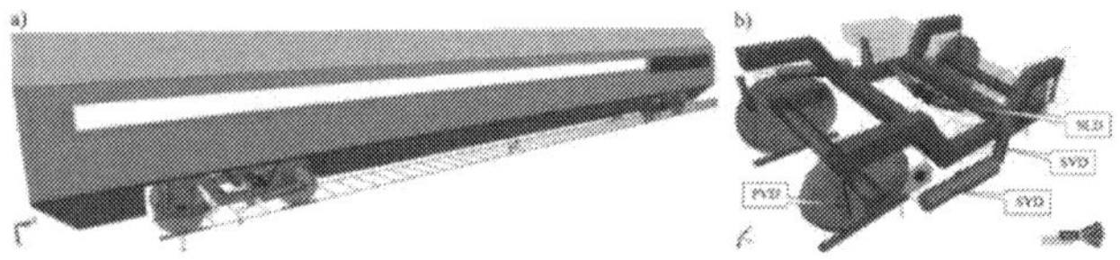

Fig. 1. A multibody model of a passenger wagon a) and a multibody model of a bogie b) created in the Simpack software: PVD – a primary vertical damper, SVD – a secondary vertical damper, SLD – a secondary lateral damper, SYD – a secondary yaw damper.

2.2 A railway track model

Simulations of railway vehicle running are usually performed for a rigid track formulation. However, there are technical applications, in which there is necessary to take into account, that the track base is not rigid, but it is flexible [9, 13-15]. Such a model of a track is extended with additional bodies, kinematic and mechanical joints with defined stiffness-damping parameters. It causes increasing of degrees of freedom. The scheme of the flexible track is shown in Fig. 2. Stiffness-damping characteristics are defined in principle in two ways: as constant values and as functional dependences on travelled distance. In our research we have used both ways.

Results of simulation computations are differentiated using characteristic words as following:

• Rigid track formulation – "Rigid",

• Flexible track formulation with constant stiffness-damping parameters – "Flex_const"with following characteristics: total vertical track stiffness $k_z = 1.65 \cdot 10^8$ N·m^{-1}, total vertical track damping $b_z = 1.90 \cdot 10^5$ N·s·m^{-1}, total lateral track stiffness $k_y = 4.15 \cdot 10^7$ N·m^{-1}, total lateral track damping $b_y = 1 \cdot 10^5$ N·s·m^{-1},

• Flexible track formulation with variable parameters – "Flex_sin". In this case, stiffness-damping parameters are defined by sinusoidal functions, where track frequency of the sinusoidal function depends on travelled distance.

Simulation computations have been performed on a real track geometry including right and left curves with various diameters. Moreover, the track model contains track irregularities. The passenger wagon was running at the speed of 60 km·h^{-1}, 90 km·h^{-1} and 110 km·h^{-1}.

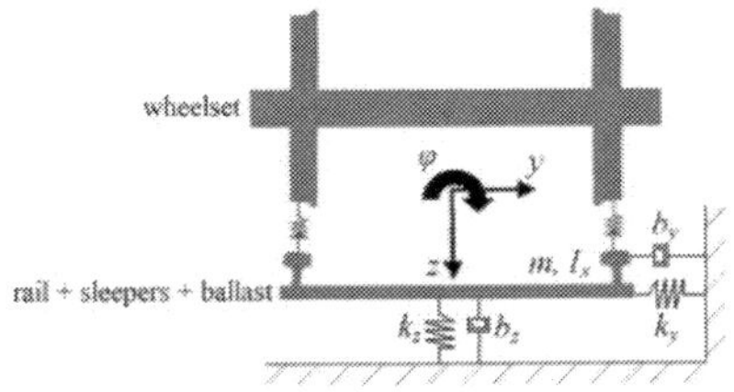

Fig. 2. A dynamical scheme of a flexible railway track.

3 Results of simulation computations

Results of dynamic analyses of tested passenger wagon with defined input parameters are introduced in this section.

As the main criteria, passenger ride comfort has been evaluated. It is represented by means of column graphs, which characterise the level passenger ride comfort in selected points. Passenger ride comfort is assessed based on distribution of the N_{MV} indices, i. e. the average ride comfort on the wagon body floor [5, 16]. For calculation of this index, there is necessary to know accelerations in each point [17-19]. The acceleration signal is processed using corresponding filters in compliance with the EN 12299:2009 standard [5, 17, 19].

Reached values of passenger ride comfort indices depending on various stiffness-damping characteristics of the track and for various running speeds are shown in Fig. 3, Fig. 4 and Fig. 5.

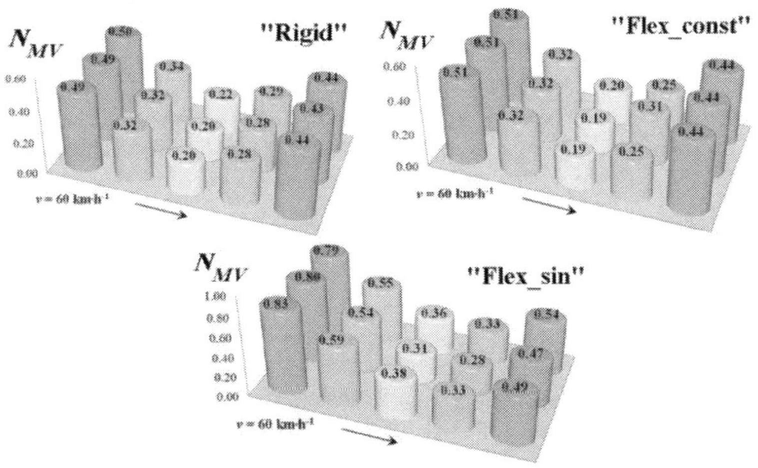

Fig. 3. Values of N_{MV} index for "Rigid", "Flex_const" and "Flex_sin" formulation, speed of 60 km·h^{-1}.

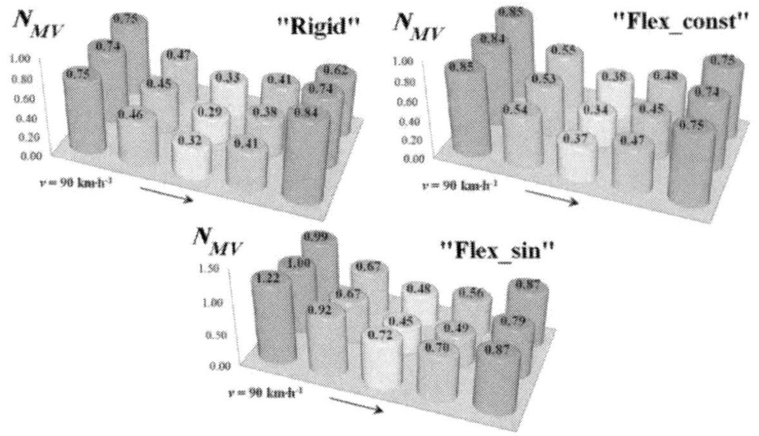

Fig. 4. Values of N_{MV} index for "Rigid", "Flex_const" and "Flex_sin" formulation, speed of 90 km·h^{-1}.

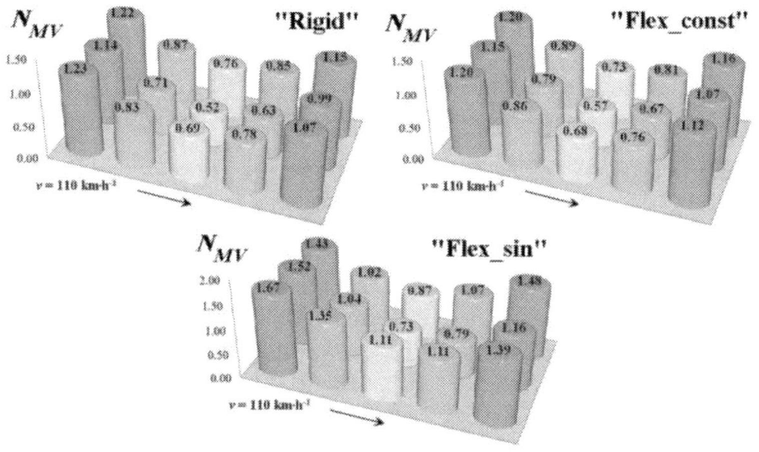

Fig. 5. Values of N_{MV} index for "Rigid", "Flex_const" and "Flex_sin" formulation, speed of 110 km·h^{-1}.

Values of ride comfort indices N_{MV} obtained from simulation computations for the speed of 60 km·h^{-1} show (Fig. 3), that definition of constant values of stiffness-damping characteristics of the track ("Flex_const") only weakly affects the ride comfort in comparison with the rigid track formulation ("Rigid"). However, the flexibility of the track is determined by means of variable stiffness-damping characteristics ("Flex_sin"), values of ride comfort are greater. Therefore, the same passenger car is evaluated as less comfortable.

Graphs in Fig. 4 show the distribution of N_{MV} indices for the speed of 90 km·h^{-1}. We can see, that higher values of the speed cause greater values of N_{MV} indices. Moreover, the formulation of the flexible track with constant values of stiffness-damping characteristics ("Flex_const") is more pronounced, while variable values of the flexible track formulation ("Flex_sin") again indicate the greatest values of N_{MV} indices.

Finally, the case of the wagon running at the speed of 110 km·h^{-1} is evaluated (Fig. 5). As expected, the ride in the wagon at this speed is least comfortable. In more detailed examination in this case, the formulation of track flexibility with constant values of stiffness-damping characteristics ("Flex_const") leads in rear part of the wagon to lower values of N_{MV} indices in comparison with the rigid track formulation ("Rigid"). However, deviations are negligible. The wagon is also in this case evaluated as very comfortable except of two points in rear part of the body, in which the wagon is evaluated as comfortable.

A railway vehicle is also assessed in term of running safety. One of the main evaluated quantities is the derailment quotient *Y/Q*, which is given as the ratio of the lateral wheel forces *Y* and the vertical wheel forces *Q*.

Figure 6 shows waveforms of the derailment quotient for the speed of 60 $km{\cdot}h^{-1}$. Waveforms belongs to the first wheelset in the running direction, while blue curve is for the right wheel and the green curve is for the left wheel.

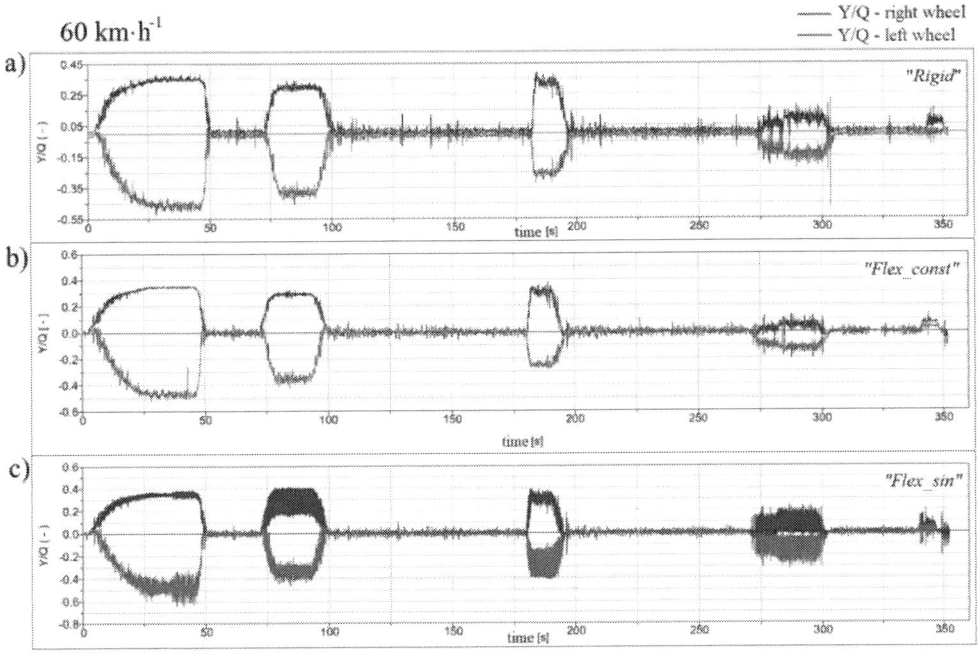

Fig. 6. Waveforms of the derailment quotient for the speed of 60 $km{\cdot}h^{-1}$: a)"Rigid" track formulation, b) "Flex_const" track formulation, c) "Flex_sin" track formulation.

When evaluating the monitored waveforms of the derailment quotient for various track formulations we can observe, that using the flexible track model with constant stiffness-damping characteristics ("Flex_const") partially results in damping the dynamic effects of the rail vehicle. From results of simulation computations reached for the flexible track formulation with variable values of stiffness-damping characteristics ("Flex_sin") we can see, the amplitudes of derailment quotient are significantly greater in comparison with the rigid track formulation ("Rigid"). It can be stated, that these greater amplitudes are due to varying parameters along the track, and the formulation used in the model does not sufficiently reflect the characteristics of the real track.

4 Conclusion

The main objective of the article was to analysed the influence of formulation of flexible railway track on output parameters. In our case we have found out, the extension of the railway vehicle multibody model by flexible track formulation has not influenced values of N_{MV} indices significantly. However, process time for defined conditions has risen for the constant values of stiffness-damping characteristics approx. of 15-20 % and for variable values of stiffness-damping parameters even of 45-50 %. Therefore, the user has to decide, which approach for modelling and simulation of a railway vehicle running on a railway track will use.

This work was supported by the Cultural and Educational Grant Agency of the Ministry of Education of the Slovak Republic in the project No. KEGA 023ŽU-4/2020: Development of advanced virtual models for studying and investigation of transport means operation characteristics.

The work was supported by the Slovak Research and Development Agency of the Ministry of Education, science, Research and Sport of the Slovak Republic in Educational Grant Agency of the Ministry of Education of the Slovak Republic in the project No. VEGA 1/0558/18: Research of the interaction of a braked railway wheelset and track in simulated operational conditions of a vehicle running in a track on the test bench.

References

1. V. Hauser, O. Nozhenko, K. Kravchenko, M. Loulova, J. Gerlici, T. Lack, Procedia Engineering **192** (2017)
2. J. Gerlici, M. Gorbunov, K. Kravchenko, O. Prosvirova, T. Lack, MATEC Web of Conferences **157** (2018)
3. J. Gerlici, V. Sakhno, A. Yefymenko, V. Verbitskii, K. Kravchenko, K. Kravchenko, MATEC Web of Conferences **157** (2018)
4. J. Soukup, J. Skocilas, B. Skocilasova, Mechanical Systems and Signal Processing **89** (2017)
5. EN 12299:2009. Railway applications. Ride comfort for passengers. Measurement and evaluation (2009)
6. R. Melnik, B. Sowinski, VSDIA 2014, **14** (2014)
7. A. Chudzikiewicz, R. Melnik, VSDIA 2012, **13** (2012)
8. A. S. Leblebici, S. Turkay, MED 2019, **27** (2019)
9. J. Sadeghi, S. Rabiee, A. Khajehdezfuly, International Journal of Structural Stability and Dynamics **19**, 6 (2019)
10. J. Harusinec, A. Suchanek, M. Loulova, P. Strazovec, MATEC Web of Conferences **157** (2018)
11. P. Stastniak, J. Harusinec, Komunikacie **15**, 4 (2013)
12. P. Stastniak, Manufacturing Technology **15**, 5 (2015)
13. J. Soukup, J. Skocilas, B. Skocilasova, L. Rychlikova, Journal of Applied Nonlinear Dynamics **3**, 4 (2014)
14. L. Izvolt, J. Harusinec, M. Smalo, Communications – Scientific Letters of the University of Zilina **20**, 3 (2018)
15. J. Gerlici, T. Lack, J. Harusinec, Communications – Scientific Letters of the University of Zilina **16**, 3a (2014)
16. J. Harusinec, M. Manurova, A. Suchanek, Manufacturing Technology **16**, 5 (2016)
17. J. Gerlici, T. Lack, Z. Ondrova, Komunikacie **9**, 4 (2007)
18. T. Lack, J. Gerlici, Komunikacie **10**, 3 (2008)
19. D. Madalina, D. I. Stanica, Materials Science Forum **957** (2019)

MATEC Web of Conferences **318**, 01004 (2020) https://doi.org/10.1051/matecconf/202031801004
ICMMEN 2020

Design of a prototype of an engine mechanism with rotating cylinders

Miroslav Blatnický[1] and *Ján* Dižo[1,*]

[1]University of Žilina, Faculty of Mechanical Engineering, Department of Transport and Handling Machines, Univerzitná 8215/1, 010 26 Žilina, Slovak Republic

Abstract. In this article, authors focus on the design and construction of a real prototype of an engine mechanism with rotating cylinders and its using mainly in piston combustion engines. It is assumed, that the normal force of a piston will be completely eliminated, because the swing angle of a connecting rod will equal to zero during the whole working cycle, since the connecting arm of the piston moves just the cylinder axis. It will by allowed by the conceptual design of the mechanism presented in this article. As rotating blocks of cylinders concurrently act as a flywheel, it is proposed, that in this way there is possible to save the mass of additional flywheels. Moreover, liquid cooling system is not necessary, because the rotating cylinders sufficiently transfer heat to ambient air. In addition, the output of torque will be reached without necessity of gear transmission, which results to decreasing of needs of mechanism lubrication. Other advance of the designed mechanism are two outputs. The first output is low-speed and it goes out from rotating cylinders, i. e. from the slider-crank mechanism with revolutions n_1. The other output is high-speed, from the crankshaft with revolutions n_2. Because of more favourable properties of the mechanism, authors have decided to create a real device to confirm all mentioned advantages of the mechanism by the suitable way.

1 Introduction

Kinematics deals with the mechanical movement of bodies, it examines the change in the relative positions of the bodies regardless of the force effect in space and time. It forms a part of classical mechanics. In kinematics, the causes of movement are ignored, i.e. force actions are not examined. In kinematics, therefore, only geometric properties (not mass) of points or bodies are considered. We will apply this knowledge to the 3-D model of the kinematic mechanism with rotating cylinders with the help of a CAD computer programme.

The disadvantage of all piston combustion engines is the fact that the larger the angle of the connecting rod β, the more loaded by side pressure the piston and other engine parts, and thus, the torque is created by less pressure force. Another disadvantage of these engines is the fact that the pistons perform a reciprocating motion relative to the cylinders, which puts increased demands on the smooth running of the engine by the distributed set of flywheel extra weights. The above-described fact encourages the creation of the

* Corresponding author: jan.dizo@fstroj.uniza.sk

construction of such a kinematic structure of the machine especially for the piston engines, which would eliminate the above-mentioned drawbacks of the piston engine designs [1-3].

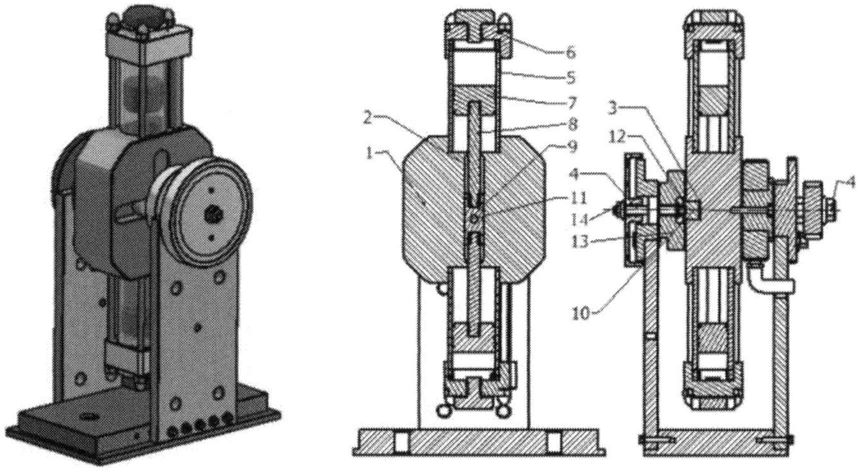

Fig. 1. A 3D model of engine design with rotating cylinders (left) and section through this model (right).

As a result of this effort, a kinematic arrangement of the mechanism with rotating cylinders will be proposed and used especially in piston combustion engines in the modification of a spark-ignition or diesel engine and a two or four stroke engine; compressors and hydraulic pump constructions as well as in compressed air engines.

2 Basic idea of the design and its methodology

The design of a kinematic arrangement of the mechanism with rotating cylinders (Fig. 1) consists of a Scotch yoke (slotted link mechanism) formed by a plate (1) with a skew line crossing arrangement of the connecting rod slot (2) and auxiliary slot (3), in whose apparent "intersection" an axis of rotation of the scotch yoke mechanism is located (4). In the axis of rotation of the mechanism (4) a shaft is located in the bearing and the bearing is mounted in the frame or chassis of the device. Radially to the axis of rotation (4) of the mechanism, two opposed cylinders (5), each ended by the cylinder head (6), are attached to the scotch yoke mechanism. The pistons (7), which are interconnected by a common connecting rod (8) passing through the connecting rod slot (2), are mounted in the cylinders (5), with the connecting rod slot nut (9) being located in the centre of the connecting rod (8). An auxiliary nut (10) is located in the auxiliary slot (3). The first pin (11) of the crankshaft (13) fits into the connecting rod nut hole (9), and the second pin (12) of the crankshaft (13) fits into the auxiliary nut hole (10). A rotation axis (14) of the crankshaft (13) is located between the first (11) and second (12) crankshaft pins. The crankshaft is mounted in its axis of rotation in a bearing which is mounted in a frame or chassis of the device. The axis of rotation (4) of the scotch yoke and the axis of rotation (14) of the crankshaft are not coaxial. The crankshaft (13) radius r is equal to ¼ of the piston (7) stroke. The axis of rotation (4) of the scotch yoke mechanism is output with the revolutions n_1. The axis of rotation (14) of the crankshaft is output with revolutions (1) [4]:

$$n_2 = 2 \cdot n_1 \tag{1}$$

Such a kinematic arrangement of the mechanism with rotating cylinders is useful for combustion two- and four-stroke engines in the modification of spark-ignition or diesel engine driven by compressed air as well as steam. The use of the kinematic arrangement of the mechanism with rotating cylinders of this design is also applicable for compressors and pumps. The advantages of the kinematic arrangement of the mechanism with rotating cylinders are evident from its effects, which are manifested externally. The effects consist essentially in the complete eliminating the side (normal) force on the piston since the angle β of the connecting rod has zero value. This is due to the piston connecting rod moving only in the cylinder axis. Furthermore, the concept of this solution implies that the rotating blocks of cylinders and pistons act as a flywheel. This means that the engine will weigh at least 15 kg less compared to the above-mentioned concepts (no need to use additional flywheels). There is also no need for a liquid cooling circuit because the rotating cylinders remove heat to the ambient air in which they are moving. By this, sufficient air circulation by increasing the Grossman factor for removing heat from the rotating cylinders is ensured. The torque output from the rotating cylinders is achieved without gearing, which results in reducing mechanism lubrication demand and also increasing the efficiency of the mechanism due to the friction reduction.

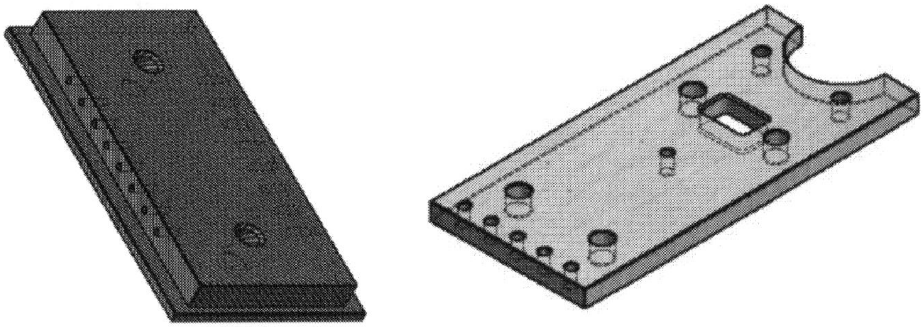

Fig. 2. A mechanism base plate (left) and its foot (right).

The first component created was a base plate (Fig. 2 left). Two through holes Ø12 mm used for fixing the plate to the frame, and threaded holes M5x15 mm for fixing the plate to the front and rear feet were drilled in. Subsequently, the chamfers 1x45° were created at the whole plate and chamfers 2x45° were created in the holes Ø12 mm. The second component is formed by two feet (Fig. 2 right), which hold the entire rotating mechanism. Both have the same size, but the holes and slots are spaced differently. The holes used for fixing to the base plate are M5x10 mm and the holes for mounting gripping are M7x10 mm. Finally, an opening for the air intake valve has been milled into the front foot.

The third component is formed by the front and rear mounting (Fig. 3 left) for the rotation of the mechanism. Counterbored holes were drilled in the mounting for its gripping to the feet by socket head cap screws. Other holes were also drilled into them, which were then further processed to the required values under the ball bearing. Finally, the outer and inner edges were chamfered to 1x45°.

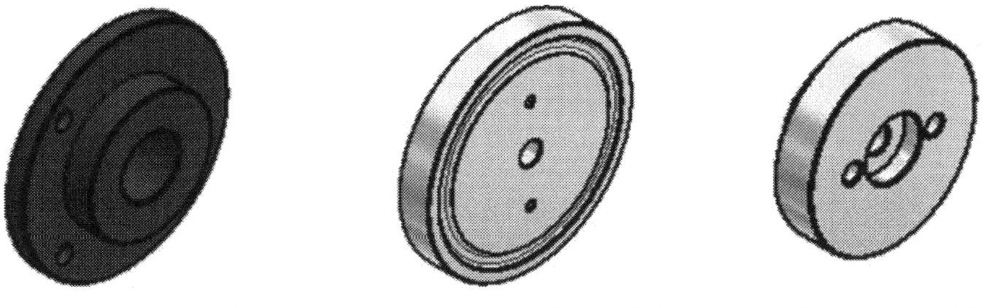

Fig. 3. Front and rear mounting (left), a rear wheel (middle) and the crank mechanism wheel (right).

The other components (Fig. 3 middle and right) were the rear wheel (middle) and the crank mechanism wheel (right), which were connected to each other with mounting by the M6x50 mm bolt and an M6 flange nut. The rear wheel has a circular groove and two holes for additional mechanism gripping on the front surface. The crank mechanism was made as a cylinder with three holes on the front surface. The middle one serves for mounting by means of a connecting screw and two for inserting Ø4 mm pins which guide the sliding keys in the grooves of the central switchgear.

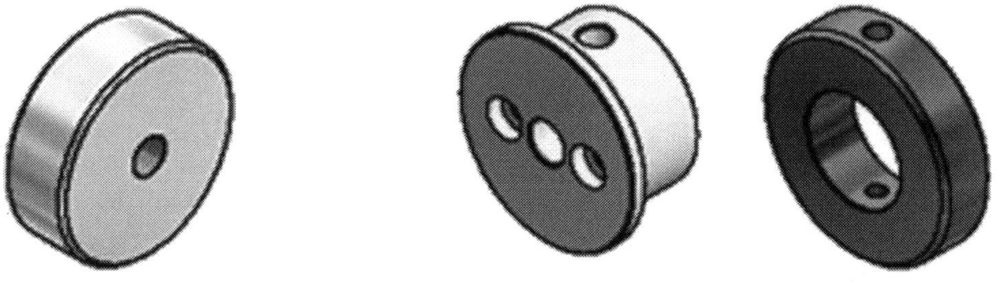

Fig. 4. Front wheel (left) and the central switchgear components (right).

Another section of the components was the creation of a front wheel (Fig. 4) with a drilled hole for M10x65 mm connecting screw that passes through the front mounting and connects two components of the central switchgear. The inner part of the switchgear was firmly connected to the central switchgear by means of two socket head cap screws and a hole for the compressed driving air has been drilled into its cylindrical circumference. The outer part can rotate freely around the inner circumference and there were also two holes for the distribution of the driving air drilled into it. An air inlet valve was placed into one of the holes.

Fig. 5. A central switchgear (left), a cylinder (in the middle) and a connecting rod (right).

Another component is a cube shaped central switchgear (Fig. 5 left), in which the holes for countersinking the cylinders and the key groove are milled. The deeper hole controls the piston movement and the other – shallower groove controls the rotation of the entire mechanism. M4x10 mm holes were also drilled into the switchgear to screw in the retaining bolts to grip the cylinder heads. The last 10x10 mm hole was used to fit the supply pipe.

Next, a piston cylinder (Fig. 5 in the centre), was made of a glass cylinder and shortened to the desired length. The glass cylinder was chosen to provide the visibility of the pistons movement. This was followed by the formation of a common connecting rod (Fig. 5 right) by means of two connecting rods. These were connected at one end by an inner key (Fig. 6 left) and at the other end with a piston (Fig. 6 in the centre).

Fig. 6. An inner key (left), a piston (in the middle) and a cylinder head (right).

The last proposed part of the rotating cylinder mechanism was the cylinder head (Fig. 6 right). The cylinder heads were made to a given shape and connected to the cylinder and the rest of the mechanism by means of fastening screws. Subsequently, a plug was screwed into the centre hole to ensure pressure in the cylinder. The manufactured mechanism (Fig. 7) works on compressed air to ensure purity. In the case of an internal combustion engine, the ignition device would be screwed in place of the plug.

Fig. 7. Image of a real prototype of a rotating cylinder mechanism captured during its performance test.

The future research activities related with this special design of a mechanism will be focused on analysis of individual part of the mechanism in term of stress distribution during its operation process. It will be analysed using FE computational method [5, 6]. Further, rotational movement of the device causes certain level of noise and vibration. These dynamics effects will be studies by means of MBS software [7, 8]. Implementing flexible bodies created in a FE software to the MBS model will allow to evaluate individual parts of the mechanism under various load conditions.

3 Conclusion

The aim of the article was the design of all members of the kinematic arrangement of the mechanism with rotating cylinders. From the figures and texts above, it can be stated that this goal has been met. The advantages of the kinematic arrangement of the mechanism with rotating cylinders are evident from its effects, which are manifested externally. The effects consist essentially in the complete eliminating the side (normal) force on the piston since the angle β of the connecting rod has zero value. This is due to the piston connecting rod moving only in the cylinder axis. Furthermore, the concept of this solution implies that the rotating blocks of cylinders and pistons act as a flywheel. This means that the engine will weigh at least 15 kg less compared to the above-mentioned concepts (no need to use additional flywheels. There is also no need for a liquid cooling circuit because the rotating cylinders remove heat to the ambient air in which they are moving. The torque output from the rotating cylinders is achieved without gearing, which results in reducing mechanism lubrication demand and also increasing the efficiency of the mechanism due to the friction reduction. The mechanism with rotating cylinders has also the advantage of providing two outputs at the engine modifications. The first output is low-speed, from the rotating cylinders, i.e., from a scotch yoke mechanism with revolutions n_1. The second output is a high-speed, from crankshaft with revolutions n_2. Another substantial advantage of the kinematic arrangement of the mechanism with rotating cylinders is the cylinders reciprocating motion removal. This is due to the offset of the crankshaft axis relative to the axis of the cylinder's rotation and due to the fact that crankshaft radius r is equal to ¼ of the piston stroke. This causes the piston to move to the top dead centre by approaching along the circular path with simultaneous rotation of the cylinder.

The advantages of the kinematic arrangement of the mechanism with rotating cylinders can emerge by its application in combustion engines by adding the relevant components associated with the fuel supply, its spark and exhausted gas routing. Another application is offered for steam-driven engines. The mechanism is complemented by steam supply components. Another application can be in compressors. The mechanism is complemented by gas supply and compressed exhausted gas components as well as components for output torque of the scotch yoke mechanism or crankshaft. Another application is available for hydraulic pumps by completing the mechanism with liquid suction and discharge components as well as torque output components in the scotch yoke mechanism or crankshaft.

Presented technical solution is currently given a utility model no. SK 24-2017 U1.

This work was supported by the Cultural and Educational Grant Agency of the Ministry of Education of the Slovak Republic in the project No. KEGA 023ŽU-4/2020: Development of advanced virtual models for studying and investigation of transport means operation characteristics.

References

1. D. Dardalis, M. Hall, R. Matthews, A. Basu, T. Ozel, SAE Technical Papers 2019
2. A. Gerlach, H. Rottengruber, R. Leidhold, IECON 2018, **44** (2018)
3. P. Kukuca, D. Barta, R. Labuda, T. Gechev, MATEC Web of Conferences **244** (2018)
4. Utility model no. SK 24-2017, Kinematic layout of a mechanism with rotating cylinders
5. P. Fabian, J. Gerlici, J. Masek, P. Marton, Komunikacie **15**, 7 (2013)
6. P. Šťastniak, M. Moravčík, P. Baran, L. Smetanka, MATEC Web of Conferences **157** (2018)
7. V. Hauser, O. Nozhenko, K. Kravchenko, M. Loulova, J. Gerlici, T. Lack, Procedia Engineering **192** (2017)
8. V. Hauser, O. Nozhenko, K. Kravchenko, M. Loulová, J. Gerlici, T. Lack, Manufacturing Technology **17**, 3 (2017)

The Study of a Control Signal's Phase Shift Influence on the Efficiency of a System for Active Vibration Damping Based on MFC Piezoelectric Transducers

Marek Płaczek[1,*]

[1]Silesian University of Technology, Department Sciences, Editorial Department, 44-100 Gliwice, Poland

Abstract. Active vibration damping of mechanical systems based on applications of smart materials has a large application potential and is getting more and more popular. In active vibration systems the fast response of actuators to the signals generated by sensors is one of the most important element that decides of the system's efficiency because the idea is to generate force by active elements that will suppress the vibrations detected by the sensors. In this paper results of laboratory tests of a control signal's phase shift influence on the efficiency of a system for active vibration damping based on application of Macro Fiber Composite (MFC) piezoelectric transducers are presented. MFCs are modern piezoelectric composite transducers produced as a thin, elastic films and can be easily installed on the surface of the mechanical subsystem or laminated in composite structures. The impact of the phase shifting between signals generated to power the actuator on the damping efficiency was verified and analysed. It was verified in what phase angle the damping of vibration has the best efficiency and if the shift of the signals causes the linear loss of the system efficiency. It was also verified whether it causes the same effects in both directions of shifting (advance or delay in the phase of the signal supplying the damper relative to the signal generated by the beam's vibration).

1 Active vibration damping using piezoelectric transducers

The damping of mechanical vibrations is an extremely important issue that is the subject of many works and scientific research [1-6]. Active vibration damping of mechanical systems based on applications of smart materials has a large application potential and is getting more and more popular. The efficiency of such systems is getting better at the same time because of development of sensors and actuators that are used as well as control systems [7-9]. One of important stages on the path of piezoelectric transducer development was NASA's development of Macro Fiber Composite transducers (MFC) [10]. The use of piezoelectric transducers is an increasingly important and popular element of modern

* Corresponding author: marek.placzek@polsl.pl

mechatronic systems that combine mechanical, electrical engineering and advanced control systems [11-13]. Piezoelectric transducers enable easy and effective conversion of electrical energy into mechanical energy or in the opposite direction.

Piezoelectric transducers are successfully used in passive vibration reduction systems, in which, together with external, passive electrical circuits, enable the dissipation of mechanical vibration energy. Systems of this type are characterized by simplicity of design and no need to introduce an additional source of energy into the system [14-18]. On the other hand, it is not possible to achieve such vibration damping efficiency as in the case of active systems. In active systems, by using a more complex system consisting of sensors that provide information on current vibration parameters to the control system and actuators controlled by this control system, it is possible to effectively damp vibration. In addition to the greater system complication, these systems also require additional energy to be supplied to the system to power both the control system and the actuators.

Active vibration damping finds a lot of applications in aviation. The example of active vibration control in aviation solutions was the Active Controls eXpert company research, which prepared a system for the damping of a tail of a military aircraft. Currently, they are also used to control aircraft wing flaps. In addition to aviation, piezoelectric systems are also used as vibration dampers for other vehicles, snowboards or baseball bats, for active damping of noise [19]. In paper [20] review of researches in the field of active vibration control and description of the most important methods of their implementation. Nonadaptive and adaptive systems (feedforward, feedback and hybrid control structures, single and multiple channel case) with adaptive algorithms are outlined. Influence and modelling of secondary path is presented. Major applications and directions for further research are indicated. A practical example of vibration damping using piezoelectric transducers is discussed in the article [21]. The article discusses the problem of self-excited vibrations in machining and proposed constructions of a vibration eliminator using a piezo actuator, thanks to which it is possible to reduce vibrations. A control system based on a Gaussian linear-square regulator ensuring high efficiency of the vibration eliminator, which has been confirmed experimentally for various cutting parameters is also presented. The experiment described in this article showed that the use of a piezo actuator allowed to modify the dynamic properties of the workpiece and reduce the level of vibration. The advantages and disadvantages of the presented solution were also indicated. The main ones include the fact that piezo elements provide greater efficiency when working at higher frequencies and are smaller in size than electromagnetic systems. The use of the discussed vibration eliminator does not reduce the machining efficiency, which is a common disadvantage of various types of active and passive methods of vibration elimination during machining. The presented solution also has an advantage over passive solutions because the system retains its efficiency in the event of changes in machining parameters. The only significant disadvantage is the high cost of the actuator and high voltage power amplifier. Also, in paper [22] active vibration control of a plate like smart flexible structure is presented. This plate is equipped with several thin piezoelectric patches. Some of them are used as sensors and the others as actuators. They are optimally positioned and not collocated. The main goal of control is to reduce the most energetic vibrating modes.

In active vibration systems the fast response of actuators to the signals generated by sensors is one of the most important element that decides of the system's efficiency because the idea is to generate force by active elements that will suppress the vibrations detected by the sensors. In this paper results and their analysis of laboratory tests of a control signal's phase shift influence on the efficiency of a system for active vibration damping based on application of Macro Fiber Composite (MFC) piezoelectric transducers are presented. MFCs are modern piezoelectric composite transducers produced as a thin, elastic films and can be easily installed on the surface of the mechanical subsystem or laminated in composite structures. They can be use both as sensors or as actuators based on direct and reverse piezoelectric effects. Therefore, the potential of their applications in variety types of mechatronic systems is very high, including systems for active vibration damping. Presented laboratory tests were conducted using a mechanical subsystem in the form of cantilever beam excited with the first natural frequency of vibration by the MFC actuator. The impact of the phase shifting between signals generated to power the actuator and damper on the damping efficiency was verified and analysed. It was verified in what phase angle the damping of vibration has the best efficiency and if the shift of the signals causes the linear loss of the system efficiency. It was also verified whether it causes the same effects in both directions of shifting (advance or delay in the phase of the signal supplying the damper relative to the signal generating the beam's vibration).

2 Laboratory stand for testing the influence of phase shift angle of signals in the system of active damping of beam's vibrations

In order to verify the impact of the phase shift angle on the efficiency of the active vibration damping system based on the use of MFC piezoelectric transducers, a laboratory stand was built, which consists of:

- a clamp with the possibility of mounting the tested piece,
- a mechanical subsystem (a cantilever beam made of an aluminium alloy),
- a generator of signals with required frequency WW5064 by Tabor Electronics,
- two high voltage amplifiers HVA 1500/50 by Smart Material,
- an oscilloscope RIGOL DS1064B
- MFC M8514-P1 and M8507-P1 piezoelectric transducers.

The created stand for laboratory tests is shown in Fig. 1.

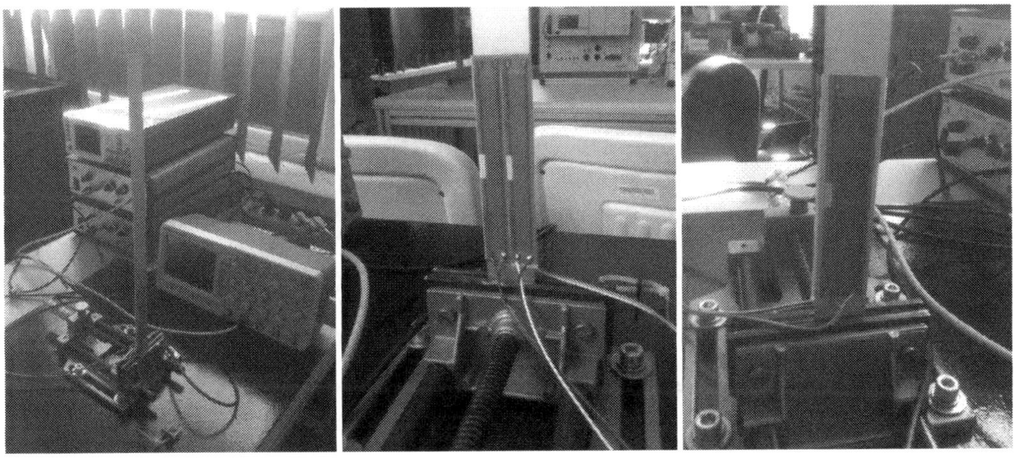

Fig. 1. The created laboratory stand (*a*), piezoelectric MFC actuator and damper (*b*) and the MFC transducer used as sensor (*c*).

Presented laboratory tests were conducted using a mechanical subsystem in the form of cantilever beam excited with the first natural frequency of vibration by the MFC actuator. Three MFC piezoelectric transducers were glued on the surface of the tested beam. The first one was used to generate vibration of the beam, while the second one was used as a damper. The third MFC was used to measure the beam's vibrations and to analyse of the damping efficiency. The vibration generator and damper were powered using laboratory signal generator and high voltage amplifiers. The signal generated to drive the vibration damper was phase shifted relative to the actuator supply signal and efficiency of the vibration damping was measured.

The system was excited with a frequency equal to the first natural frequency, which in the case of the considered beam was 8 Hz. Such action was aimed at obtaining maximum values of the amplitude of vibration of the mechanical system. The MFC M8514-P1 transducer was used to indirectly measure the amplitude of vibration of the mechanical subsystem. The value of voltage generated by this transducer depends on its deformation, resulting from the beam deflection and the frequency of deformations, which was constant during all tests. At a constant value of the frequency of excitation of beam vibrations, the value of voltage generated by the MFC sensor was therefore influenced only by the amplitude of the system vibrations.

3 Measurements results

The tests were carried out at five different values of the phase shift angle between the signal supplying the transducer exciting vibrations and the signal supplying the piezoelectric damper. Values from $0{,}8\pi$ to $1{,}2\pi$ with a gradient of $0{,}1\pi$ were assumed. This corresponds to the situation in which the signal supplying the active vibration damper in phase is faster than the signal supplying the vibration generator (when $\varphi < \pi$), is in the opposite phase ($\varphi = \pi$) or is delayed in phase relative to the vibration excitation signal ($\varphi > \pi$). The tests were carried out in such a way that vibrations of the mechanical system were forced, and then after some time the power supply of the vibration damper was turned on. After stabilization of the damped oscillation amplitude, the damper was turned off while the actuator of vibrations was still powered. The tests were repeated five times for each phase shift angle value, simultaneously the signal generated by the MFC transducer being the measuring element was recorded.

Fig. 2 shows the waveforms of the electric voltage as a function of time. It is a signal generated by the MFC transducer, which is a measurement element of the effectiveness of the active vibration damping system. The vertical lines in black indicate the moment of activation of the system of active vibration damping, while the vertical lines in red indicate the moment of system deactivation, while the vibrations of the mechanical system were still excited. The waveforms are presented at different values of the phase shift angle φ of the signal, which fed the transducer acting as a vibration damper in relation to the signal supplying the actuator which excite system vibration.

Table 1 summarizes the values of the average maximum amplitude of the signal generated by the measuring transducer after stabilization of the system vibration amplitude during the operation of the active vibration damping system.

Table 1. Average value of the voltage amplitude generated by the measuring transducer.

Signals phase shift angle [rad]	V_{max} [*V*]
0,8 π	2,42
0,9 π	1,31
1,0 π	0,35
1,1 π	0,62
1,2 π	1,62

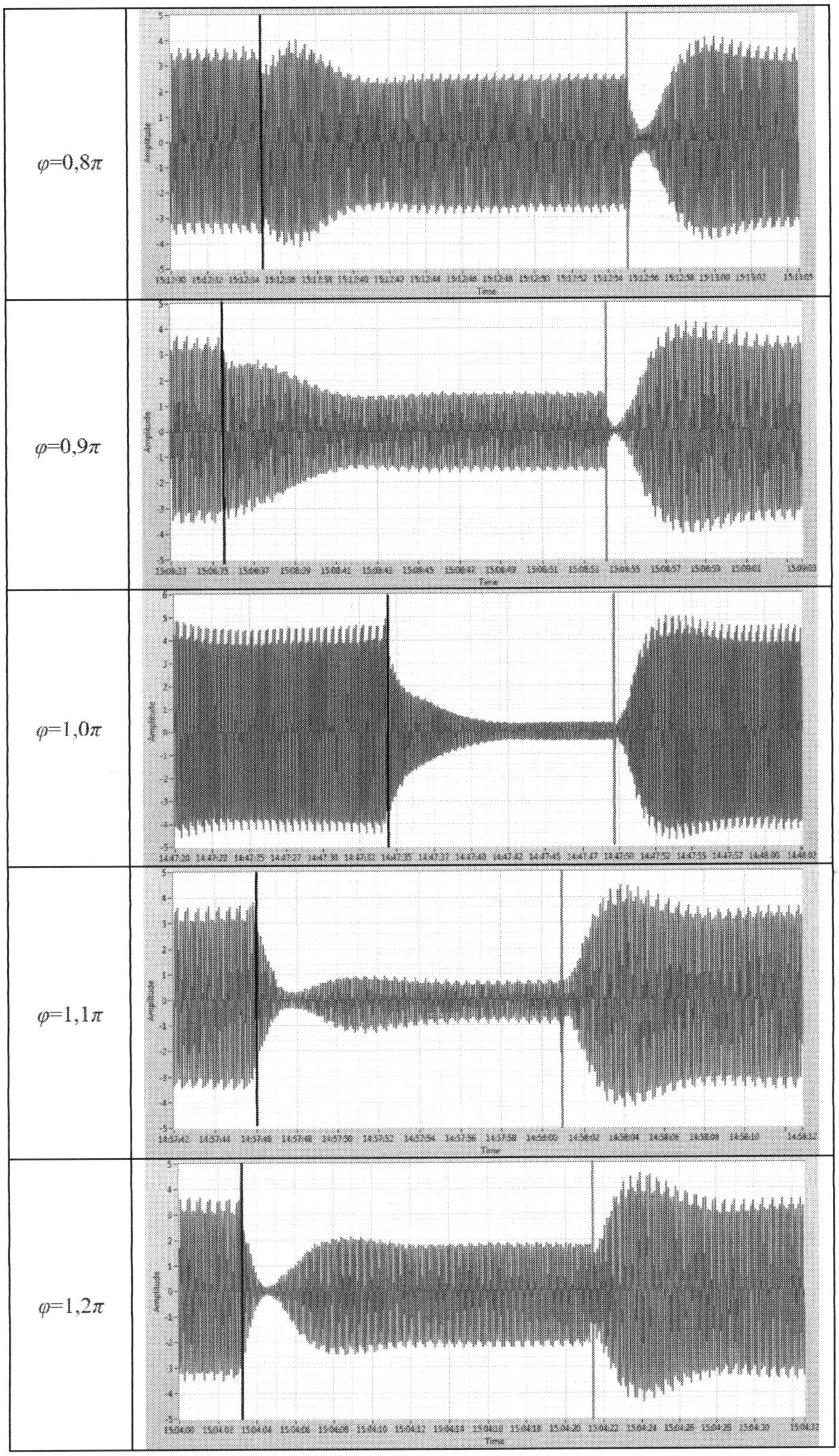

Fig. 2. Time waveforms of the voltage signal generated by the MFC transducer acting as a measuring element with different angles of phase shift of the signals.

4 Conclusions

Laboratory measurements of the effectiveness of the active vibration damping system at variable phase shift angle values between the excitation signal of the mechanical system and the signal supplying the piezoelectric damper have shown its high significance for the efficiency of vibration damping of the system. To achieve maximum vibration damping efficiency, these signals must be in the opposite phase (shifted by 1.0 π radians). Any deviation from this value causes a decrease in the effectiveness of the system. It should be emphasized that in the event of a delay in the phase of the signal supplying the piezoelectric damper relative to the excitation signal ($\varphi > \pi$), the decrease in system performance is smaller than in the case when the signal supplying the piezoelectric damper is faster in the phase than the beam deformation resulting from its excitation ($\varphi < \pi$). It should also be noted that in the event that $\varphi > \pi$ after switching on active vibration reduction, there is a transition period in which the vibration damping efficiency is very high, followed by a decrease in efficiency and stabilization of the vibration amplitude at a certain level. If $\varphi < \pi$, this phenomenon occurs when the active vibration reduction system is turned off. When designing active vibration reduction systems based on the use of piezoelectric transducers, these phenomena should be considered and strive to develop fast control algorithms in which it will be possible to achieve maximum efficiency and effectiveness of vibration damping.

References

1. K. Białas, Solid State Phenomena **164** (2010)
2. K. Białas, J. of Vibroengineering **14** (2012)
3. A. Dymarek, T. Dzitkowski, J. of Materials Processing Technology **164-165** (2005)
4. A. Dymarek, T. Dzitkowski, J of Vibroengineering **15**, 3 (2013)
5. A. Wróbel, J. of Vibroengineering **14**, 2 (2012)
6. S. Zolkiewski, Int. J. of Bifurcation and Chaos **21**, 10 (2011)
7. T. Bartel, S. Herold, D. Mayer, T. Melz, *Proceedings of the 6th ECCOMAS Conference on Smart Structures and Materials* (2013)
8. S. Herold, D. Mayer, Actuators **5**, 7 (2016)
9. L. Sui, X. X. Shi, Physics Procedia **25** (2012)
10. M. Płaczek, G. Kokot, Sensors **19**, 2 (2019)
11. W. Banaś, A. Sękala, K. Foit, et al., IOP Conf. Series: Mat. Sc. and Eng. **95** (2015)
12. A. Gwiazda, K. Herbuś, G. Kost, P. Ociepka, Solid State Phenomena **220-221** (2015)
13. A. Gwiazda, A. Sękala, Z. Monica, W. Banaś, Adv. Mat. Res. B (2014)
14. A. Buchacz, M. Płaczek, A. Wróbel, Eksploatacja i niezawodnosc – Maintenance and Reliability **16**, 2 (2014)
15. M. Płaczek, Advanced Material Research **1036** (2014)
16. A. Buchacz, M. Płaczek, Solid State Phenomena **147-149** (2009)
17. K. Marakakis, G. Tairidis, P. Koutsianitis, G. Stavroulakis, Front. Built Environ. **5** (2019)
18. Z. Skup, J. of Theoretical and Applied Mechanics **44**, 1 (2003)
19. S. Yoshikawa, A. Bogue, B. Degon, *Proceedings of the 11th IEEE International Symposium on Applications of Ferroelectrics* (1998)
20. D. Milijkovic, *MIPRO 2009 Conference Proceedings* (2009)
21. A. Parua, K. Marchelek, S. Domek, M. Hoffmann, Modelowanie Inzynierskie – Modelling in Engineering **39** (2010)
22. S. Tliba, H. Abou-Kandil, C. Prieur, IFAC Elsevier Publications, 16th Triennial World Congres, Prague, Czech Republic (2005)

Progress and recent trends in generative design

Ioannis Ntintakis[1,2] and *George E.* Stavroulakis[1,*]

[1]School of Production Engineering and Management, Computational Mechanics and Optimization Laboratory, Technical University of Crete GR-73100 Chania, Greece email: gestavr@dpem.tuc.gr, www.comeco.tuc.gr
[2] School of Engineer, Department of Mechanical Engineering, Hellenic Mediterranean University, Estauromenos, 71004 Heraklion, Greece e-mail: ntintakis@hmu.gr

Abstract. Due to recent developments in the field of additive manufacturing enormous advantages have become in product design and manufacturing process. Before the appearance of additive manufacturing, developing very complex or light weight structures was difficult to manufacture. The development of artificial intelligent technology helps to develop new collaborative tools and algorithms. Generative design approach is one of them. The outcome model from a generative design study is not depending only from designer/engineer experience or his knowledge. Designers can react with sophisticated algorithms through CAD programs to specify the shape and the topology of the model. A significant tool on a generative design system is topology optimization which is able to generate different solutions. The changes in design process are significant. A rough conceptual design (sketch) or a 3d model is first prepared. Then, boundary conditions, safety factor, manufacturing limitations and materials properties are defined. The generative design system generates potential solutions. It's up to the designer to find the design that best fits to his need. In this paper the review covers the limitations of current systems through the study of specific design cases using commercial generative design systems.

1 INTRODUCTION

During the design process designers/engineers has to take decision for product shape, ergonomic design, material selection, product stiffness, durability and so on. The combination of these parameters create a complex problem. Designer has to take a decision very fast and in accurate. In this effort advanced systems like generative design, topology optimization and additive manufacturing are very helpful.

Generative Design (GD) has become more and more popular in recent years. In general GD mimics nature by using algorithms inspired, for example, by the way in which bones grow in animals. A basic outcome of this system is to add material where is need to be and to remove from the domain where it doesn't. GD is used today in mechanical design, architecture and civil engineering, microstructural design, smart structures, micromechanics, multiphysics and other [1-2]. As a term is the repetition of process of potential outcomes that meet certain constrains, boundary conditions. Designer can change the inputs in a way to

* Corresponding author: gestavr@dpem.tuc.gr

choose the best solution. It is therefore a fast method of exploring design possibilities that is used in various design fields. As a process generative design can be analysed as a linear diagram [3]. The process starts with an idea which designer set up in the system. Mainly the idea described in a general form with a rough 3d model or a sketch. Then the rules are defined and an algorithm starts to formalize the rules in parameters in order to output potential solutions. Each time the algorithm offers a number of different solutions. Designer studies the solutions and modifies the inputs rules so to take more accurate solutions. The process ends when the designer judges that the potential solution is the appropriate. The algorithmic method transform requirements into design and product.

2 GENERATIVE DESIGN IN INDUSTRY

In 2019 the company Kartell produced the first commercial chair made using generative design in collaboration with Autodesk and with famous designer Philippe Starck. As Starck said they ask from intelligent system to carry the body chair with the least amount of material. In aerospace industry generative design is a well know technology. Also, in RUAG Sentil satellite the designers used topology optimization and Additive Manufacturing (AM) for the antenna bracket with EOS Aluminium AlSi10Mg which has high strength and resistance to dynamic stress. The optimization goal wasn't only the light weight structure but also the minimization of undesired vibration from rocket high speed. The antenna bracket weight reduction was over than 40 %, from 1.6 kg to 0.94 kg [4].

In automotive sector in 2018 General Motors starts to develop new car components. One of them is a new seat bracket which is be 40% lighter and 20 % stiffer than the previous one. The potential solutions was chosen from about 150 different models [5]. Audi used AM to produce the water connection for the W12 engine and Porsche makes the gearstick for the 959 model. BMW producing more over than 25,000 prototypes helping the design process [6]. Also, in Ford there is no need for tooling for prototypes saving a huge cost in product development process [7-8].

In aviation industry the Airbus group has run out an analytically study about the reducing of CO_2 emission about 40% from weight reduction. For this purpose they redesign with TO toll the Airbus A320 nacelle hinge Bracket using the direct metal laser sintering AM technology rather than traditional casting methods [9].

3 Soft Robotics

Traditional rigid robots are characterized by their incredibly powerful and precise performance abilities. A typical robot construction is made of rigid materials and rigid joints. There is no ability for elastic deformation therefore in case the robot acts on rigid bodies it cannot adapt its shape. The case arises, for example, during interactions with human body or environmental constrains which can also be defined as lack of multi-functionality [10].

In the domain of healthcare, robots have to adapt their functionality in human body requirements and to be more sensitive, compliant and adaptive. Next generation robots have to be more deformable and their shape can be inspired from nature. In the design and control of new age robots biological principles are applied, the so-called soft robots. Soft robots are expected to have better adaptability on human constrains with safe and smooth interaction with them [11].

A typical structure of a soft robot consists from a number of actuators which are usually manufactured from deformable material. Usually actuators activated with compressed air thus generate motion [12]. In other cases fluid motion creates mechanical action [13]. Other researcher [14] gives a specific spatial architecture to the bladder as a bellow. Or other use thin films for the bladder structure [15]. The film flexibility and the appropriate plane stiffness conduce to transform the structure in the right way. The use of materials with different stiffness values can easily be used in order to add 'programming' to the behaviour of the bladder. The use of stiffer material in only one side of the actuator is another possibility [16]. Other actuators allow faster actuation with lower volume change. Along the surface of each finger flexible tubes are established, so allowing faster actuation. A soft robot is composed of three parts, an actuator, a sensor(s) and a controller.

In this study we are going to test an actuator which is powered by air. The aim of this series of experiments was to evaluate the inflating behaviour of a soft robotic actuator using materials with different stiffness so on to add a programming behavior to the actuator. Soft robots mimic the nature and especially animal capabilities and they are at the crossroad of engineering and mechatronics, chemistry and plastics [17]. Also we are going to use a generative design platform in order to design a rigid body to reinforce a soft robot actuator. The new design stability will be tested by using a FEA analysis which is the appropriate tool to test a wide range of structures [18]. Results related to the design of controlled flexible structures can be founr in [19-20]

4 Generative Design (GD) and Topology Optimization (TO)

Topology Optimization is a basic tool in generative design. In GD potential solutions refer to different size and shape of the product. TO results combine shape and size in order to satisfy the objective function which usually wants to minimize the volume [21]. So designers using topology optimization are able to select a design solution that satisfy design requirements. Very often the selected solution is not manufactured or it is economically unprofitable with subtracted methods.

In particular TO has the following benefits in the design: a) creating light structures b) generation of a ready-to-build part/assembly c) minimize the amount of raw material d) energy saving e) less need for natural prototypes f) reduction of physical testing g) reduced entry time [22]. Elastic properties of a material compared with their density may vary in model area. The material can be permanently removed from model space. During the TO study models have to be applied all necessary boundary conditions. Two methods have been developed for TO study. The first one is truss based and second is volume based.

The meaning of optimization problem is a combination of many parameters. The target is to find the appropriate structural design solution in a domain. To solve the optimization problem restrict parameters as set up, such as, the boundary conditions, the initial volume and the specific area which will remain void or will be filled with material. Potential solutions for the shape or the size of the model are unknown [23]. Mathematical formulation is specified from minimize of the objective function specified as below:

$$\text{Specify } \chi = \begin{Bmatrix} \chi_1 \\ \chi_2 \\ \cdot \\ \cdot \\ x_n \end{Bmatrix} \text{ which minimize f(x)} \qquad (1)$$

$$where\ gi\ (x) \leq 0,\ i=1,2,\ldots.,m$$
$$hj\ (x) = 0,\ j=1,2,\ldots,n$$

The domain area is an n-distance Cartesian area in which each axis matched in a parameter. Where x is the vector of the variables that the design must meet in order to be acceptable. There are essentially two categories of restrictions: a) Operational constraints refer to system behavioral constraints b) Geometric constraints refer to physical constraints. Where, f(x) is the objective function or the cost function, essentially a criterion for comparing acceptable solutions to select the optimal design. The objective is to minimize f (x). The functions gi (x) and hj (x) are called constraints of inequality and equality that determine the constraints of the problem. Points on the constraint surface with gi (x) = 0 satisfy the constraints marginally, so the design area is divided into two distinct spaces. The first space refers to marginally acceptable and acceptable points where gi (x) <0. The second space refers to unacceptable points and applies gi (x)> 0.

Extensions of topology optimization in order to take into account wished deformability patterns, and thus instead of a structure to deliver a flexible mechanism, are also possible [23-24]. These results can be used for the design of innovative metamaterials with optimal microstructures, like auxetics [25].

5 Experimental Validation

We are going to test a common shape of a soft robot actuator in order to predict its behavior using a finite element analysis. In the first stage we are going to present a prediction of the actuator's behaviour in a specific domain using a set of boundary conditions. After that, using a generative design study we are create a stiffer structure which will improve the actuator's behaviour. Inflating behaviour mainly dependent on the material flexibility. We use the Ecoflex 30 which is a flexible material with a tensile strength of 120psi (0.827 MPa). The actuator diameter is 10mm and is based on a rigid base. The thickness of air chamber to 0.1 mm. The outcomes from the first FE analysis shows that the maximum displacement of the soft robot actuator is about 9mm and von Mises stress is 3.83 MPa (figure 1). The total displacement of the actuator is quite big.

In order to control the actuator behavior we are going to use a reinforcement body on the outer actuator surface, as a result of generative design. Potential solutions are outcome from the analysis. Based on one of the potential solution we design the reinforcement element (figure 2).

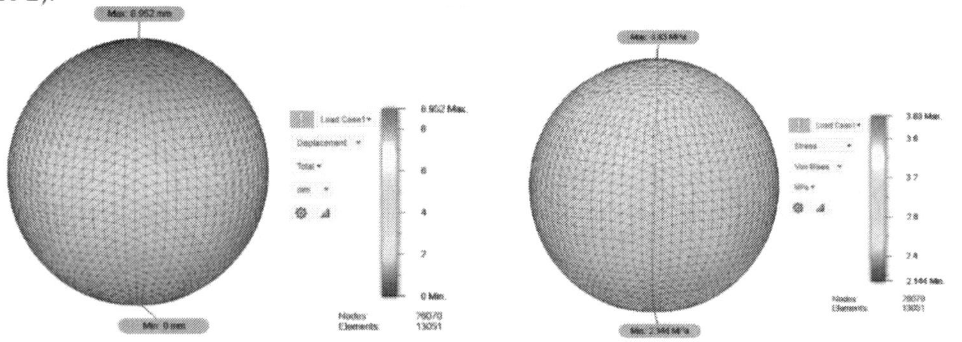

Fig. 1. Total displacement and stress results of soft robot actuator

The selected material for reinforcement element is Polylactic Acid (PLA) which is a bioplastic and common material for 3d printing process. PLA can be used in soft robot actuator for surgery use. The actuator behavior changes in height during inflation t at large mechanical strains. Generative design analysis starts from an initial sphere of 13mm diameter and 3mm thickness (figure 2).

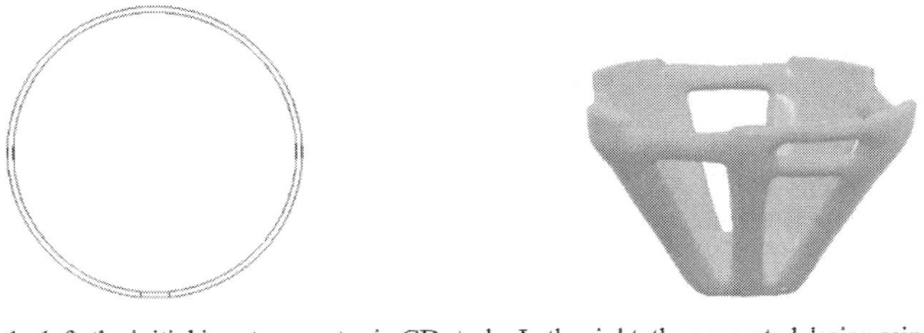

Fig. 2. In the left, the initial input geometry in GD study. In the right, the generated design reinforcement element from bioplastic material

Now the air is incoming from the base of the outer body and passes through a hole to the main actuator surface. After the generative design study is completed we verify the behavior of the reinforced structure using FEA. The tangent surfaces between the actuator and the reinforcement element are bonded. All other boundary conditions are the same as the first FE analysis. As its presented in fig. 3 the displacement of the actuator is limited, is not bigger than 2 mm and von Mises stress is now less than 1.733E+08 Pa.

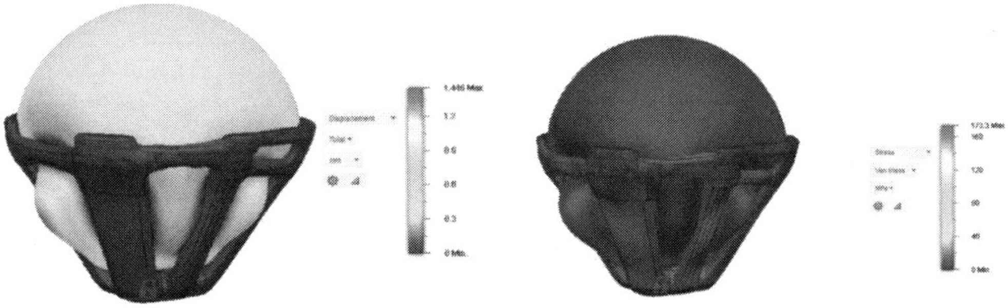

Fig. 3. Total displacement and stress results of soft robot actuator bonded with reinforcement element

6 CONCLUSION

Generative design is a powerful tool for designers and engineers. As methodology GD is not new, but in combination with computer raising and intelligent systems usage they become a well-developed and reliable design tool. The results of GD study are characterized by organic shapes which try to mimic the nature. Moreover, the results of a GD study can not be predicted but not exceed operational and geometric constrains. In most cases the GD 3d models are very difficult or can't be machined with traditional manufacturing methods. Additive manufacturing technology is the appropriate production method for generative design models. The generated model of soft robot actuator can be produced with a two printed head, i.e. additive manufacturing using two materials.

In the presented example, it has been demonstrated that the regenerated model limits the displacement of the soft robot actuator. The generated model is stiffer than the initial one. Further study could take into account variable (unilateral) contact effects and contact force optimization.

References

1. E. Tyflopoulos, D. Tollnes Flem, M. Steinert and A. Oslen. *NordDesign*, 1-24 (2018)

2. X. Chen, Y. Tao, G. Wang, R. Kang, T. Grossman, S. Coros and S. Hudson. *CHI,* 1-12 (2018)
3. H. Bohnacker, B. Grob, J. Laub, and C. Lazzeroni, *Generative Gestaltung* (2009)
4. EOS Aerospace: https://www.eos.info/en/3d-printing-examples-applicationsinnovation-stories/ruag-aerospace-3d-printed-satellite-components (2018)
5. GMmotors.https://media.gm.com/media/us/en/gm/news.detail.html/content/Pages/news/us/en/2018/may/0503-lightweighting.html (2018)
6. Stratasys. https://3dprinting.trimech.com/case-study-fdm-bmw (2014)
7. S. Jauhar, K. Asthankar and A. Kuthe AMEA. **2** 181 (2012)
8. Center FM Ford. https://www.plasticstoday.com/content/ford-3d-printed-auto-parts-save-millions-boost-quality/97214356120060 (2014)
9. M. Tomlin and J. Meyer *7th Altair CAE technology*, 1–9 (2011)
10. C. Majidi, soft robotics **1,** 5-11(2013)
11. R. Pfeifer, H. Gravato and F. Iida *Twenty-Third international joint conference on Artificial Intelligence* 1-9 (2013)
12. A. Calderón, J. Ugalde, Z. Joakin, C. Juan and N. Perez-Arancibia *IEEE International Conference on Robotics and Biomimetics*, 31-38 (2016)
13. F. Giorgio-Serchi and Weymouth G.D. *Soft Robotics: Trends, Applications and Challenges* **17**, 37-46 (2017)
14. K. Digumarti, A. Conn and J. Rossiter *IEEE Robotics and Automation Letters*, **2**, 2302, (2017)
15. R. Niiyama, X. Sun, C. Sung, B. An, D. Rus and S. Kim Pouch *Soft Robotics,* **2**, 59-70 (2015)
16. P. Polygerinos, Z. Wang. K.C. Galloway, R. J. Wood and C. J. Walsh, RAS 73, 135-143 (2015)
17. F. Ilievski, A. Mazzeo, R. Shepherd, X. Chen and G. M. Whitesides. Angewandte C., 50, (2011)
18. I. Ntintakis, V. Iakovakis, G. Ntalos and J. Kechagias, *Current Issues in Global Furniture,* 14-21 (2013)
19. G. Tairidis, Phd Thesis, https://www.didaktorika.gr/eadd/handle/10442/38484 (2016)
20. G. Tairidis, A. D. Muradova and G.E. Stavroulakis, Frontiers in Built Environment J. **5**, (2019)
21. I. Ntintakis, G.E. Stavroulakis and N. Plakia, *SIM-AM 2019,* 225-236 (2019)
22. A. W. Gebisa and H. G. Lemu, *MESIC 2017*, 724-731 (2017)
23. M.P. Bendsoe and O. Sigmund, *Topology Optimization* (2013)
24. N. Kaminakis, https://www.didaktorika.gr/eadd/handle/10442/36368 (2015)
25. N. Kaminakis, T. Drosopoulos and G.E. Stavroulakis, Arch Appl Mech **85**, 1289–1306 (2015). https://doi.org/10.1007/s00419-014-0970-7

Stop wasting! A plea for maintenance

Felicia Veronica Banciu[1,*], *Eugen* Pămîntaş[1]

[1]IMF Department, Mechanical Faculty, Politehnica University of Timişoara, Bd. Mihai Viteazu No. 1, 300222, Timişoara, Romania

Abstract. It is produced, it is worn out, it is thrown away, it is becoming more and more polluted, it is recovering, but it is not completely away, and therefore the material reserves are diminished and some have become even deficient. The handiest solution, at least to slow down the dynamics of pollution and draining of supplies of materials and materials is maintenance. In the following we will explore this solution to make it as efficient and available as possible in industrial application.

1 Clarification Concepts

We are surrounded by products, of all kinds, tangible and intangible, made by man to cover various needs and which, like everything on earth, suffer physical and moral degradation of functions and characteristics.

To clarify the concepts (notions), the United Nations' International Standard Industrial Classification standard, the Statistical Classification of Economic Activities in the European Community (NACE) in the EU and similar systems elsewhere, have a first-level hierarchy that reflects whether the economic goods are tangible or intangible.

When the product no longer meets the needs or no longer fulfils its functions, it is considered to have ended its so-called life cycle. The specialized literature records dozens of graphic representations of the “product life cycle”, some more detailed, others more brief and to make us understand but without entering into controversies with other authors, is presented in figure 1 our own short version, which has basically an analogy with the "life cycle" of a living thing on Earth. The essence of the requirements regardless of the system adopted for the representation and operation of the life cycle of a product, service, etc. is a good design and easy maintenance. These have been and will remain ideal goals, but in some cases achievable. In this paper, we mainly consider the products resulting from a manufacturing process.

In manufacturing (secondary sector), products are bought as raw materials (primary sector) and then sold as finished goods. A service (not as tertiary sector, but with reference to maintenance and service for manufactured products – authors' note) are also regarded to as a type of product [1]. “It is sometimes hard to define whether a given company is part and parcel of the secondary or tertiary sector [2].

The term of circular economy is often met - where the lifecycle of products is extended to reduce the use of raw materials and the production of waste through recycling

* Corresponding author: felicia.banciu@upt.ro

of what can be recycled in order to reinsert these raw materials at the beginning of this new, endless cycle.

However, with dramatic cost reduction and speed and reliability improvements in the transportation of people and the communication of information, the maintenance sector now includes some of the most intensive international competition.

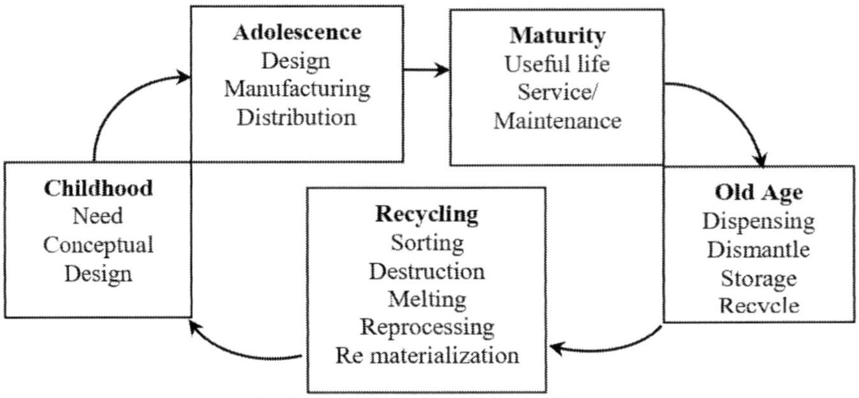

Fig. 1. A vision of product life cycle.

2 About the manufactured product

Making a product requires determining the needs or requirements of customers, developing the product concept as a result of the marketing study, developing the technical specifications of the product, product design and manufacturing technologies, the actual manufacturing, distribution and service in warranty and post -guarantee for the products sold, and finally the elimination of the product.

From the above we can only refer to two notions:

a) The notion of service - it must be specified that, on the one hand in this case, the term refers to a tangible product, result of a manufacturing process, and on the other hand, that it in its content includes not only the activities of maintenance organized by the manufacturer for that product but also those that must be performed by the user in order to maintain its operating characteristics.

b) The notion of disposal of the product at the end of the life cycle - we believe that, closer to the reality of the content, would be the use of the terms of withdrawal of the product that has become useless and the recovery of the materials from which it is made. This is because the product cannot be disposed of, literally, i.e. after the removal action there are no traces and no consequences, as if it did not exist. In addition, the result of elimination is at least devoid of positive economic effects, while the reintroduction of recovered materials into the economic circuit is undoubtedly profitable both for the industrial field and for the terrestrial ecosystem.

Recycling and disposal activities provide information on components, materials and other resources, but also on product maintenance, repair and dismissal support systems. A PLM environment enables collaboration between – and informed decision making by – various stakeholders of a product over its lifecycle [3]

To see the interest in minimizing waste and facilitating product maintenance, let's briefly review the phase theory of a product design process [4]

- The first phase is to clarify the task (clarification and planning of tasks). The result of this phase is the initial description of the product expressed in the form of a list of functions and features that must be fulfilled by the product, a set of constraints and possibly objectives related to cost and time to market.

- The conceptual design phase (conceptual study) is seen as an iterative, evolutionary and top-down process, which ultimately leads to a solution of principle or a concept for a product..

- In the embodiment design phase,

-In the last phase, detailed design (detailed conception) are elaborated all the documents regarding the production. All aspects of ergonomics, safety, assembly, operation, maintenance, recycling, costs, aspects that are progressively discovered and suffer from iterations must be taken into account.

Even if the lists of requirements for the future product require the importance of maintenance, it is usually defined briefly, incompletely, insisting only that it should be easy, which in fact has other implications materialized in design rules for carrying out inspection, service and repair - maintenance such as: prefer self-balancing and self-adjusting solutions; aim at simplicity and few parts; use standard components, allow easy access, provide for easy disassembly; apply modular principles, use few and similar service and inspection tools.

At a simple analysis of these rules we find that ensuring easy access is a requirement imposed by ergonomics, ensuring a good view of the defective area, locations for service, inspection, repair must be easily accessible, the tools used must be ergonomic, the location of the defect must be done easily, the change of components must be done easily and last but not least to have clear instructions, procedures and support activities.

3 About maintenance

In a very broad sense, maintenance means all the operations of organization, planning, supervision and monitoring of useful functions operation, maintenance, reconditioning and repair by which a product - equipment, machinery or building - is maintained throughout its life cycle so that it operates within the designed parameters.

For practitioners, maintenance means any activities such as inspections, tests, measurements, replacements, adjustments. An enterprise maintenance service may also be determined to participate in studies to improve the industrial process and must take into account numerous restrictions such as quality, safety, environment, costs, etc. According to the European standard EN 13306: 2001, maintenance is the set of all technical, administrative and management actions, during the life cycle of an equipment / system, designed to maintain or restore it in a state in which it can fulfil its required function [5].

NATO's AAP-6 Standardization Agency, in a Glossary of Terms and Definitions [6] provides a broad definition: “Maintenance is the set of measures taken to put or restore a system / equipment to a specified state, until the end of its use, including inspection, testing, maintenance, modification, repair, restoration, reconditioning, reuse, recovery and removal of parts from products to repair other damaged equipment and machinery."

Theorists in the field distinguish several possible types of maintenance: corrective, palliative, systematic preventive, conditional preventive, preventive-planned, predictive, total productive (TPM, which, once chosen, decisively make their mark on a company. Qualitative and quantitative assessment of maintenance is done through a series of indicators, such as: maintainability, average repair time (MTR), average operating times (MTF), failure rate, etc. and also by the volume of maintenance costs.

Maintenance costs include the costs involved in the operation of the maintenance department, which cover the costs of: labor, subcontracting to third parties, spare parts and consumables, equipment and tools, transport and storage of spare parts, reconstruction, automation and self-equipment, training and staff training, etc. Equally significant but more difficult to quantify are the costs of maintenance inefficiency which include: serious personal injury to staff due to insecurity, unavailability of equipment, and the cost of replacement measures taken, contractual penalties, intermediate stocks of surplus equipment, non-quality attributable to maintenance non-compliant products, reprocessing, scrap, reclassification of non-compliant products, returns, etc.).

Maintenance activities use a series of methods and tools for: identifying potential product defects (FMEA), developing the most appropriate maintenance strategy in terms of efficiency and costs (RCM - Reliability Centred Maintenance, for which the MIL- STD-2173 [7] provides a guide for RCM analysis), analysis of the chain of defects (Analysis of the fault tree - see standard SR EN 61025: 2007), rapid restoration of the operating condition of the product (Maintenance Manual – e.g. British Standard BS EN 13460: 2002 [8], analysis adapted to assess the complex structures of systems and complex maintenance and repair strategies (Markov Analysis).

Maintenance is an important part of life cycle management, its main purpose being to increase the eco-efficiency of the product life cycle [9], [10]. In this sense it aligns with the desideratum that The goal of manufacturing is no longer to produce products in an efficient way, but rather to provide the functions needed by society while minimizing material and energy consumption in order to make our society sustainable through reducing the level of pollution and depletion of materials and energy, maintaining an ecological balance [11].

Also proper maintenance is essential to reduce energy losses. Energy saving through maintenance can be obtained by housekeeping and simple maintenance; condition based maintenance and advanced condition monitoring. [12]. The general rule is to reduce the possibility of failure and failure of products / equipment, reduce maintenance and eliminate it if possible. In other words, instead of analysing the potential failures of the systems, solutions should be analysed for products / equipment that require minimal, inexpensive maintenance and that are easy to repair when they fall.

Several maintainability prediction procedures have been developed, the goal being, instead of predicting how the system will fail, to make products that require minimal maintenance and are easy and inexpensive to fix when they fail [13]. System failure is inevitable at some point, no matter how reliable it is designed to be, it is therefore important to make sure the machine can be quickly restored. Maintainability is an important aspect in overall system continuous improvements efforts, along with reliability, safety, and other factors vital to overall product viability. In fact, many times the term RAMS (Reliability, Availability, Maintainability and Safety) or "reliability and maintainability" will be used [14].

4 About waste

Regarding the consumption of industrial products in Europe [15] it is stated that there is an increase in demand and corresponding supply of products so that the use of raw materials and waste production increases. The effects on the environment are obvious. Products and systems are becoming more complex, environmental restrictions, resources that are not inexhaustible, product recycling and ensuring a sustainable design are elements without which products can no longer be manufactured. Specialized knowledge and multidisciplinary teams are needed and the concept no longer exists without the EcoDesign side.

Recent initiatives include reducing pollution, waste management, and protecting the environment. These initiatives should not be seen as solutions negative effects already done, but there should be preventive actions to remove or limit the negative effects on the environment. Thus, all actions regarding the life cycle of a product must be treated in an integrated manner not only from a technical point of view but also integrated from an environmental point of view - which is done with the help of ECoDEsign and Design for sustainability. Each of the stages of the life cycle has an impact on the environment: raw materials, manufacturing, distribution, and disposal. The impact can be more or less significant in the long or short term and can occur at regional / local or global level [16].

Here are some of the best known ways to reduce environmental impact [17]:

- Waste reduction: reducing material use through better utilization and by reducing waste during production involves a design with the possibility of extended durability - this being seen as an aspect of quality, involving a decrease in product weight, so that the environment means a decrease in the volume, storage and transport capacity required.

- Materials management: materials recovery, their reuse (recovery must be done considering minimizing all processes related to energy consumption), recycling materials by reusing or reprocessing production waste, products and parts of products, substituting materials for those becoming rare and expensive

- Management of toxic materials:

Pre-processing and reconditioning make significant contributions to effective recycling. Reuse involves retaining the product shape whenever possible. This type of recycling represents a high level of utilization and should therefore be aimed for. Determining the recycling strategy, i.e. if the product or materials from which it is made is recycled, must be the one that will dictate the entire design process.

5 Conclusions

If we reduce the maintenance activity strictly to the interventions to keep the product in working order, it would be reduced to having access / being able to disassemble / repair-recondition-replace / reassemble-test. However, its direct dependence on the way in which the product is constructively thought leads to aspects related to the whole process, from the constructive concept to assembly and disassembly and further to reuse and recycling. At the same time, through its involvement in the possibility of repairing / reconditioning certain parts / modules and even products themselves, maintenance is important in terms of ensuring a design for the environment by extending the life of products, reducing costs and reducing the waste of raw materials, materials and energy.

It should be noted that in all the theoretical description of the design process of a product, in no phase is attention drawn to the need to give it other attributes, such as; repairability, accessibility to troubleshooting, disassembly in order to reuse components or selective recycling. Moreover, the notion of maintenance appears only in the last phase, which will cause delays in the process if there are changes to be made to the final project.

But who uses all the theoretical concepts and prescriptions in the current design of new products? An overview of the new products that have appeared on the market in the last decade reveals that they are increasingly "boxed" in terms of not only the functions but also the actual construction. Most of the new products are either irreparable or usually not removable, forcing the user, in case of failure of one of the functions of the product, to dispense with the entire product. The predominant use of non-removable joints or cast metal or composite materials and injected plastics that have inserts of various metallic or non-metallic materials makes it impossible to disassemble for reconditioning or repair and particularly difficult, or economically inefficient, the effort to separate materials for recovery and reuse. In order to avoid the above cases, at least one new rule should be introduced in the design of new products, namely, the priority use of removable joints based on the classic screw-nut and intelligent coupling devices, coupling based, for example, on materials used elastic characteristics.

On the other hand, the challenges posed by changes in the labor market must be taken into account, where, among other things, the lack of qualified staff for various occupations is already a serious threat and even more so in the field of maintenance where the specific job of the maintenance operator requires as he knows from all technical branches and therefore are perfectly difficult to train and practical experience is a very important factor.

Engineering faculties must train designers with both an open mind and practical skills for repair and reuse. The product design activity must be done with maintenance in mind - both as a design / conception requirement and as an objective to be achieved. Good design and easy maintenance help, on the one hand, to make a product last longer, and on the other hand, it makes it easier to disassemble and sort materials in order to repair, recondition, reuse and finally recycle materials, minimizing both the efforts and the costs and especially the waste of any kind from the resources of humanity-Terra.

References

1. P. Kotler, G. Armstrong, L. Brown, and S. Adam. *Marketing*, 7th Ed. Pearson Education Australia/Prentice Hall, (2006)
2. Definition by the European Foundation for the Improvement of Living and Working Conditions Archived July 20, 2014, at the Wayback Machine
3. S. Terzi et al, Int. J. Product Lifecycle Management **4**, No. 4, (2010)
4. G Pahl. et al. *Engineering Design: A Systematic Approach* (Elsevier P. House, 2007)
5. Standard EN 13306:2001 *Terminologie de la maintenance, (2001)*
6. North Atlantic Treaty Organization, Nato Standardization Agency (NSA) AAP-6- *Glossary of terms and definitions*, p. 43
7. MIL--HDBK--2173 *Department of Defense Handbook: Reliability-Centered Maintenance Requirements for Naval, Aircraft, Weapons systems and support equipment*, (1998)
8. Standard BS EN 13460:2002 *Maintenance- Documents for Maintenance*, (2002)
9. K Takamura, KP Lok, R Wittlinger, *Microsurfacing for preventive maintenance: eco-efficient strategy*, Proceedings of the Forty-Seventh Annual Conference of the Canadian Technical Asphalt Association (CTAA): Calgary, Alberta, (2002)
10. P. Salling et al., Int. J. of LCA, **7(4)**, 203-218, DOI: 10.1007/BF02978875, (2002)
11. S. Takata, F. Kimura et.al. Maintenance: Changing Role in Life Cycle Management Annals of the CIRP, **53 (2)** , 643-655, (2004)
12. S. Backlund, P. Thollander, M. Ottosson and J. Palm, Energy Policy **(51)**, 392-396., (2012)
13. L. Chen J Cai**,** Reliab. Eng. Syst. Saf. **81/2,** 147-154, https://doi.org/10.1016/S0951-8320(03)00075-9, (2003)
14. https://www.relyence.com/2019/07/24/guide-maintainability-prediction/ , Accesed 15 february, 2020
15. L. M. Agudelo Gutierrez, *Decision making support through the consideration of environmental criteria in preliminary design*, ph. D. Thesis, (2016)
16. ISO / TR 14062: 2004 Environmental Management - Integrating Environmental Aspects Into Product Design and Development, (2004)
17. S. Anders S., *Global Life Cycle Impact Assessments of Material Shifts: The Example of a Lead Free Electronics Industry*, DOI 10.1007/978-1-84882-661-8, ISBN 978-1-84882-661-8, (2010)

Effect of machine hammer peening on the surface integrity of a ZnAl-based corrosion protective coating

Alina Timmermann[1,*], *Mohamed* Abdulgader[2], *Leif* Hagen[2], *Alexander* Koch[3], *Philipp* Wittke[3], *Dirk* Biermann[1], *Wolfgang* Tillmann[2], and *Frank* Walther[3]

[1]Institute of Machining Technology (ISF), TU Dortmund University, Baroper Str. 303, 44227 Dortmund, Germany
[2]Institute of Materials Engineering (LWT), TU Dortmund University, Leonhard-Euler-Str.2, 44227 Dortmund, Germany
[3]Department of Materials Test Engineering (WPT), TU Dortmund University, Baroper Str. 303, 44227 Dortmund, Germany

Abstract. Thermally sprayed protective coatings are applied onto many mechanically stressed components such as support structures, shafts, turbine blades or heat exchangers. In addition to the static or cyclic load, a superimposition with corrosion processes occurs in many cases. Thermal sprayed ZnAl coatings are known for their performant corrosion protection properties. Within this context, the potential of ZnAl-based layer systems was analyzed regarding corrosion fatigue behavior. Therefore, a time- and cost-efficient testing strategy based on a corrosion-superimposed load increase procedure was used to estimate the effects of a corrosive attack during cyclic loading. The investigated coating systems were thermally sprayed and partially post-processed with a Machine Hammer Peening (MHP) operation. This treatment was identified as an appropriate technique for compressing and smoothing coated surfaces. The inter-relationships between the parametrization of the MHP process, the resulting surface integrity, and the estimated corrosion fatigue properties were analyzed. The investigations indicate a positive effect of MHP post-processing operations on the surface properties of the ZnAl-based coating system.

1 Introduction

Offshore structures such as offshore wind plants are exposed to strong corrosive influences (see Fig. 1) with difficult conditions for repair and maintenance. In addition to cathodic corrosion protection (protective current), the application of corrosion protection coatings is an option [1-2]. For the design of optimized protective coatings in offshore applications, the consideration of mechanical resistance is necessary due to the influence of the foundation's weight, the water and wind force [3]. Therefore, appropriate measurement techniques are required to enhance the mechanical properties of the corrosion protected structures.

* Corresponding author: alina.timmermann@tu-dortmund.de

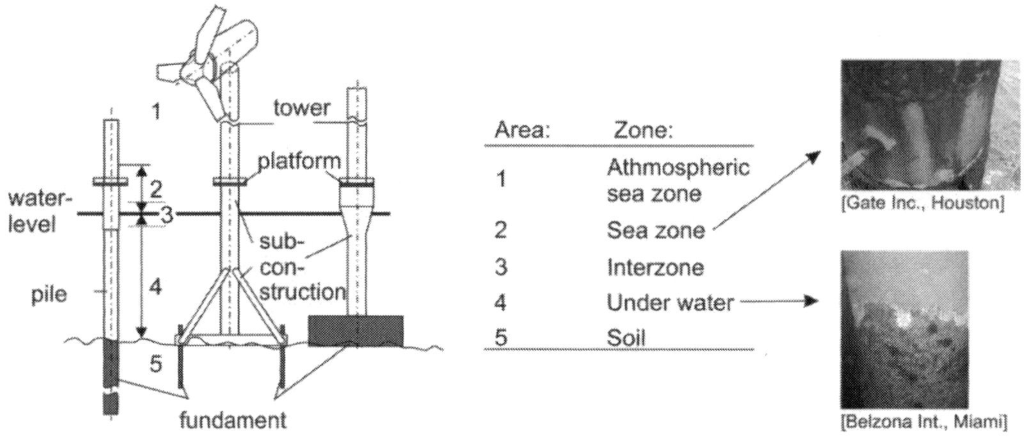

Fig. 1 Corrosion zones of an offshore wind energy plant [1]

ZnAl-based coatings have become an established option to protect steel structures, buildings and offshore wind turbines against corrosion. The superior corrosion protection properties of ZnAl based coatings are mainly assigned to the formation of hydrozincite or simonkolleite, as well as alumina as a resulting corrosion product [1]. ZnAl-based coatings are commonly deposited using the twin wire arc spraying (TWAS) process [4-5]. , TWAS-sprayed coatings are characterized by a heterogeneous lamellar microstructure with an increased amount of pores and rough coating surface. In the as-sprayed state, the obtained coatings also exhibit a high surface roughness, which reduces its resistance against corrosion under cycling loading.

An improvement of the corrosion resistance requires the use of appropriate post-treatments, such as Machine Hammer Peening (MHP). This procedure has demonstrated the potential for densifying the microstructure and smoothening surface asperities of thermally sprayed coatings [6]. In the field of surface engineering using thermal spray technology, MHP presents a novel technological approach for modifying the surface integrity of the coating.

Nevertheless, MHP became an established post-processing method for increasing the mechanical resistance of thermally sprayed wear protection coatings [6-7]. MHP involves hammering against the surface of e.g. a metallic work piece with an impact tool in defined path distances. The hammering induces kinetic energy and thus generates densification and smoothening effects of the coated surface and compressive residual stresses as well as strain hardening effects due to plastic deformation [7-8].

In order to estimate the corrosion fatigue behavior in a time and cost-efficient way, corrosion superimposed load increase tests were used, see [9]. Therefore, different measurement techniques were applied in order to measure the material response on the stepwise increased load and the superimposed corrosion. The material response can give additional information about the fatigue progress, so that a detailed estimation of the fatigue properties, especially the fatigue strength as well as the whole fatigue progress gets possible [10-11].

In this research project, a novel, interdisciplinary approach is investigated to post-process arc-sprayed ZnAl corrosion protection coatings by using MHP. Surface quality and corrosive-mechanical resistance of the generated hybrid material are the main objects of study.

2 Material and experimental procedure

The process chain, which is applied for the experiments (Fig. 2), included the application of the corrosion protection layer, the MHP post-treatment as well as corrosion fatigue tests for the evaluation of surface quality and corrosive-mechanical resistance. Flat specimens were processed for evaluating the surface integrity, while round specimens were used for the fatigue tests.

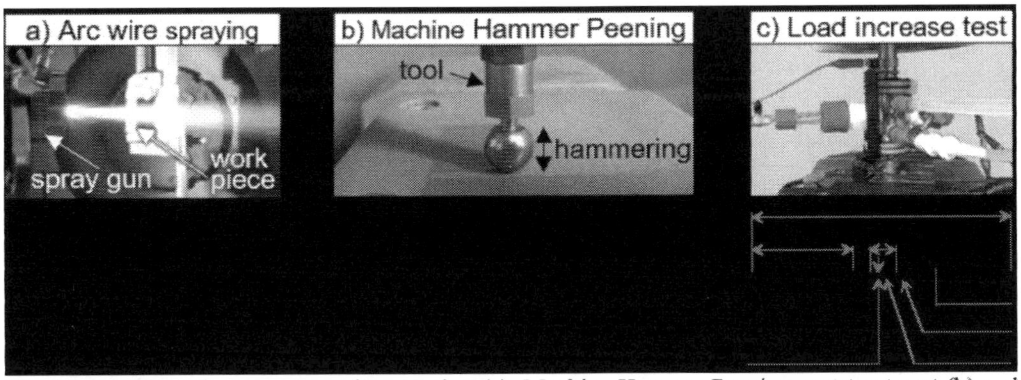

Fig. 2 Experimental set-up: Arc wire spraying (a), Machine Hammer Peening post treatment (b) and load increase test under corrosive conditions with specimen's geometry (c)

2.1 Layer application and post-treatment

The sandblasting pre-treatment of planar C45 (AISI 1045) steel was carried out using alumina oxide powder with EKF 24, a blasting pressure of 4 bar, 100 mm stand-off distance, and a blasting angle of 45°. The TWAS-coating was done using Durum Durspray 450 by Durum GmbH, Zn5Al solid wire as feedstock, and an industrial robot system type IRB 4600 (ABB Ltd) which was used to allow the processing of steady coating thickness.

The pneumatically operated FORGEFix Air tool (3S-engineering GmbH) was applied for the MHP process. The MHP tool with a ball diameter of d_b = 10 mm consisting of cemented carbide, was integrated into the Deckel Maho DMU 50 eVolution CNC machine tool. The process parameters tool feed velocity v_f and the applied tool working pressure p were each varied in two stages.

2.2 Surface measurement and cyclic testing

Shape and roughness of the ZnAl-coated, MHP post-treated surface characteristics were measured using the 3D profilometer InfiniteFocus (Alicona). Cross-sections of the specimens were examined including metallographic shape, porosity and microhardness properties. For this purpose, a confocal white light microscope µsurf (NanoFocus) and the Nanoindenter G200 (Agilent) were applied.

For the investigation of the fatigue behavior under consideration of the corrosive influence, instrumented load increase tests were carried out using a Schenck PSB100 with Instron 8800 controller servohydraulic testing system equipped with a load cell with a load capacity of 100 kN under corrosive conditions in a 3.5 % NaCl solution as well as air for comparison reasons. The specimens were tested using a stress ratio of R = 0.1 and a test frequency of f = 10 Hz with a triangular load-time function at room temperature. Additional measurement techniques such as an extensometer, a thermocouple and a three-electrode arrangement were used to follow the materials responses.

3 Results and Discussion

3.1 Treatment of the ZnAl layer with MHP

Surface appearance as well as averaged roughness depth values Rz are shown in Figure 3 for the as-sprayed condition and the MHP-treated coating with MHP-tool feed velocities of $v_f = 250$ mm·min^{-1} and $v_f = 1000$ mm·min^{-1}. Apparently, profile peaks, total profile height and thus the surface roughness decreased in comparison to the sprayed condition, particularly for a low feed velocity of $v_f = 250$ mm·min^{-1}, probably due to the extended contact time. This tendency could also be observed in the context of the MHP treatment of the wear protection layers [6, 12]. For $v_f = 1000$ mm·min^{-1}, defects in form of delamination of the coating occurred. These results indicate that MHP is suitable for the treatment of the ZnAl coating system, if appropriate parameters are applied.

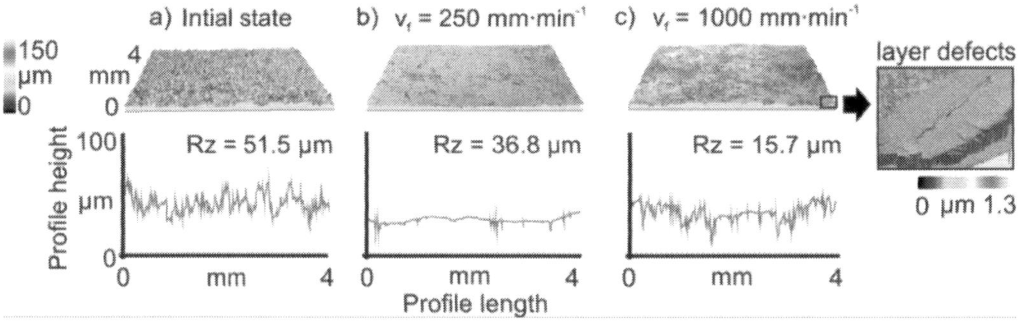

Fig. 3 Surface shape and roughness of sprayed condition (a), sprayed and MHP-treated with feed velocity of $v_f = 250$ mm·min^{-1} (b) and $v_f = 1000$ mm·min^{-1} with a section of layer damage (c)

In Figure 4, cross-sections of the sprayed condition and the MHP-treated ZnAl layer systems under varied tool pressure of $p = 4$ and 6 bar are shown. At low working pressure of $p = 4$ bar, the porosity in the sub-surface was significantly reduced from about 7 to 1 % and the microhardness increased from 47 to 57 HV0.01. The observed compression of the microstructure and increase in hardness is similar to MHP post-treatment results of wear protection layers [7]. At a higher working pressure of $p = 6$ bar, the effective compression depth in the coating layer increased, although the porosity and microhardness is similar.

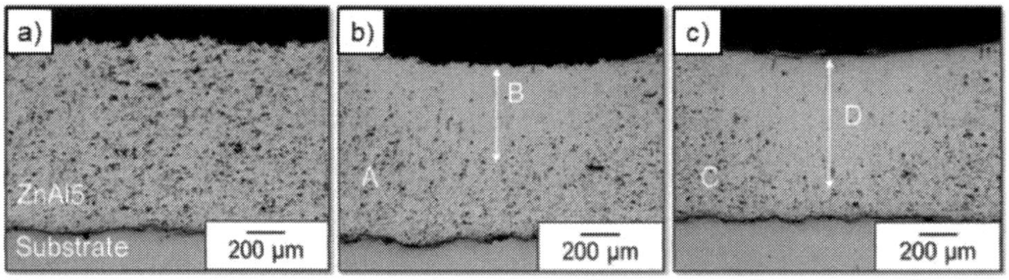

Area	A	B	C	D
Porosity (Po) in %	7.1 ±0.5	0.7 ±0.1	6.2 ±0.6	0.8 ±0.1
Microhardness (MH) in HV0.01	47.6 ±1.4	56.9 ±4.7	47.6 ±2.5	56.1 ±2.4

Fig. 4 Light microscopic cross-section analysis of sprayed condition (a), MHP-treated specimen with low working pressure $p = 4$ bar (b) and high working pressure $p = 6$ bar (c)

3.2 Testing of the corrosive-mechanical resistance

Figure 5a shows the corresponding load increase test (LIT) in a 3.5 % NaCl solution as a function of the number of load cycles N for the ZnAl-coated and MHP-treated specimen with a feed velocity of $v_f = 1000$ m·min^{-1} and a working pressure of $p = 6$ bar. While the plastic strain amplitude, determined from stress-strain hysteresis measurements, represents the purely mechanical part of the corrosion fatigue test, the change in the open circuit potential shows the relationship between corrosion processes and strain behavior, as it refers to the area of active iron dissolution [13].

The stress amplitude was increased in intervals of $\Delta N = 10^4$ cycles by $\Delta\sigma_a = 10$ MPa, starting from $\sigma_{a,start} = 100$ MPa, until specimen failure. The plastic strain amplitude shows values close to zero up to about $N = 28 \cdot 10^4$ cycles, followed by an exponential increase until specimen failure at $N_f = 38 \cdot 10^4$ cycles or a fracture stress amplitude of $\sigma_{a,f} = 470$ MPa. The fatigue strength is estimated based on the first changes in plastic strain amplitude (see [9]) at $\sigma_{a,E} = 370$ MPa. The increase of the open circuit potential at the beginning of the test shows that the passivation properties of the specimen's surface change. The transition to the decreasing section suggests that a critical stress or strain has been reached at which crack initiation in the corrosion protective coating occurs, so that the C45 substrate gets in contact with the electrolyte. The exponential increase towards the end of the test can be explained by the fact that softening mechanisms predominate and macro-crack formation is present in the base material.

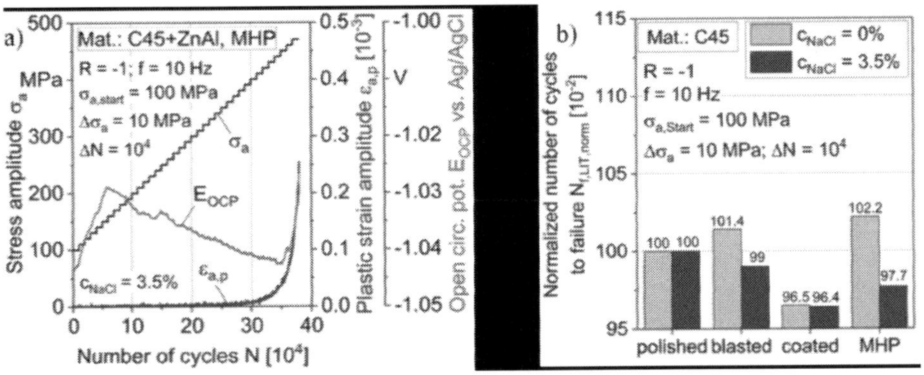

Fig. 5 Load increase test (LIT) of a ZnAl-coated MHP-treated specimen in a 3.5 % NaCl solution (a) and comparison of the lifetime reached in LIT for different treated specimens (b)

Figure 5b compares the normalized number of cycles to failure $N_{f,LIT,norm}$ reached in the load increase tests under different test conditions for uncoated polished or blasted, as well as coated non-treated and MHP-treated specimens. Blasted and MHP specimens in air show slightly increased numbers of cycles to failure compared to the polished state. This can be explained by induced compressive residual stresses of the processes into the specimen, which have a positive effect on the fatigue properties [14]. Regarding tests with superimposed corrosion, all specimens achieve lower numbers of cycles to failure compared to the results in air. The electrolyte thus has an influence on the fatigue properties. The coated specimen fails first, analogous to the investigations in air. This can be explained by the roughness of the coating, which adds possible crack initiation sites to the surface. However, since both values of the coated specimens are at the same level, a protective effect against corrosive attack by the coating can be assumed. In contrast, the lifetime of the blasted and the MHP-treated specimen in the electrolyte decrease in comparison to the investigations in air.

4 Conclusions

In this research, the potential of Machine Hammer Peening (MHP) for post-processing ZnAl corrosion protection coatings with the purpose of improving the corrosion fatigue behavior under mechanical loading was examined. The focus of the investigations was on the created surface quality as well as the corrosive-mechanical properties of the hybrid material evaluated by load increase tests under corrosive conditions.

The test results showed that the MHP post-treatment of the corrosion protection layer ZnAl is applicable without the occurrence of surface defects if appropriate parameters are applied. Reduced surface roughness, less porosity and higher hardness are potential outcomes of MHP. The load increase test for the evaluation of the mechanical behavior with superimposed corrosion showed, based on the measurement of the open circuit potential, an accumulation of microcracks in the coating surface during the test. Due to the increasing electrolyte contact with the substrate, the test specimen subsequently fails earlier. MHP-treatment of the ZnAl layer system increased the resistance to mechanical as well as to corrosive loads in relation to the untreated ZnAl layer. The ZnAl-coated specimens, MHP-treated and non-treated, demonstrated a higher resistance under mechanical, and a lower resistance under corrosive-mechanical load compared to the polished and the sandblasted reference state.

The results described were obtained within the project DFG 426365081. The authors would like to thank the German Research Foundation (DFG) for supporting the research project.

References

1. A. Momber, Materials and Corrosion, **62**, p. 391-404 (2011)
2. VGB PowerTech e.V., Bundesanstalt für Wasserbau, *VGB-S-021-04-2018-07-DE* (2018)
3. A. Hildebrandt, A. Stahlmann, T. Schlurmann, *Hansa*, **2**, p. 35-36 (2010)
4. F. Prenger, J. Spriestersbach, *International Thermal Spray Conference and Exposition*, Houston, p. 104-109 (2012)
5. D. J. Varacelle, D. P. Zeek, V. Zanchuck, E. Sampson, K. W. Couch, D. Benson, G. S. Cox, J. Therm. Spray Technol., **7(4)**, p. 513-520 (1998)
6. J. Wied, Ph.D. Dissertation, Technical University of Darmstadt (2011)
7. D. Biermann, P. Kersting, S. Rausch, *FORM + Werkzeug* **3**, p. 22-23 (2015)
8. F. Lienert, J. Hoffmeister, A. Erz, V. Schulze, *Proceedings of the 12th International Conference on Shot Peening* (2014)
9. P. Starke, F. Walther, D. Eifler, Mater. Test., **60**, 7-8, p. 669-677 (2018)
10. M. Klein, F. Walther, *Materials Science Forum*, **941**, p. 1728-1733 (2018)
11. A. Koch, P. Wittke, F. Walther, Materials, **12(15)**, 2372 (2019)
12. S. Rausch, P. Wiederkehr, D. Biermann, A. Zabel, U. Selvadurai, L. Hagen, W. Tillmann, *Procedia CIRP* **45**, p. 275-278 (2016)
13. J. Riskin, Electrocorrosion and Protection of Metals, **2**, p. 213-223 (2019)
14. G. Luo, Y. Lio, Mar. Struct., **58**, p. 367-381 (2018)

3D-Printed Composite Bone Bricks For Large Bone Tissue Applications

Evangelos Daskalakis[1], *Fengyuan* Liu[1], *Anil* A. Acar[2], *Edera-Elena* Dinea[1], *Glen* Cooper[1], *Andrew* Weightman[1], *Bahattin* Koç[2], *Gordon* Blunn[3], and *Paulo* Jorge Bártolo[1,*]

[1]School of Mechanical, Aerospace and Civil Engineering, University of Manchester, Manchester, UK
[2]Department of Manufacturing Engineering, Sabanci University, Instabul, Turkey
[3]School of Pharmacy and Biomedical Sciences, University of Portsmouth, Portsmouth, UK

Abstract. This study investigates the use of low cost, customizable, biodegradable, polymer-ceramic composite porous structures (bone bricks) for large bone tissue regeneration. Different ceramic materials (hydroxyapatite (HA), β-tri-calcium phosphate (TCP) and Bioglass (45S5) were mixed with poly-ε-caprolactone (PCL). Bone bricks with different material compositions were produced using an extrusion-based additive manufacturing system. Produced bone bricks were morphologically and mechanically assessed. Results allowed to establish a correlation between scaffolds architecture and material composition and scaffolds performance. Reinforced scaffolds showed improved mechanical properties. Best mechanical properties were obtained with PCL/TCP bone bricks and topologies based on 38 double zig zag filaments and 14 spirals.

1 Introduction

Additive manufacturing is a very promising fabrication strategy for the production of three-dimensional (3D) porous scaffods for the regeneration of damaged bones and other tissues [1-5]. These scaffolds must be infection resistant, biocompatible, biodegradable, custom made, cost-effective and modular, contributing to the establishment of a proper biomechanical environment that promotes tissue regeneration [6-8]. These characteristics can be obtained through a proper combination of scaffold design and suitable materials providing mechanical stability and promoting cell attachment, differentiation and proliferation [9-11].

Under an EPSRC/GCRF (Engineering and Physical Sciences Research Council/Global Challenges Re-search Fund) project entitled "Bone Bricks: Low cost effective modular osseointegration prosthetics for large bone loss surgical procedures" we are developing a novel low cost osseointegrated modular prosthetic solution to treat large bone loss injuries (20cm) to enable limb salvage. The immediate application is to treat Syrian refugees who have been displaced to Turkey, but it can be used for injuries made from conflict or natural disasters. The project proposes to build on the current treatment of external fixation but with the addition of an engineered internal prosthetic implant to improve patient outcomes, avoid painful limb lengthening and reduce recovery time. A patient specific prosthetic to fill the bone lost due to injury is being developed using biodegradable and biocompatible modular

[*]Corresponding author: paulojorge.dasilvabartolo@manchester.ac.uk

pieces (bone bricks) from a pallet of shapes and sizes that fit together in a lego-like way to form the prosthesis. The assembled prosthesis will create a hollow cage which will be filled with an infection prevention paste containing calcium sulphate and polymeric microbeads encapsulating antibiotics (Fig. 1). This paper presents preliminary results considering anatomical designed bone bricks produced using different architectures and material composition. Produced structures are morphologically and mechanically investigated.

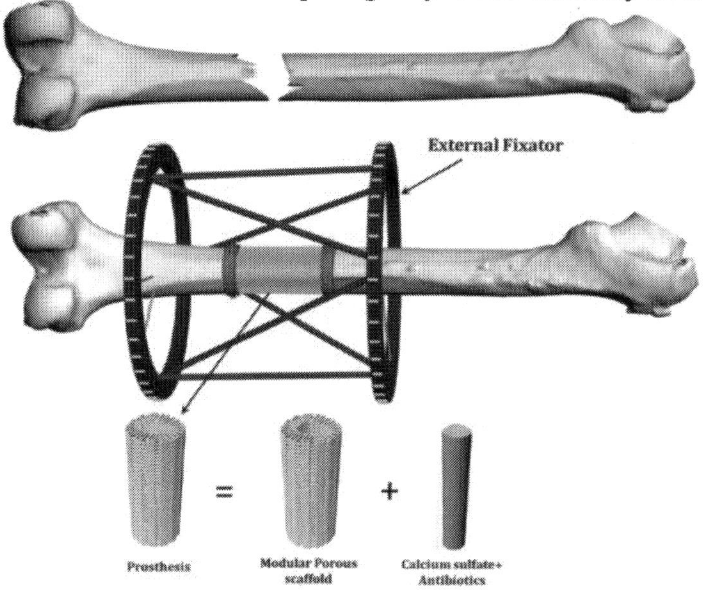

Fig. 1. Bone bricks approach for large bone loss treatment.

2 Materials and Methods

2.1 Materials

Polycaprolactone (PCL) (CAPA 6500, Mw=50000Da), a semi-crystalline linear aliphatic biopolymer, was supplied by Perstorp Caprolactones (Cheshire, UK) in the form of pellets. Hydroxyapatite (HA) (Mw= 502.31 r/mol, MP=1100 °C) was supplied by Sigma-Aldrich (St. Loius, USA) in a nanopowder form (< 200mm particles size), β-tri-calcium phosphate (TCP) (Mw= 310.18 r/mol, MP= 1391 °C) was supplied by Sigma-Aldrich (St. Loius, USA) in a nanopowder form (< 200mm particles size) and Bioglass (45S5), with a composition of 45wt% SiO_2, 24.5wt% CaO, 24.5wt% Na_2O, and 6wt% P_2O_5, was supplied by CeraDynamics Ltd. James Kent Group (Stoke, UK) in nanopowder form (< 0.005mm particles size). PCL composite blends containing different bioceramic contents (20wt% HA, 20wt% TCP and 20wt% Bioglass) were produced by melt blending. Briefly, PCL pellets were melted at 150 °C in a porcelain bowl before adding the ceramic material. Composite materials were mixed around 1 hour to ensure a uniform distribution of the ceramics in the polymer mix.

2.2 Scaffolds Production

Bone bricks were produced using the screw-assisted extrusion based additive manufacturing 3D Discovery system (RegenHU, Switcherland). A computational geometry-based algorithm, with data collected from anthropometric measurements by surgeons in Turkey was used to create a continuous path planning algorithm, using zig-zag (25 and 38 double filaments) and spiral like patterns (9 and 14 filaments) to produce four groups of bone bricks with overall porosity varying between 52% and 68% (Table 1). The process parameters used for the production of the scaffolds were: melting temperature of 90 °C, deposition velocity of 18mm/s and screw rotation velocity of 14 rpm. The diameter of the needle was 0.33mm.

Table 1. Anthropometric-based geometries and different path planning strategies considered to produce bone bricks with different porosities.

Zig Zag Displacement		Spiral Layer
25 Filaments	38 Filaments	
Case 1 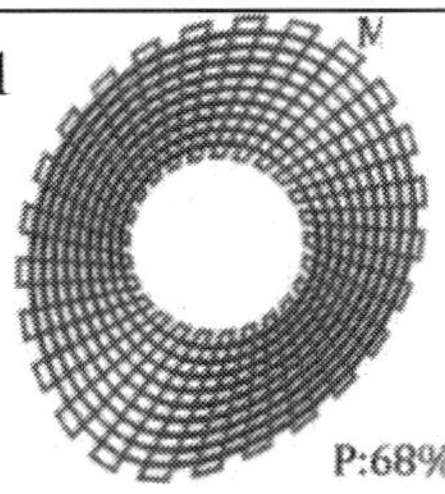	Case 2 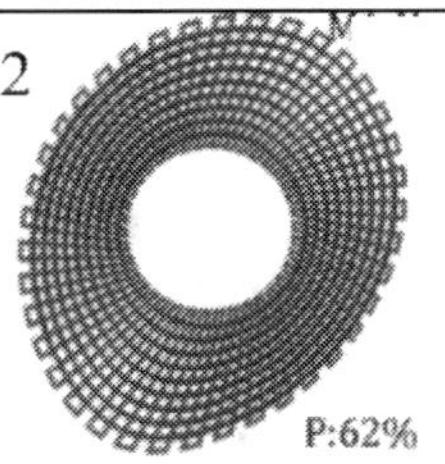	9 Filaments
Case 3 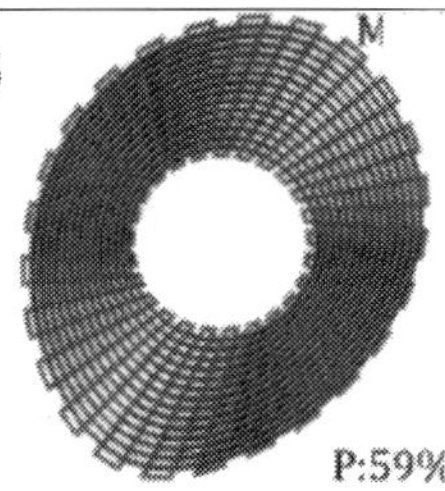	Case 4 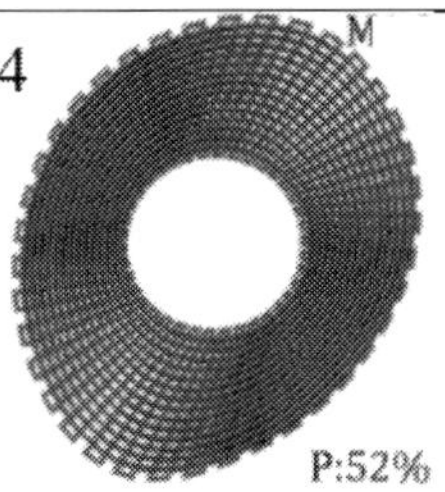	14 Filaments

2.3 Morphological Characterization

The morphology of printed scaffolds was investigated using the Scanning Electron Microscopy (SEM) FEI ESEM Quanta 250 (FEI Company, United States) at an accelerated voltage of 15kV. EMITECH K550X sputter coater (Quorum Technologies, UK) was used for coating the structures (gold coating) prior imaging. The obtained images were processed by ImageJ (NIH, USA) allowing to determine the pore size (PS) and the filament width (FW). For each scaffold 10 measurements were considered to obtain the average and standard deviation.

2.4 Mechanical Characterization

Compression tests were performed on the INSTRON 3344 (Instron, UK) in dry state with a 2kN load cell and a displacement rate of 0.5 mm/min, according to the ASTM D695-15. The Bluehill Universal software (Instron, UK) was used to collect the data and to determine compression modulus.

3 Results

https://doi.org/10.1051/matecconf/202031801009

3.1 Morphological Analysis

Fig. 2 shows high magnification SEM images of both top view and cross-section view of printed bone bricks considering as an example case 2 and all material compositions. Pore size and filament width values are presented in Table 2. Results show that pore size decreases and filament width increase by increasing the bioceramic content. Moreover, for the same configuration and level of reinforcement filament width is higher (lower pore size) in HA bone bricks than in TCP and Bioglass scaffolds. The pore size also decreases by increasing the number of spiral filaments.

Table 2. Morphological characteristics of bone bricks considering different material compositions.

[redacted]

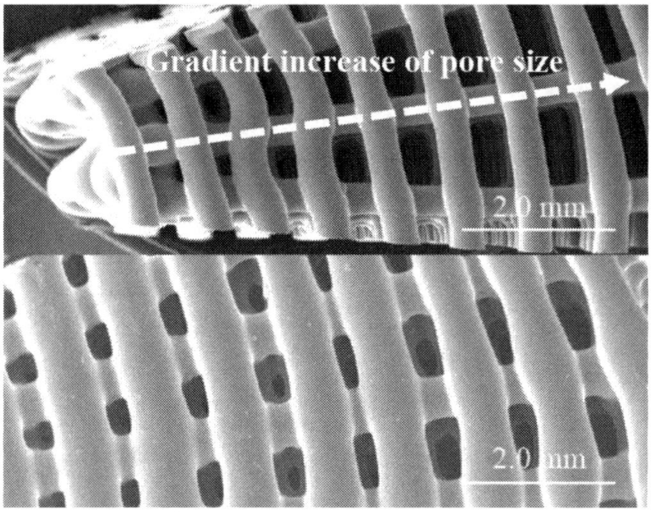

Fig. 2. SEM images of the top view of 20wt% HA scaffolds with different architectures (A) case 1 and (B) case 4.

3.2 Mechanical Analysis

As shown in Fig. 3, the mechanical behavior of the bone bricks strongly depends on the architecture and ceramic content. For the same architecture, the compressive modulus increases by increasing the ceramic content and for the same level of reinforcement the compressive modulus is higher in bone bricks containing TCP than HA and Bioglass. For scaffolds with the same material composition and the same number of double layers, compressive modulus increases by increasing the number of spiral filaments. This can be explained by the overall decrease in porosity. Results also show that by controlling the number of double and spiral filaments it is possible to create scaffolds with compressive modulus in the trabecular region and presenting much higher values than previously reported for standard regular square scaffolds [12].

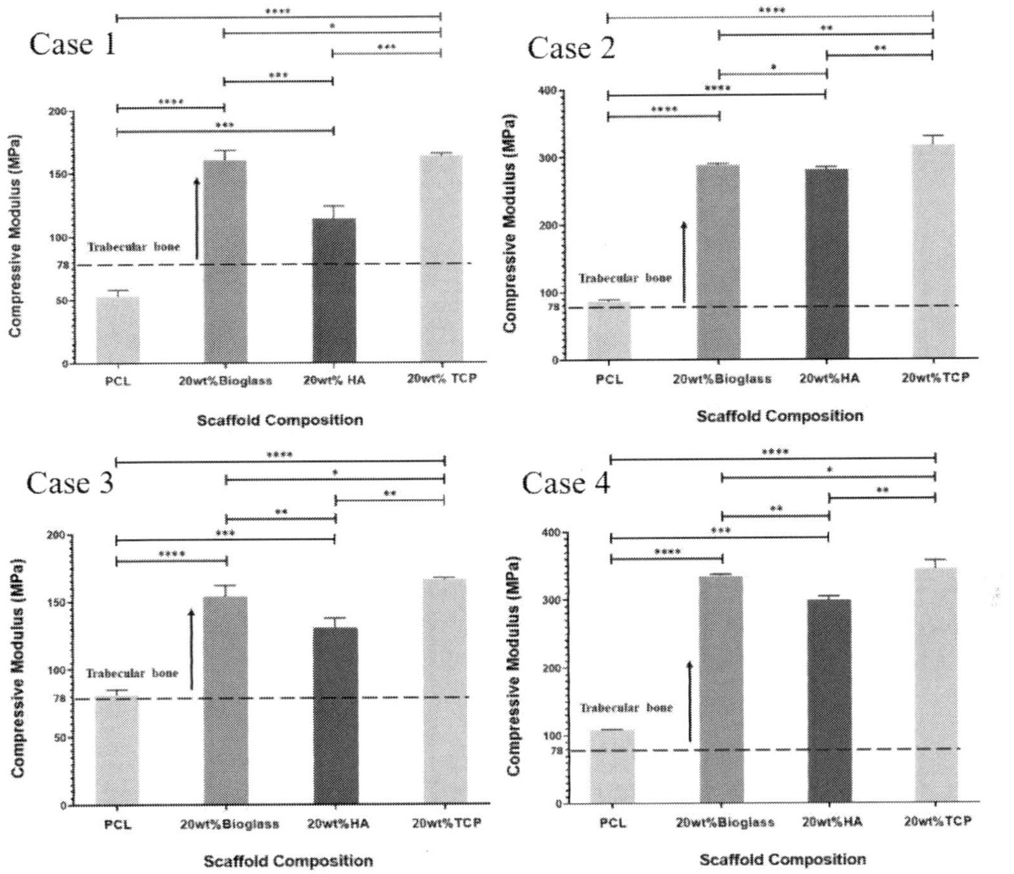

Fig. 3. Compressive modulus as a function of bone brick architecture and material composition. *Statistical evidence ($p < 0.05$) analyzed by one-way ANOVA, and Tukey post-test.

4 Conclusion

This paper investigates the effects of bone bricks architectures and materials composition on the morphological and mechanical properties of printed structures, which were fabricated according to anthropometric measurements. The results show that scaffolds with complex architectures mimicking the patient bone structure were successfully produced using the screw-assisted extrusion based additive manufacturing. Moreover, results show that pore size decreases and filament width increases by adding ceramics into the polymeric matrix, while for the same configuration and level of reinforcement, filament width is higher (lower pore size) in PCL/HA scaffolds than in PCL/TCP and PCL/Bioglass scaffolds. Results also show that scaffolds produced using architecture 2 show improved mechanical properties compared to the other architectures. The addition of ceramics (Bioglass, HA and TCP) enhances the mechanical properties of all bone bricks configurations. Mechanical proprieties increase, for the same concentration of ceramic reinforcements, as follows: PCL/TCP, PCL/Bioglass and PCL/HA.

This project has been supported by the University of Manchester and the Engineering and Physical Sciences Research Council (EPSRC) of the UK, the Global Challenges Research Fund (CRF), grant number EP/R01513/1.

References

1. F. Melchels, M. Domingos, T. Klein, J. Malda, P. Bartolo, D. Hutmacher, Pr. in Pol. Sc., **37**, 1079 (2012)
2. R. Katari, A. Peloso, G. Orlando, Adv. in Sur., **48**, 137 (2014)
3. A. Mao, D. Mooney, Proc. of the Nat. Acad. of Sc., **112**, 14452 (2015)
4. J. Lee, W. Ng, W. Yeong, App. Phys. Rev. **6**, 011307 (2019)
5. W. Ng, C. Chua, Y. Shen, Progr. In Pol. Sc., **97**, 101145 (2019)
6. P. Bartolo, J. Kruth, J. Silva, G. Levy, A. Malshe, K. Rajurkar, M. Mitsuishi, J. Ciurana, M. Leu, CIRP AN, **61**, 635 (2012)
7. F. Liu, B. Huang, S. Hinduja, P. Bartolo, Biocer. And Biocomp.: From Res. To Clin. Pract., 17 (2019)
8. P. Lichte, , H. Pape, T. Pufe, P. Kobbe, H. Fischer, Inj., **42**, 569 (2011)
9. C. Vyas, R. Pereira, B. Huang, F. Liu, W. Wang, P. Bartolo, Cur. Opin. in Biom..Eng., **2**, 1 (2017)
10. B.Guillotin, F. Guillemot, Tren. in Biot., **29**, 183 (2011)
11. S. Murphy, A. Atala, Nat. Biot., **32**, 773 (2014)
12. B. Huang, G. Caetano, C. Vyas, J. Blaker, C. Diver, P. Bártolo, Mat., **11**, 129 (2018)

Preliminary studies on the suitability of PETG for 4D printing applications

Mohamed H.Hassan[1,*] , *Abdalla* M. Omar[2], *Evangelos* Daskalakis[3], *Fengyuan* Liu[4], and *Paulo* Bartolo[5]

[1]Department of Mechanical, Aerospace and Civil Engineering, University of Manchester. mohamed.hassan@manchester.ac.uk
[2]Department of Mechanical, Aerospace and Civil Engineering, University of Manchester. abdalla.omar@manchester.ac.uk
[3]Department of Mechanical, Aerospace and Civil Engineering, University of Manchester. evangelos.daskalakis@manchester.ac.uk
[4]Department of Mechanical, Aerospace and Civil Engineering, University of Manchester. fengyuan.liu@manchester.ac.uk
[5]Department of Mechanical, Aerospace and Civil Engineering, University of Manchester. paulojorge.dasilvabartolo@manchester.ac.uk

Abstract: Shape memory polymers represent a class of polymers that can recover its original shape under stimulus. This paper presents the preliminary results of a broader research programme that investigates the potential use of polyethylene terephthalate glycol (PETG), a glycol modified version of polyethylene terephthalate (PET), for the fabrication of smart bone tissue engineering scaffolds. PETG. A full characterization analysis is carried out, including chemical analysis, mechanical analysis and shape recovery characteristics. Chemical analysis show the presence of terephthalic acid (TPA), ethylene glycol (EG) and cyclohexanedimethanol (CHDM) in the polymer structure. Mechanical analysis, considering a quasi-static tensile test, shows that PETG presents better than PET. Shape recovery is assessed using a cyclic thermomechanical experiment where stress and temperature are controlled during the programming and recovery phases and demonstrates that PETG is able to change and recover its initial shape.

1 Introduction

Smart or intelligent materials (e.g. shape memory alloys and polymers, electro-rheological fluids, piezoelectric materials) represent a special class of materials able to change some of their properties in response to changes in their environmental conditions [1]. External stimuli include changes in temperature, pressure, moisture, pH, electrical and magnetic fields or acoustic stimulation [2, 3]. Due to their potential sensing, actuation, self-cleaning, self-healing, self-diagnosis and shock absorber properties they are promising materials for a wide range of applications such as construction, aerospace, robotics, biotechnology and tissue engineering [2, 4].

*Corresponding author: mohamed.hassan@manchester.ac.uk

In the field of tissue engineering, smart materials present significant potential to design smart three-dimensional (3D) biocompatible and biodegradable porous scaffolds (support structures that provide a proper biomechanical environment for cell attachment, differentiation and proliferation, promoting the formation of a new tissue) providing a dynamic change during tissue regeneration and delivering specific stimulations to cells [5]. Scaffolds have been produced using different 3D Printing techniques and a wide range of biocompatible and biodegradable polymers, ceramics and polymer/ceramic composites [6]. The use of 3D Printing to process smart materials, a novel approach known as 4D Printing (3D fabrication of a product that can be programmed to produce a specific response to a stimulus; time and/or functionality is considered as the fourth dimension) [7], is an emerging topic of research in the field of tissue engineering [8]. However, most studies focused on soft active materials and hydrogels and, consequently, limited to scaffolds for non-load bearing application. In order to address this problem, our group is investigating the potential use of polyethylene terephthalate glycol (PETG) for 4D Printing aiming to develop more effective bone scaffolds. Preliminary results are presented in this paper, which investigates the chemical composition, mechanical properties and 4D behaviour of PETG.

2 Materials and methods

2.1 Materials

PETG with molecular weight of 300 g/mol and printing temperature ranging between 195 °C and 220°C was purchased from RS components (UK).
Chloroform (HPLC grade), phenol, and 2, 5 Dihydroxybenzoic acid (DHB), materials required for sample preparation for matrix assisted laser desorption/ionisation time of flight mass spectroscopy (Maldi-ToF-Ms), were purchased from Sigma Aldrich (UK).

2.2 Fabrication

PETG dog bone structures were designed using Type I specimen dimensions (Figure 1) according to the ASTM D638-14 standard and produced using a filament-based extrusion 3D printer (Flashforge Creator pro, China) with 100% infill. Considered processing conditions are shown in Table 1.

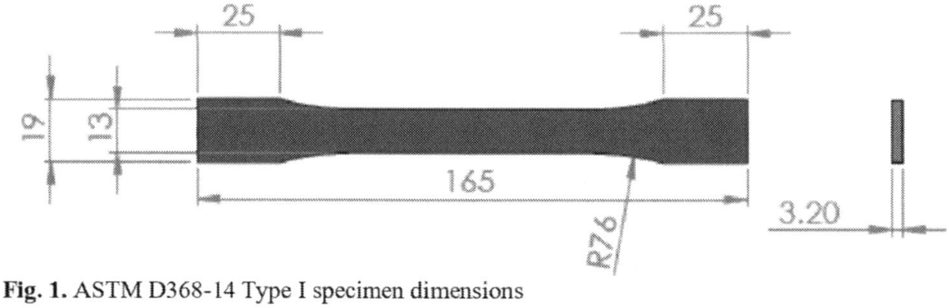

Fig. 1. ASTM D368-14 Type I specimen dimensions

Table 1. Processing conditions

Parameters	Values
Layer thickness (mm)	0.15
Filament diameter (mm)	0.40

Nozzle Temperature (°C)	230
Bed temperature (°C)	60
Printing speed (mm/s)	60

2.3 Chemical characterisation

MALDI-TOF MS, used to characterise the chemical structure of PETG, is based on the use of short, high intensity laser pulses to produce gas ions. In this technique, the analyte molecules are not directly exposed to the laser shots. They are mixed with a matrix of small organic molecules presenting strong absorption of the laser wavelength. The matrix compound, usually weak organic acids, must have high electronic absorption for the used laser wavelength, stable under vacuum, low vapour pressure, good solubility in solvents that dissolve the analyte, and good miscibility with the analyte in the solid state. Examples of commonly used matrices for synthetic polymers are 2,5- dihydroxybenzoic acid, dithranol (1,8,9-trihydroxyanthracene (DHB), 3-β-indoleacrylc acid (IAA), and 2,(-4-hydroxyphenylazo)benzoic acid (HABA) [9, 10].

In this study, PETG was dissolved in phenol/chloroform, 1/1 (v/v), and DHB was selected for the matrix. A dried droplet sample was prepared by putting 1µL of polymer in phenol/chloroform, 1/1 (v/v), and 1µL of a mixture of 1ml of DHB in phenol/chloroform, 1/1 (v/v), as a matrix on the target plate. The test was conducted using the Axima Confidence mass spectrometer Plus MALDI-TOF/TOFMS analyser (Shimadzu Biotech, Japan) equipped with a Nitrogen laser emitting at 337.1 nm allowing data acquisition in positive ion mode. The acceleration voltage used was 20KV and the extraction delay time used was 15000 in reflector MS mode. Data acquisition and data processing were carried out using the Kratos analyser software (Kratos, UK). All mass spectra were collected by averaging the signal of 1620 laser shots.

2.4 Mechanical characterisation

Uniaxial tensile tests were carried out using the SAUTER TVL machine (Sauter GmbH, UK) in dry state with a 500N load cell and two AC 01R fine point clamps fitted to the platform and the loadcell. Five tensile samples were tested up to ~2% strain which is the yield point.

2.5 Shape recovery

Shape recovery was evaluated by heating the printed structure above the glass transition temperature (T_g), apply a force to deform the structure, cooling down the structure below T_g to keep its temporary shape, remove the force, and finally heating up the structure again over T_g allowing the structure to regain its permanent shape [11]. This procedure was repeated considering 5 and 10 cycles to investigate the effect of the cycle number on the shape recovery process. The steps followed to deform the structure and keeping its temporary shape is considered as programming and the following step of reheating the structure to regain its permanent shape is considered as the recovery. The PETG has a glass transition temperature of 75°C, and both programming and recovery were carried out at 80°C [12].

 https://doi.org/10.1051/matecconf/202031801010

3 Results and discussion

3.1 Chemical analysis

The MALDI-TOF MS spectrum of PETG is presented in Figure 2 showing a broad peak in the high mass range, and consequently the structural information cannot be obtained. Results show that PETG has a TPA backbone which is common in other polyesters with different diols which produce different properties. It is also possible to observe the presence of CHDM (not present in PET), which improves the printability of PETG and reduces its crystallinity [9, 10]. In Figure 2 two lines were drawn, a black line corresponding TPA+EG and a blue line corresponding to TPA+CHDM [13, 14]. Besides the TPA+EG and TPA+CHDM peaks the MALDI spectrum contains other peaks that correspond to fragment ions.

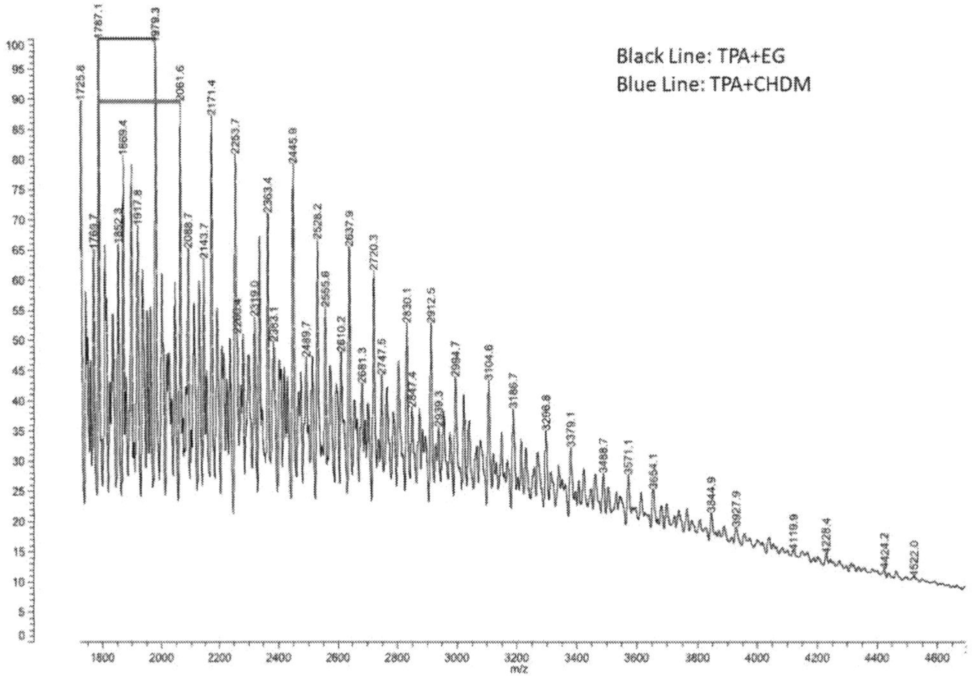

Fig 2. PETG MALDI-TOF MS spectrum where the Y-axis is relative abundance and X-axis is the mass per charge considered as mass in mass spectrometry

Table 2 highlights the constituent compounds that make up PETG. Molecular weight results were determined considering the condensation reaction that produces the oligomers.

Table 2. PETG Chemical components

Compound	Molecular Formula	Molecular weight
Terephthalic acid (TPA)	$C_8H_6O_4$	166.13
Ethylene Glycol (EG)	CH_2OH_2	62.07
Cyclohexanedimethanol (CHDM)	$C_8H_{16}O_2$	144.21

3.2 Mechanical Tensile test

The mechanical tensile test results obtained for the five tested samples are presented in Table 3. As observed the overall elastic modulus is around 1756 MPa, lower to PET due to the reduced crystallinity level, but higher than other polymers commonly used for bone scaffolds such as PCL, and PLA [15, 16].

Table 3. Mechanical properties of PETG

Test	Force (N)	Displacement (mm)	Stress (MPa)	Strain	Modulus (MPa)
Test 1	399.70	1.53	49.94	0.02	1749.41
Test 2	404.30	1.56	42.42	0.02	1726.87
Test 3	397.50	1.58	41.71	0.02	1676.34
Test 4	404.90	1.46	42.49	0.02	1847.89
Test 5	400.10	1.49	41.98	0.02	1789.21
Average	401.30	1.52	42.11	0.02	1756.20

3.3 Shape recovery results

Figure 3 shows the shape recovery of a PETG printed structure in response to heat changes. Figure 3(A) shows the original printed structure before any programming or recovery takes place, Figure 3(B) shows the recovered shape after one cycle of programming and recovery, Figure 3(C) shows the recovered shape after five cycles of programming and recovery, and Figure 3(D) shows the recovered shape after ten cycles of programming and recovery. All cycles were given a period of 10s to recover.

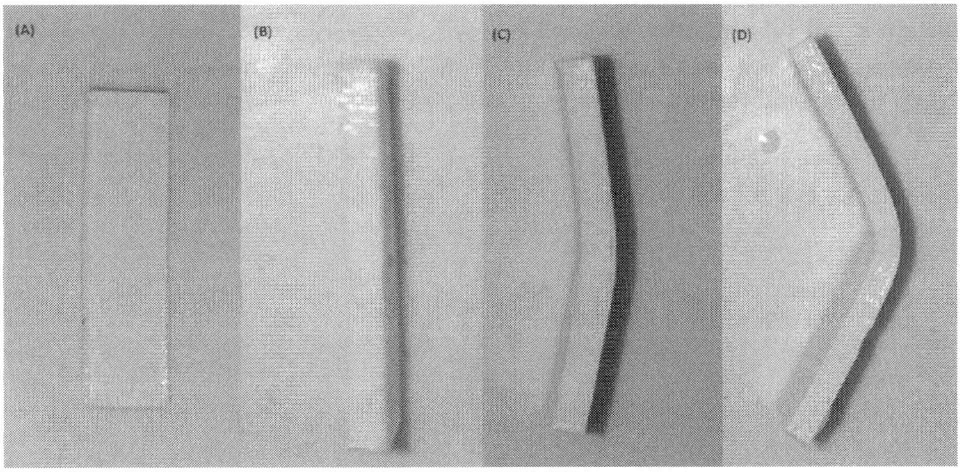

Fig 3. Shaper recovery for PETG using different recovery cycles, (A) original printed structure, (B) shape recovered after 1 cycle, (C)shape recovered after 5 cycles, (D) shape recovered after 10 cycles.

The crystalline and amorphous phases in SMPs determine the shape memory behaviour. The amorphous phase is related to shape fixity and the crystalline phase is related to shape recovery. As previously reported, PETG has a 70%: 30% ratio of amorphous to crystalline phases which explains the deteriorating shape recovery performance as the number of cycles increase [17, 18].

4 Conclusion

This paper shows that PETG is a suitable material for 4D Printing applications. In the case of bone tissue engineering applications, PETG exhibit higher mechanical properties than commonly used polymers such as PCL and PLA [15, 16]. Maldi-ToF MS confirmed the presence of CHDM that is responsible for decreasing PET crystallinity, hence increasing its printability. Further investigation needs to be carried out on PETG considering different factors such as morphological changes with time, potential shape change upon other external stimuli rather than temperature, degradability and biological behaviour.

Reference:

1. J. Kim, Biopolymer Composites in Electronics, p. 311-331 (2017)
2. Z. Zhang, K.G. Demir, and G.X. Gu, Int J Smart Nano Mater,**10**(3): p. 205-224 (2019)
3. M. Bashir, C.F. Lee, and P. Rajendran,J Eng Appl Sci, **12**: p. 50-56 (2017)
4. C.N.Bowman, Polymer,**23**(55): p. 5847-5848 (2014)
5. P. Bártolo, and B. Bidanda, Bio-materials and prototyping applications in medicine,p. 15-26 (2008)
6. R. Dwivedi, S. Kumar, R. Pandey, A.Mahajan, D. Nandana, D.S. Katti, and D. Mehrotra,J.Oral Biol Craniofac Res.**10**(1),9.381-386 (2020)
7. C. Muehlendfeld, S.A. Roberts ,3D and 4D printing in biomedical applications: process engineering and additive manufacturing,pp. 1-23 (2019)
8. E. Pei,and G.H. Loh, Technological considerations for 4D printing: an overview. Prog. in Addit. Manuf.,**3**(1-2): p. 95-107 (2018)
9. G. Montaudo, F. Samperi,and M. Montaudo, Prog. in Poly. Sci. **31**(3),p. 277-357 (2006)
10. T. Wu, H.L. Hu, INT J POLYM ANAL CH, **19**(5): p. 441-452 (2014)
11. M. Heuchel, J. Cui, K. Kratz, H. Kosmella, A. Lendlein,Adv. in Poly. Sci. **51** (26),p.6212-6218 (2010)
12. B. Yang, W. Huang,C. Li,and L. Li , Poly., **47**(4): p. 1348-1356 (2006)
13. S.Turner,J. of Poly. Sci. Part A: Poly. Ch.,**42**(23): p. 5847-5852 (2004)
14. L. Charles, Mass Spec. Rev.,**33**(6),p.523-543 (2013)
15. H.S. Abdo, A.A. Elzatahry, H.F. Alharbi, and K.A. Khalil, Biopolymer Composites in Electronics, p. 13-25 (2017)
16. G. Narayanan,V.N. Vernekar, L.E.Kuyinu, and T.C. Laurencin, Adv. Drug Deliv. Rev.,**107**, p. 247-276 (2016)
17. T. Chen,G. Jiang, G.Li, Z. Wu, and J. Zhang, RSC Advances, **5**(74): p. 60570-60580 (2015)
18. S. Abdullah, A. Jumahat, N. Abdullah,and L. Frormann, Procedia Engineering, **41**, p.1641-1646 (2012)

The Influence of the AWJM Working Parameters on Manufactured Surfaces Microgeometry

Iulian Patirnac[1,*], *Razvan George* Ripeanu[1], and *Eugen* Laudacescu[1]

[1]University Petroleum-Gas of Ploiesti, Mechanical Engineering Department, 100680 Bucharest Blvd., Romania

Abstract. The aim of this paper is to make a research in order to analyze the modification of the microgeometry parameters of AWJ machined surfaces for a ductile material used in oilfield and petrochemical equipment. Experimental research was performed on waterjet cutting machine model YCWJ-380-1520 using pre-establish working conditions. The SURTRONIC 3+ device was used in order to measure the microgeometry characteristics of the machined surfaces. Experimental tests were performed regarding some values for: material thickness, working feed-rate and standoff distance. The obtained outcomes highlighted that there are three zones on cutting surfaces for thick material and only two zones for thin materials. There are a lot of profile parameters provided by the software and there were taken into account only the average roughness (Ra) and average waviness (Wa). For each of those two parameters were accomplished graphical representation for every working parameters in order to establish an analytical dependence between them. The outcomes which were obtained are polynomial four order equations of average roughness and waviness according to standoff distance. The analytical relations obtained give us the possibility to estimate the average roughness of the surfaces related to thickness, feed-rate and standoff distance.

1 Description of AWJ machine and working parameters

The aim of this paper is the study of microgeometry obtained on the kerf surfaces generated by the AWJ process, machining a soft alloyed steel used in oilfield and petrochemical equipment. The study of kerf microgeometry is made in the framework of the experiments using AWJ machine WUXI YCWJ-380-1520, presented in **fig.1**. This machine is made up by the follows: machine table (known as machine bath), CNC controller using CAD/CAM software, high pressure system, cutting head and abrasive particules controling system.

The cutting head of AWJ machine is made up by high pressure nozzle (also called water nozzle), mixing tube and main cutting head body. Regarding the types of the machine active parts, the flowing parameters used in the AWJ prosess are shown in Table 1 and the abrasive material used by WJ machine is Garnet 80Mesh. According to the experiment by measuring the volume of water filled in a specified period of time, it was obtained that the average velocity at the outlet of water nozzle is about 622.35 m/s.

* Corresponding author: ipatirnac@gmail.com

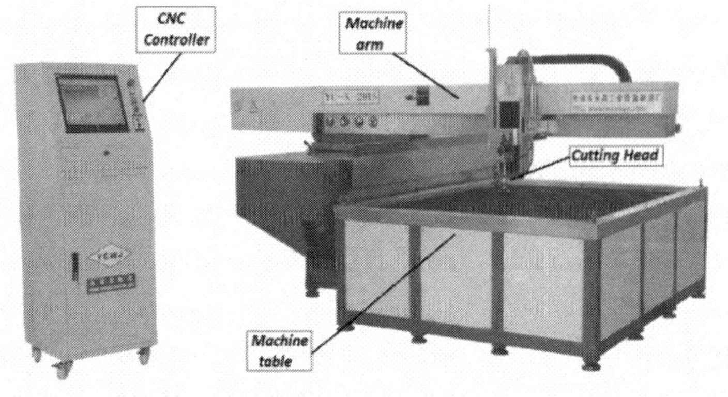

Fig. 1. AWJ machine WUXI YCWJ-380-1520, [1].

At the outlet of the focusing tube, the maximum value of water velocity obtained by CFD simulation on the active parts of AWJ machine is about 187.5m/s. According to some authors [2, 3], the water velocity at the end of the mixing tube is between 120 and 400 m/s, depending on the flow parameters.

Table 1. Flowing parameters used in AWJ process

Water nozzle diameter, [mm]:	0.25
Mixing tube diameter, [mm]:	0.76
Water flow rate, [l/min]:	1.83
Abrasive flow rate, [g/s]:	3.543

The geometry of the material sample subjected to AWJ machining it is presented in **fig.2**. That shape was been adopted taking into account the capabilities of AWJ Machine and the mechanical characteristics of the material subjected to the cutting. The initial raw material was steel plate having 40mm thickness and the two other dimensions were obtained by using cutting conventional machining (milling) to the dimensions L_p x l_p = 280mm x 240mm. In order to highlight the waterjet influence on the material, it was adopted four stages of piece thickness, but due to limitation of material width, it was adopted that every thickness stage to be 58mm width (l_{tr}).

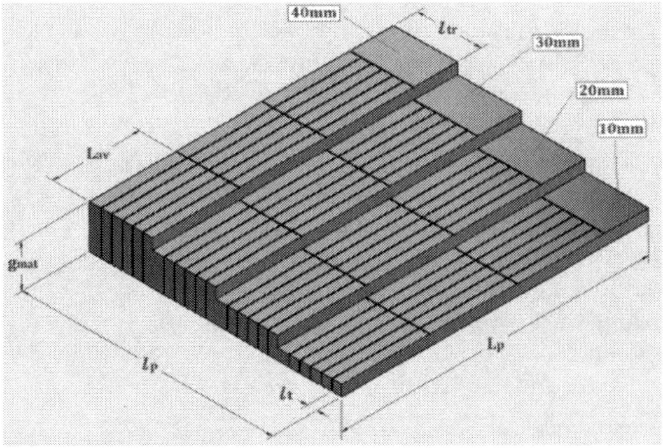

Fig. 2. Schematic of the workpiece machined on AWJ.

In order to emphasize the effects of AWJ technological parameters on the material characteristics and microgeometry of kerfs surfaces, it was adopted the working parameters presented in Table 2. In order that jet of water has no influence on the neighboring kerfs, it was made that distance between them (l_t) to be at least 9mm and the length L_{av} to machine one feed rate stage is set to 80mm.

Table 2. Working parameters used for experiments

Stage thickness, *h* [mm]:	10 ; 20 ; 30 ; 40
Standoff distance δ, [mm]:	1.0 ; 1.5 ; 2.0 ; 2.5 ; 3.0
Working feedrate *f*, [mm/min]:	4 ; 6 ; 9

The material used inside of the experiment is P275NL2, that means a soft alloyed steel used for high pressure equipment, containing 0.17%C and the yield strength is about 275MPa.

2 Investigation of surface microgeometry

It was used SURTRONIC 3+ profilometer in order to measure the kerf microgeometry and the obtained values are processed using TalyProfile Lite 2.1 software. The profile diagram of microgeometry resulted after the program process it is shown in **fig.3**. The 2D parameters can be calculated respectively on the raw profile (P), or after filtering, on the roughness profile (R) or on the waviness profile (W). For each parameter, the type of filtering and the cut-off to be used are defined. According to ISO 4287 standard, some parameters can be defined on sampling length and others on assessment length. The sample length is set on 2.5mm for each measurement.

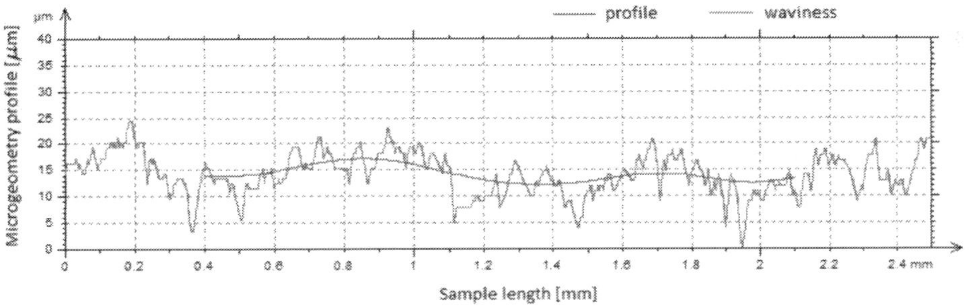

Fig. 3. Profile diagram generated by the software.

There are a lot of parameters which can define a surface microgeometry. In this study are processed only three of them as it follows below, for roughness profile (R) and waviness profile (W). It is made the sense that X prefix is to be replaced by P, R or W according to the needs [4-6].

Xa: arithmetic mean of the assessed profile. It is defined by the expression (1) and the result is expressed in length unit of the Z axis.

$$Xa = \frac{1}{lb}\sum_{0}^{lb}|Z(x)| \tag{1}$$

In the above relation, Z(x) represents the height measured on the vertical axis [µm]; *lb* represents the sampling length.

Xt: total height of the profile on the evaluation length. It has the inconvenience of being very sensible at abnormal points (deep holes or extreme peaks).

Xz: maximum height of the profile within a sampling length. It is defined as: Xz=Xp+Xv, where: Xp represents the maximum profile peak height within the sampling length, and Xv represents the maximum profile valley depth within a sampling length.

In the case of small thicknesses there are two zones on the kerf surface: the superior zone (S) situated at the entrance of waterjet to the material, and the inferior zone (J) situated at the waterjet exit to the material, as can be shown in **fig.4**. At the high thicknesses (over 20mm), it can be distinguished a central zone (C) placed between the two others. According to some authors [7], there are two mechanisms of wear which rule the AWJ process along jet trajectory through the material. So that, the upper zone is characterized by cutting wear and the lower zone by deformation wear.

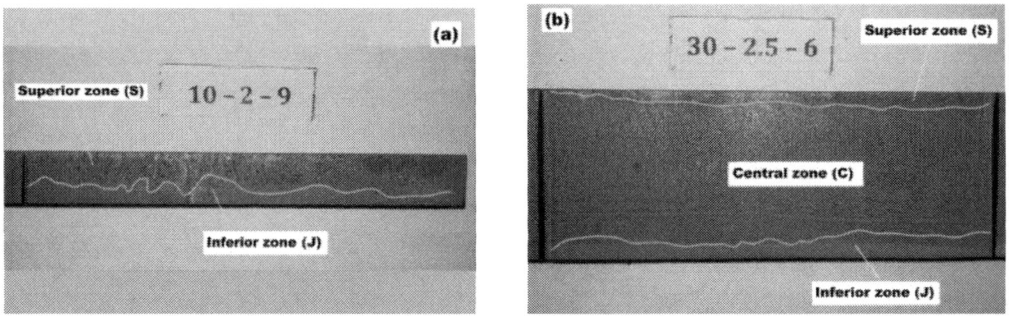

Fig. 4. Kerf surface zones. (a) 10mm thickness; (b) 30mm thickness.

In example, in **fig.5** it is presented the graphical representation of average roughness (Ra) for the upper zone of the kerf surfaces machined with 4mm/min. It can be noticed that all four graphical forms present a waveshape of average roughness of surfaces along to the regarded standoff distances. In this case, the maxim value of average roughness is 5.46 μm and it is obtained cutting the 30mm material thickness at 3mm standoff distance. The lower value is 1.31μm obtained at 1.5mm of standoff distance cutting the same material thickness. The highest values of (Ra) are obtained for 10mm of material thickness and the lowest for 30mm, exception for 3mm standoff where was obtained the maxim value of this case.

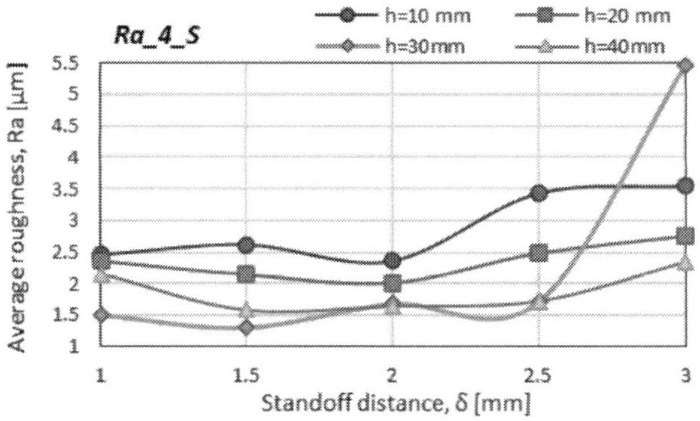

Fig. 5. Roughness (Ra), feed-rate 4mm/min, superior zone (S).

3 Analytical investigation of kerf profile

Due to average roughness (Ra) is indicated on ISO standard drawings, and for a shorter period of lapping (in case of machine parts are manufactured through such a procedure), it is recommended that Ra to be (1.5…2) Wa. It was searched to find some laws to express these two parameters on the entire height of the kerf. Due to there are many zones with different profiles on kerf surface, it was made the average of Ra and Wa using relation (2), where Xa represents the average value of either rouhness or waviness, Xa_i is the roughness or waviness measured in the specified zone on the kerf, n is the number of zones.

$$\mathrm{Xa} = \frac{\sum Xa_i}{n} \tag{2}$$

From above, results that there are 24 charts and equation to debate in that study. In order to simplify the analysis, it is debated only roughness (Ra) and waviness (Wa) it is debated in this paper, both in case of 10mm material thickness and 4mm/min working feed rate, in order to simplify the analysis. The others have a similarity to these two cases. As can be observed in **fig.6**, the average roughness along the height of the kerf has an alternat variation showing two peaks at 1.5mm and 3mm standoff distance. The maximum value in the case of roughness is 2.99μm and it is obtained at 3mm standoff distance. The minimum value is about 2.54μm and it is obtained at 2mm of standoff.

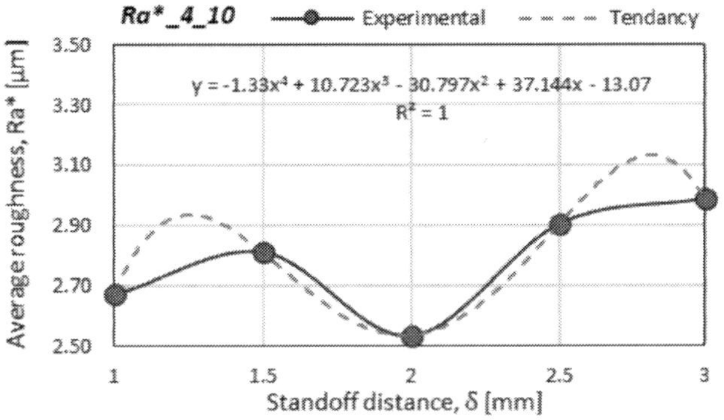

Fig. 6. Roughness distribution for (Ra) 4mm/min working feed-rate and 10mm material thickness.

Any resulted equation was obtained using interpolation method and the factor $R^2=1$ for one another. That means any experimental point is fitted on the trend curve for any standoff distance regarded. As the result of the analyses, both average roughness Ra and average waviness Wa, can be expressed as an equation, generally, written in the form of fourth order equation given by expression (3), for each material thickness and feed rate range.

$$\mathrm{Xa} = \mathrm{A}\cdot\delta^4 + \mathrm{B}\cdot\delta^3 + \mathrm{C}\cdot\delta^2 + \mathrm{D}\cdot\delta + \mathrm{E} \tag{3}$$

In the above expression, Xa represents either average roughness or waviness, δ represents the standoff distance and coefficients *A, B, C, D, E* are given in Table 3, only for 10mm working piece thickness and each feed rate.

Table 3. Coefficients of expression (3) for 10mm material thickness

Feed-rate [mm/min]	Parameter [μm]	Coefficient				
		A	B	C	D	E
4	Ra	-1.33	10.723	-30.797	37.144	-13.07
	Wa	-1.278	10.419	-31.037	40.151	-17.39
6	Ra	-0.5833	5.1567	-16.074	20.891	-6.785
	Wa	-2.4177	19.55	-56.544	68.458	-27.098
9	Ra	0.8333	-6.1267	16.142	-17.883	9.68
	Wa	3.25	-25.583	71.467	-83.134	35.635

Similar to the above table were obtained the coefficients for the other stages like 20mm, 30mm and 40mm. Knowing the laws which rule the surface microgeometry it is created a software application in order to estimate either roughness (Ra) or waviness (Wa). The interface of the program is shown in **fig.7** and was created using Visual Studio 2019.

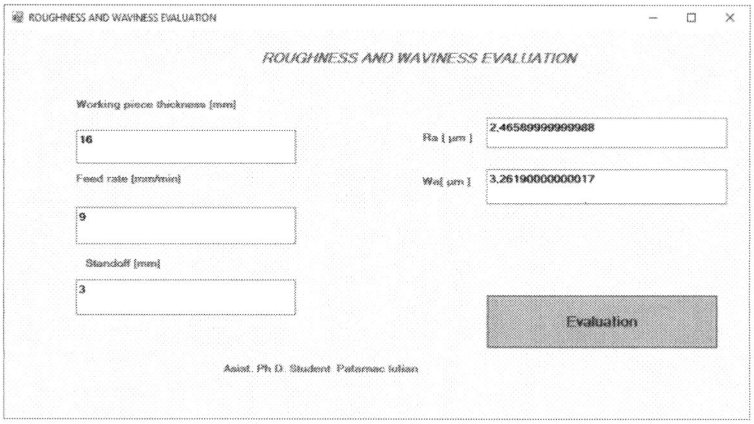

Fig. 7. The panel of the software application.

4 Conclusion

For thin metallic materials (up to 15 or 20mm) machined by AWJ process there are two distinguished zones on the kerf surface, along the jet trajectory. Over 20mm thickness, can be noticed that there are three zones and their areas have a random variation. The maximum values of average roughness (Ra) are obtained on the inferior zone and the minimum values are obtained on the superior zone. According to some authors, the upper zone is featured by a smooth area and the lower zone by a rough surface.

According to the outcomes obtained in this analysis, it can be concluded that average roughness and waviness can be estimated using de equations developed in this paper, inside of AWJ machining process used for the steel P275NL2. With the aid of these equations, it is possible to create a software program in order to calculate either roughness or waviness, for a specified thickness, feed rate or standoff distance, whose values are inside the intervals which are taken into account on this paper.

References

1. https://chinawaterjet.com/ycwaterjet/products/full-set-yc-water-jet-cutting-machine/ (accessed: may, 2019)
2. N. H. Kamarudin, A.K. Prasada, A. Azhari, IOP Conf. Ser.: Mater. Sci. Eng. **114**, 5 (2016)
3. X. Long, X. Ruan, Q. Liu, Z. Chen, S. Xue, Z. Wu, Pow. Tech. **314**, 637 (2017)
4. UNE EN ISO 4287:1999/A1:2010 Geometrical Product Specifications (GPS) - Surface texture: Profile method - Terms, definitions and surface texture parameters - Amendment 1.
5. UNE EN ISO 4288:1998: GPS - Surface texture: Profile method - Rules and procedures for the assessment of surface texture.
6. SURTRONIC 3+ technical booklet
7. G. Cosansu, C. Cogun, J. Mater. Sci. Eng. **26**, 2372 (2012)

Advanced technologies for shoe sole production

Tatjana Spahiu[1,*], *Henrique* Almeida [2,3], *Rita* M. T. Ascenso[2, 3], *Liliana* Vitorino[2], and *Anabela* Marto[2, 3, 4]

[1]Faculty of Mechanical Engineering, Textile and Fashion Department, Polytechnic University of Tirana, Albania.
[2]School of Technology and Management, Polytechnic Institute of Leiria, Leiria, Portugal.
[3] Computer Science and Communication Research Centre, Polytechnic Institute of Leiria, Leiria, Portugal.
[4] Information Systems and Computer Graphics, Institute for Systems and Computer Engineering, Technology and Science (INESC TEC), Porto, Portugal.

Abstract. Advanced technologies for modelling and production are an important part in the whole process of product manufacturing. These advancements have changed the way of product development and play an important role in customization. In the footwear industry, as in any other industry, the use of these technologies is widely spread. Footwear comfort is one of the main selection criteria for purchase. Considering this fact, a case study of different steps for shoe designing according to individual foot shape will be presented. Taking into consideration the aesthetics of the sole and in a more sustainable view, through topological optimization reducing of material wastage for sole production will be presented. By means of the topological optimization in the shoe design process, sole optimization is realized. As a part of personalization, feet's plantar pressure maps taken from 1 participant gave a better explanation of weight distribution of each foot. Following, sole personalization according the plantar pressure maps for each foot gives the possibility to obtain the best least material design according to the feet's pressure while maintaining biomechanical performance.

1 Introduction

Footwear is a garment worn by feet but with high requirements regarding fit and comfort compared to the other type of garments. The feet support us for a long time sowe need to show a great care. As the first step of footwear production, shoes last plays an important role in the fit and comfort and using the right last and materials, shoes with the right fit and comfort are produced. Insoles are one of the most important parts of the footwear as they are in direct contact with the foot. An insole helps to correct any mal-alignment problems and makes it easier to walk comfortably, run easily, and avoid injury [1]. Applications of 3D printing in the shoes industry are of great interest. A researcher conducted a study by

*Correspondingauthor:tspahiu@fim.edu.al

wearing a customized 3D-printed single-sided lateral wedge insole and demonstrated gait rehabilitation [2]. Despite the shape, different materials used for insole production play an important role by affecting ankle dorsiflexion and plantarflexion and maximum knee flexion during walking [3]. Researches done from several authors have concluded that production of anatomical insoles by 3D printing is cost effective and feasible at the industrial level [4], reducing pressure [5] and through the use of 3D scanning making it an affordable option for footwear [6]. With CAD tools designed by a group of authors it is possible to reduce time, increase sole qualityand alsofunctionality. Moreover, 3D printing technology offers new possibilities for manufacturing of complex geometries [7]. The sole is usually made of leather, vulcanized rubber, TPR (thermoplastic rubber), polyurethanes, TPU (thermoplastic polyurethanes) or EVA (ethylene vinyl acetate) [8]. Sole hardness plays an import role in footwear comfort. The tests conducted on the relationship between foot pressure distribution and sole hardness found a great impact on the shoe comfort [9]. Due to irregular geometries and complex material properties of soles, computational modelling as FEM techniques help to study the foot-footwear interaction, where personalization of insoles is more important to reduce peak plantar pressure than the stiffness of the material used for insole production [10]. By mean of reverse engineering, a produced shoe can be used to investigate and suggest improvements of material allocation within the midsole [11]. Results taken from finite element simulations of insole and midsolelayers made from various foam materials as the case of sport shoes optimize the distribution of the elastic modulus in the midsole to improve operability [12] aiming to reduce the Ground Reaction Forces [13]. Moreover, these applications help in controllingsole properties by using lattice structures [14] and through the combination of mechanical meta-material structures to create personalized soles [15] and shoes by 3D printing [16]. Lattice structures are generated by some CAD software's, butthey may present difficulties in the operationalization of the design process. To avoid this, researchers have developed an interface in Visual Basic programming language which is based on the API functions in SolidWorks. This can help to automatically generate complex models of lattice structures [17] and using finite element analysis method to investigate the mechanical properties of each configuration of the lattice structure [18] and produce by 3D printing and investment casting [19]. Another innovative technique as Constrained Free Form Deformation and 3D scanning have shown to have a great impact by accelerating the process of shoe development and production [20]. Since footwear is the interface between the foot and the floor which has relevance for each person, the investigation, understanding and management of foot biomechanics through pressure measurement is mandatory for footwear design. Foot plantar pressure is the pressure field that acts between the foot and the support surface during everyday locomotor activities. Plantar pressure assessment alone is not enough to thoroughly investigate foot biomechanics. Plantar pressure, however, has great potentialities in the field of research and footwear customization [21].

2 A Case Study – Shoe Redesign with Topological Optimization Based on Biomechanical Data

To demonstrate the potential of combining biomechanics with several digital and physical technologies, a shoe design case study is presented. It focuses on a sole model that was previously designed in Shoemaker software as depicted in figure 1.In a previous work the numerical simulation of sole models was realized. In this case without taking into considerations the feet pressure maps [22]. But considering the biomechanical factor that each person presents asymmetric behaviourbetween both sides, the biomechanical study began with the definition of the feet's plantar pressure in order to understand the differences between both the left and right foot.

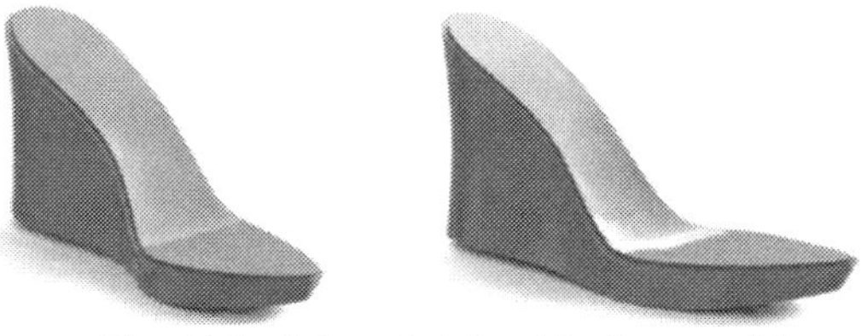

Fig. 1.The left and right side shoe models.

In figure 2 is illustrated the scanning procedure and the obtained plantar pressure plot and values. The hardware system used is a high speed 2m Hi-End footscan® system and the software used is the Footscan® 9, both from RSSscan Systems. The test was repeated 3 times for the patient.

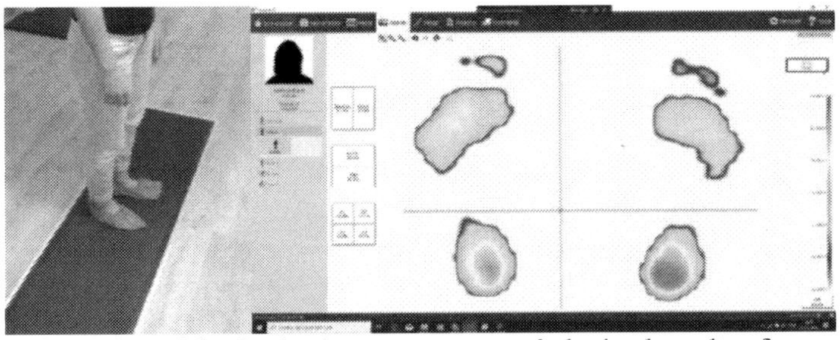

Fig. 2.Scanning of the feet's plantar pressure and obtained results ofpressure plot.

The next step consisted in creating the curves for the generation of the surfaces of the pressure areas on the soles for the definition of the loading conditions for both the structural simulations and the topological optimizations. With the aid of CAD software (SolidWorks 2019), the plantar pressure plot was imported into the software and the necessary scaling of the image was performed in order to correspond with the patient's shoe size of 37. Also, both the left and right CAD soles were orientated to correspond both feet with the plantar pressure plot. Figure 3 illustrates the orientation of both soles and the pressure curves used to create the surface areas according to the plantar pressure plot.

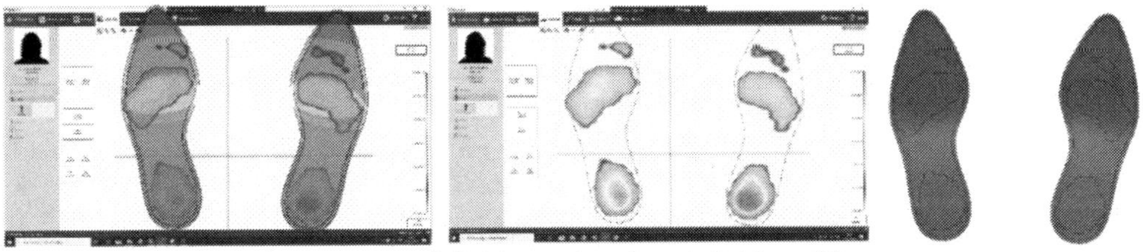

Fig.3. Image import, curve definition and surface definition of the pressure areas.

To begin the engineering design phase, the first step consisted in performing the structural simulations in order to validate the mechanical performance of the designed sole. The patient's data was used in the simulations to define the loading conditions for both the left and right side. The previous designed pressure areas along with the values of the pressure plots and distributed weight is presented in figure 4. The patient presents a weight of 51 kg with a difference of 5.6% in weight distribution between both the left and right side. The fixed constraints were defined on the lower surface of both sole models which were meshed with the same mesh size in order to perform an adequate comparison between both models. Considering that Additive Manufacturing is to be used for the fabrication of the shoe soles in an affordable system to guarantee that the price of both soles is similar when compared to other regular designed shoes in the market, the material selected was TPU 95A from Ultimaker. The results of the structural simulations are presented in figure 5 and demonstrate that the designed shoe can withstand the patient's weight. Alsoit is visible the influence of the patient's weight distribution in both the stress and displacement results.

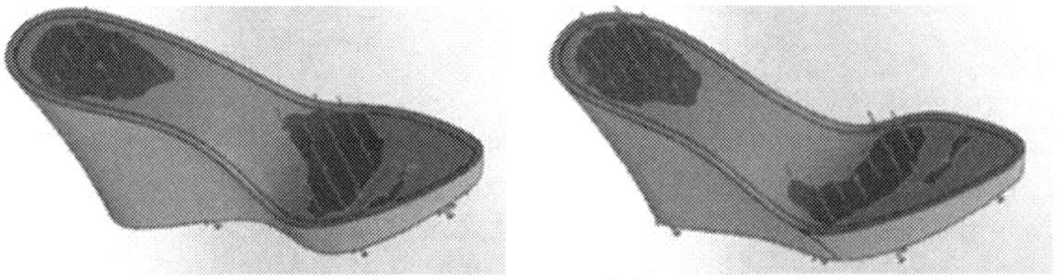

Fig. 4.Loading and constraint conditions for both shoe soles.

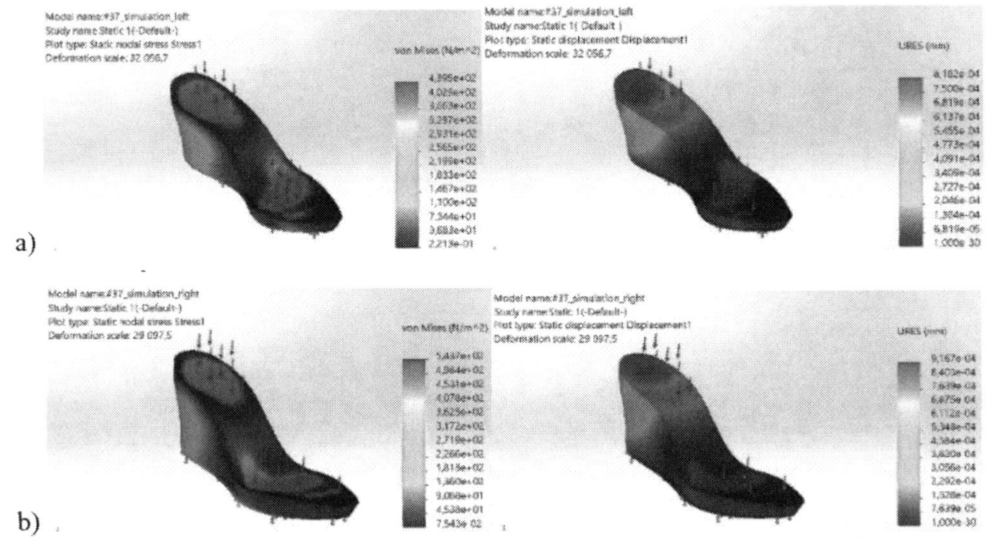

Fig. 5.Stress and displacement results for both feet: a) left and b) right side.

After performing the structural simulations and validating both the design and material selection, the topological optimisation was then performed based on a mathematical method that optimizes the material layout within a given design space, for a given set of loads, boundary conditions and constraints with the goal of maximizing the system performance. The material percentage reduction for both soles was of 25 % and the area selected in which the topological optimisation will occur is the centre core of the shoe sole. Figure 6 illustrates the initial model and then the optimised topological design of the centre cores for both the left and right sides. Similar, to the structural simulations, again the optimised designs also present differences between them.

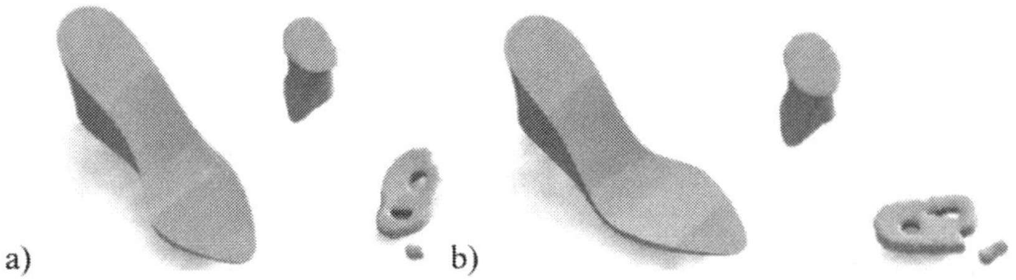

Fig. 6.Topological optimisation solutions of the centre sole for the a) left and b) right sides.

Figure 7 illustrates the final topological optimized design for both the left and right shoe soles.From the work presented, it is possible to observe that engineering design schemes may be introduced in shoe design with an additional characteristic of structural simulations in order to predict the product's mechanical performance. To end the product design cycle, the following step is to produce both designs for consumer approval, which may suffer new iterations according to the consumer's preferences.

Fig. 7.The left and right side topological optimized shoe models.

3 Conclusions

Personalisation helps to improve the efficiency and reduce costs, by offering the companies a sustainable growth with differentiated products and services in a profitable manner. This is already possible due to the advanced digital and physical manufacturing systems and distribution technologies. In this paper, a case study focused on the design of a shoe design considering the biomechanical asymmetry of the consumer. After obtaining the 3D models and the biomechanical data, engineering design schemes was implemented, namely topological design. From the framework presented, it is possible to observe that engineering design schemes can be introduced in shoe design with an additional characteristic of structured simulations in order to predict the product's mechanical performance while maintaining a tailored aesthetic appearance and most important a biomechanical performance.

References

1 Y. Urabe, N. Maeda, S. Kato, H. Shinohara and J. Sasadai, J Phys Fitness Sports Med. **3**, no. 4, 385-398, doi.org/10.7600/jpfsm.3.385, (2014).

2 H. Jin, R. Xu and J. Wang, Med Sci Monit. **25**, 25:7720-7727, doi:10.12659/MSM.919400(2019).

3 R. Ozmanevra, S. Angin, I. H. Gunal and A. Elvan, Am Podiatr Med Assoc. **108**, no. 5,390-396,doi.org/10.7547/16-175, (2018).

4 M. Davia-Aracil, J. J. Hinijo-Perez, A. Jimeno-Morenilla and H. Mora-Mora, Computers in Industry **95**, 38-53, (2018).

5 S. Telfer, J. Pallari, J. Munguia, K. Dalgarno, M. McGeough and J. Woodburn, BMC Musculoskelet Disord. **13**, no. 84, doi.org/10.1186/1471-2474-13-84, (2012).

6 A. S. Salles and D. E. Gyi, J Sports Sci. **31**, no. 4: 442-450. doi:10.1080/02640414.2012.736629, (2013).

7 M. Davia-Aracil, A. Jimeno-Morenilla and F. Slas, Int J Adv Manuf Technol. **86,** 3495–3516, doi.org/10.1007/s00170-016-8427-5, (2016).

8 S. S. Muthu, S. Limited and P. R. China, "The environmental impact of footwear and footwear materials," *Handbook of footwear design and manufacture*, A. Luximon, The

Textile Institute and Woodhead Publishing, (2013).

9 X. Xie, Leather Footwear J. **18**, no. 4, 249-258, doi: 10.24264/lfj.18.4.1, (2018).

10 J. T-M. Cheung. M. Zhang, Arch Phys Med Rehabi. **86**, no. 2, 353-358, doi: 10.1016/j.apmr.2004.03.031, (2005).

11 D. Drougkas, E. Karatsis, M. Papagiannaki, S. Chatzimoisiadis, F. Arabatzi, S. Maropoulos and A. Tsouknidas, Appl Bionics Biomech, **9**, doi.org/10.1155/2018/6520314, (2018).

12 K. Shimoyama, K. Seo, T. Nishiwaki, S. Jeong and S. Obayashi, J Sport Eng Technol. **225**, no. 4, 179-188, doi.org/10.1177/1754337111414485, (2011).

13 M. R. Shariatmadari and R. English and G. Rothwell, "Finite Element Study into the Effect of Footwear Temperature on the Forces Transmitted to the Foot during Quasi-Static Compression Loading," *IOP Conf. Series: Materials Science and Engineering*, doi.org/10.1088/1757-899X/10/1/012126, (2010).

14 D. Guoying, T. Daniel and Z. Y. Fiona, "Design of Shoe Soles Using Lattice Structures Fabricated by Additive Manufacturing," *22nd International Conference on Engineering Design*, Delft, The Netherlands, doi.org/10.1017/dsi.2019.76(2019).

15 D. J. Nogueira Amorim, T. Nachtigall and M. B. Alonso, "Exploring Mechanical Meta-Material Structures through Personalised Shoe Sole Design," *Symposium on Computational Fabrication (SCF '19)*, Pittsburgh, PA, USA, doi.org/10.1145/3328939.3329001 (2019).

16 T. R. Nachtigall, S. Wensveen, O. Tomico, P. Van Dongen, R. Wakkary and L. T. Van Noorden, "Towards Ultra Personalized 4D Printed Shoes,"*CHI Conference on Human Factors in Computing Systems*, Montreal QC Canada, doi.org/10.1145/3170427.3174369, (2018).

17 D. S. Nguyen, T. Tuan Tran, D. Kien Le and V. Than Le, "Creation of Lattice Structures for Additive Manufacturing in CAD Environment," *International Conference on Industrial Engineering and Engineering Management*, Bangkok, Thailand, doi: 10.1109/IEEM.2018.8607271 (2018).

18 D. S. Nguyen and T. A. Nguyen-Van, J. Korean Soc. Precis. Eng. **37**, no. 4, 305-318,doi.org/10.7736/JKSPE.019.100 (2020).

19 G. Maliris, I. T. Sarafis, T. Lazaridis, A. Varoutoglou and G. Tsakataras, "Random lattice structures. Modelling, manufacture and FEA of their mechanical response," *IOP Conf. Series: Materials Science and Engineering* **161**, doi:10.1088/1757-899X/161/1/012045(2016).

20 L. Orazi and B. Reggiani, Int J Adv Manuf Technol, **106**, 2295-2303, doi.org/10.1007/s00170-019-04717-8, (2020).

21 F. T. Piller &. M.Müller, Int. J. Comput. Integr. Manuf. **17**, no. 7, 583–593, doi.org/10.1080/0951192042000273140,(2004).

22 T. Spahiu, E. Piperi, A. Ehrmann, H.A.Almeida, R.M.T.Ascenso, L.C.Vitorino,"3D Printing: An Innovative Technology for Customised Shoe Manufacturing",*Progress in Digital and Physical Manufacturing,* 171-180, doi.org/10.1007/978-3-030-29041-2_22, (2019).

Functional Analysis Diagram of a Flux Modulated Magnetic Gear

Gheorghe Galmadi[1,*] and *Oana* Dodun[1]

[1]Gheorghe Asachi Technical University of Iaşi, Department of Machine Manufacturing Technology, 59A, Dimitrie Mangeron Bldv., Iaşi, 700050, România

Abstract. The reliability of a system is generally done by a functional analysis (FA), whose mission is to provide a deep understanding of the object to be designed, to discover all the valuable solutions, modes, and configurations, and finally to choose the best options for the system's design. Functional analysis is then matched with a dysfunctional analysis, such as Failure Modes and Effects Analysis, to foresee and take measures to avoid the possible dysfunctionalities of the system and to determine the most suitable model. In this paper, the functional analysis diagram (FAD) is used in the conceptual phase of the design process of a Flux Modulated Magnetic Gear.

1 Introduction

Magnetic Gear (MG) are used to transmit torque without mechanical contact between two shafts and resembles the traditional mechanical gear in geometry and function. The magnetic coupling uses a magnetic field instead of a physical-mechanical contact. The MG main feature consists of no-wear of the shaft seals or gear teeth surfaces, which means enhanced reliability of the device. More of that, the magnetic transmission does not require lubrication and has less noise in operation, the transmission has no local heating, no friction losses or risk of breaking the elements.

Mechanical gears have many industrial applications for power transmission. Even though their use is important, they have some inherent problems such as contact, overheat, friction, noise, and wear, which are often damaged in an over-torque situation and require periodic lubrication and maintenance. Magnetic gears can be solutions for these mechanical problems. Therefore, in recent years, magnetic gears (MGs) are increasingly receiving attention from both research institutions and industry; many authors have proposed several new magnetic gears to overcome these drawbacks. MGs offer lighter weight, reduced maintenance, high reliability, no wear and tear, contact-free operation, no noise and, higher efficiency.

However, magnetic gears have not attracted much attention since their initial development because of the low torque capability caused by the poor utilization of the permanent magnets [1]. In the past decade, a more efficient topology for magnetic gears was proposed in [2,3]. By adopting a coaxial arrangement of two permanent magnets

* Corresponding author: oanad@tcm.tuiasi.ro

rings, all of the permanent magnets could contribute to the torque transmission simultaneously; hence a torque density exceeding 100 kN·m/m^3 could be achieved [2].

This paper presents the concentric Magnetic Gears with a flux modulator, which can compete against certain mechanical constructions. The performances of different topologies and classification for the flux modulated magnetic gears available today are presented. The functional analysis methodology is used in order to achieve the necessary parameters of a certain magnetic gear for a certain purpose.

2 Classification, characteristics, and performances of a concentric magnetic gear

This section is dedicated to describing the most important concentric magnetic gears with flux modulator topologies that are available nowadays and present the best characteristics, have a buildable construction design, and were experimentally tested. Although there are many available topologies for which were obtained good characteristics after simulations, they do not have a practical design and applications because of difficulty or even impossibility to build them experimentally. These types of MG are much simpler to realize, considering the mechanical construction. Various aspects of this type of gear have been investigated.

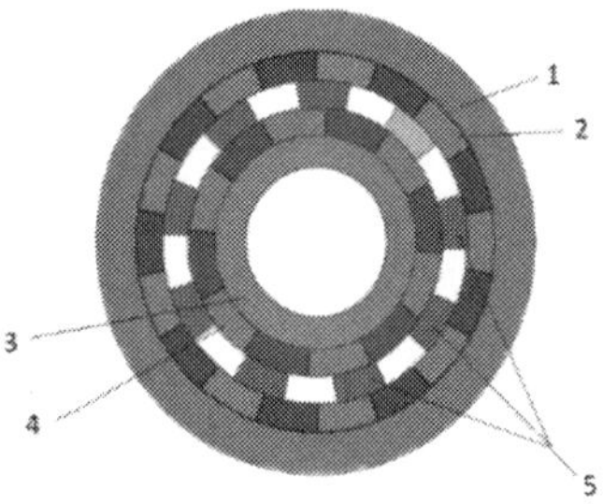

Fig. 1. Constructive scheme of a coaxial magnetic transmission with flux modulator:
1,2 - magnetic core and radial PM of the stator;
3,4 - Magnetic core and PM of the high-speed rotor;
5 - ferromagnetic pole-pieces of the low-speed rotor.

Based on the orientation of magnetic flux, concentric MGs can also be divided into three categories namely:

- radial flux magnetic gear or coaxial magnetic gear (CMG) [2,3];
- disc type or axial flux magnetic gear [4];
- and linear or transverse flux magnetic gear [5,6].

Permanent magnets (PM) play an essential role in MG construction. Earlier MG designs suffered from poor torque producing capabilities of the available PMs at that time. With the realization of high-performance rare-earth magnets in the 1980s, there was a rapid positive shift in research and development in MG technology. Different types of MGs, which employ permanent magnets, can be divided into two big categories availability:

- MG with Non-Rare earth magnets: with Aluminium-Nichel_Cobalt (AlNiCo) permanent magnets or with Ceramic (Ferrite) PMs
- MG with Rare earth magnets: with Smaaium-Cobalt (SmCo) PMs or Neodymium-Iron-Boron (NdFeB) PMs

NdFeB PMs possess high magnetic strength but have low mechanical strength [7]. Due to its high magnetic strength, NdFeB was extensively used in MG research despite its high price.

The basic operation principle of a concentric MG is that the ferromagnetic pole-pieces interposed between the inner and outer PM rotors modulate the magnetic field such that each rotor "sees" a working space harmonic corresponding to its number of poles (Figure 1). The flux modulator is the key to the operation of concentric type MGs. It operates as either a low-speed rotor (high torque) or the mechanical earth of an MG.

In the absence of an intermediate ferromagnetic wheel, magnetic systems of external and internal magnetic rotors create magnetic fields with a different number of pairs of poles, and stable power transfer between them is impossible. When an intermediate magnetic rotor (modulator) is installed between the external and internal magnetic rotors, the magnetic field created by the external magnetic rotor will change its harmonic composition due to the presence of soft magnetic material segments in the modulator.

Different researchers have realized several prototype flux-modulated gears until now. Table 1 lists some examples, with their outer diameter (Do), gear ratio (Gr), and active volume torque density (T/VA) reported. Although the majority of experiments could not exceed 100 kN·m/m^3, in recent years, a few high torque magnetic gears have been experimentally obtained.

Table 1. MG performances from prototypes elaborated by different researchers.

Authors	Outer Diameter D_0 [mm]	Gear ratio G_r	Torque density T/V_A [kNm/m^3]
Atallah [2]	140	5,5	72
Molokanov et al. [8]	232	11.5	44.4
Rasmussen et al. [1]	120	5.75	92
Frank and Toliyat [9]	120	5.5	42
Valiente-Blanco [10]	~50	10	95
Uppalapati et al. [11]	112	4.25	33
Gerber and Wang [12]	150	10.5	62
Fukuoka et al. [13]	156	10.3	49
Shah et al. [14]	120	5.5	56
Bird et al. [15]	220	4.25	239

3 Functional Analysis of a concentric magnetic gear

Functional Analysis (FA) is a fundamental tool of the design process to discover new meaningful concepts, to define their structure, and to have a deep understanding of the researched object [16]. FA is used in engineering to define the functional requirements of the new product, to map its functions to physical components by listing all the necessary components, and, where is needed, by eliminating the unnecessary ones, a method that will help designers to understand the relationship between product's components.

Functional analysis is particularly useful during the conceptual design phase, as it provides a clear understanding of functions related to energy, materials and signals and can help choose the best option among a wide range of feasible solutions discovered in the process of exploring ideas that may offer significant advantages.

The method is also used to define material limits, the operations, and functions performed by the system and its configurations. A practical way of using FA is by visually representing the decomposed system, braking its functions down into sub-functions, providing a format, and a specified set of rules to perform FA of all functions of the system [17].

Functional analysis is performed in two stages [18]: external functional analysis which shows the relations between the system and its external environment and internal functional analysis which perform breakdown of the system into elements and describes the functions of the system but also the relationships between the components and therefore the physical interactions between the different entities of the system.

Although FA is relatively easy to use and its application is similar to any system, its steps are adaptable and defined by the various aspects of the analyzed object. A common set of steeps are described in [19]: identification of system states and modes; development of the functional tree of all functions the system; identification of flow relationships across functions; identification of function interfaces; allocation of requirements to functions; identification of function time synchronization requirements.

The researches on functional analysis led to development of such representation methods as Function Tree, Data Flow Diagram, Function Structure and FAST Diagram [20].

A form-dependent method for functional representation is the function analysis diagram (FAD) which uses blocks to represent device or system structure and arrows with labels to represent functional relations between components. FAD originated in Invention Machine Goldfire software based upon TRIZ methodology [17] as a tool which visually represents the decomposed system, providing the space and a specified set of simple rules to perform a functional analysis of all functions of a system at any level [20].

FAD orders the functions in a logical order, it contributes to the clarification of the functional state of the product and it allows to order the functions identified by the designer or the analyzer, to have a good knowledge of the product or system studied thus become aware of the relative importance of elements or structures in relation to the functions they perform which results in synchronizations between independent functions.

The methodology is used to examine external and internal functions concerning a potential magnetic transmission of an electric car. External Functional Analysis displays the relation between the transmission and the systems to which it is connected, in our case, the electrical engine and the wheel, and the internal FA lists all the functions of the magnetic transmission (Figure 2).

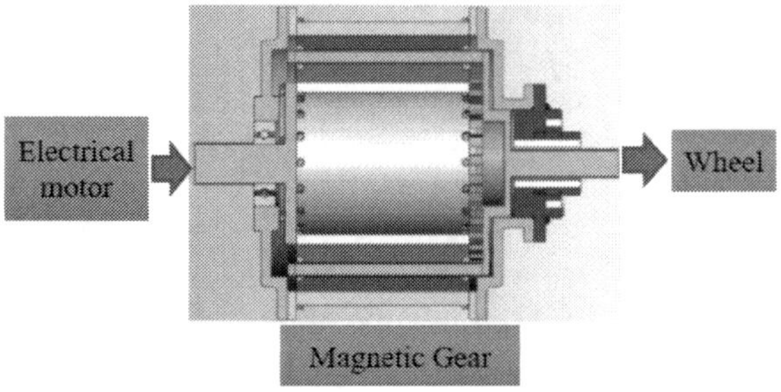

Fig. 2. Magnetic Gear of an electric car and its relation with connected systems.

Usually, Functional Analyses Diagrams (FAD) are developed for large systems that incorporate many subsystems, but here, it is analyzed a smaller assembly, such as a magnetic gear. The diagram is developed using a hierarchical structure.

The relationship between two sub-systems or system items can be represented in the diagram as relationship between: a) system function and the equipment implementing it; b) system function and the system states requiring it; c) internal interfaces between system functions; d) system external interfaces; e) system function and performance requirements [19]. The first step starts with the analysis of primary gearbox functions and their interactions with the peripherical systems, as shown in Figure 3.

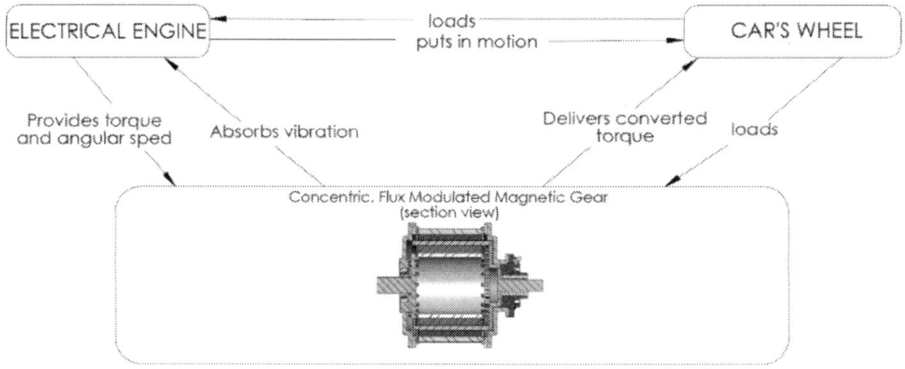

Fig. 3. Top-level functions of MG of an electric vehicle and its interactions with the electrical engine and wheel.

The product's top-level functional diagram contains its primary product functions, in our case, the primary function of the MG is to take over the torque from the engine and to deliver it to the car's wheel at a necessary angular speed and torque. In the top-level FAD model, there can also be found additional functions such as vibration absorption and noise cancelation in magnetic gear due to the nature of magnetic interaction between the rotors.

The second or middle level of the FAD examines the subassemblies of the MG and the functions that link them. Every arrow represents a function a subassembly exerts upon another and the way the entire assembly works. This type of representation will be very useful when Failure Modes and Effects Analysis are done as every function could be analyzed to evaluate the probability of failure.

All the functions represented at the level of the FAD sub-assembly work together to ensure the execution of the functions of the assembly. The representation from Figure 4 provides a general understanding of how the Magnetic Gear is assembled and how it works.

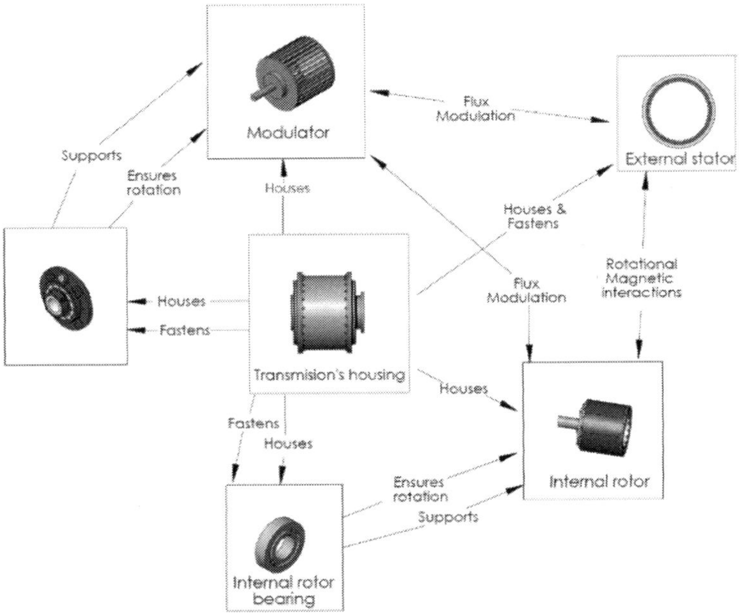

Fig. 4. Sub-assembly level of MG diagram and the functional interaction between them.

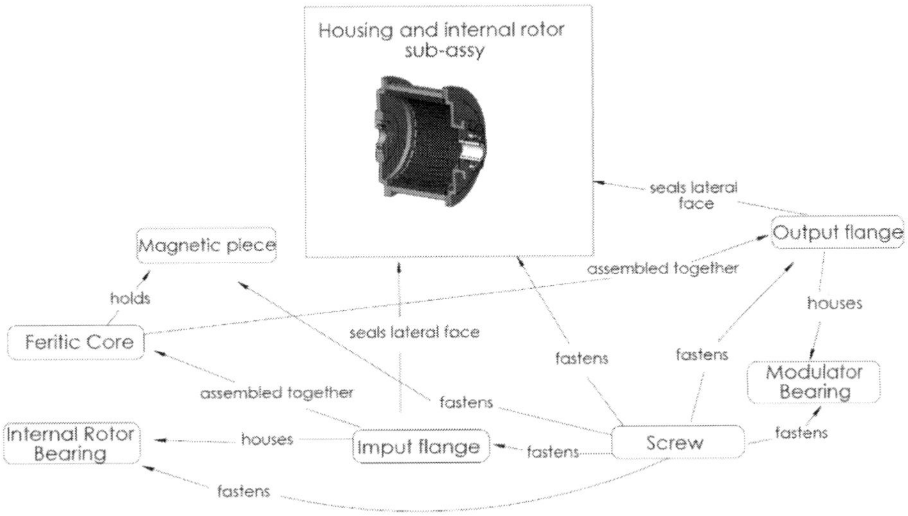

Fig. 5. Functional Diagram.

The final step is the development of the functional analyses diagram represents and explains the function of every component and its utilization in the assembly. FAD could be useful for analyzing the necessity of a certain component and a possible replacement with another one that does the same function in a better fashion, or if the function is not necessary, the component could be eliminated to simplify the assembly design (Figure 5).

4 Conclusion

In this paper, an overview of concentric magnetic gear with a flux modulator has been presented. Although the emphasis was on radial flux concentric MG, the other two types require further analysis because of the advantages they present. The growing capabilities of this type of MG in recent years due to optimized topologies and better use of PM properties make possible their serial manufacturing and practical application in diverse industries. For this reason, a Functional Analysis related to the design process of MG's has been done to facilitate the Design Process of the Magnetic Gear. The method has shown capability for optimization of the magnetic gear design and displays a possible utilization in a interdisciplinary environment due to its quality to support high levels of understanding trough functional decomposition. The future works will use the functional analyses diagrams of this type of MG to utilize it in the process of design failure mode and effect analysis (DFMEA) of such a system.

References

1. P.O. Rasmussen, T.O. Andersen, F.T. Jorgensen, O. Nielsen, IEEE Trans. Ind. Appl. **41**, 764–770 (2005)
2. K. Atallah, D. Howe, IEEE Trans. Magn. **37**, 2844–2846 (2001)
3. K. Atallah, S. Calverley, D. Howe, IEE Proc. Electr. Power Appl. **151**, 135–143 (2004)
4. S. Mezani, K. Atallah, D. Howe, J.of Appl. Phys. **99**(8):08R303-08R303-3 (2006)
5. K. Atallah, D. Howe, J. of Appl. Phys., **97**(10):10N516-10N516-3 (2005)
6. N. Niguchi, K. Hirata, M. Muramatsu, Y. Hayakawa, Proceedings of the 14th Biennial IEEE Conference on Electromagnetic Field Computation (2010)
7. M. Todorova, V. Mateev, I. Marinova, Proceedings of the 19th International Symposium on Electrical Apparatus and Technologies SIELA, 1–4 (2016)
8. O. Molokanov, P. Dergachev, V. Kiruhin, P. Kurbatov, Proceedings of 18th International Symposium on Electrical Apparatus and Technologies SIELA, (2014)
9. N.W. Frank, H.A. Toliyat, IEEE Energy Conv. Congress and Exposition ECCE, 2977–2984 (2010)
10. I. Valiente-Blanco, J.L. Perez-Diaz, E. Diez-Jimenez, Efren, in 44th Aerospace Mechanisms Symposium, 255-262 (2018)
11. K. Uppalapati, W. Bomela, J. Bird, M. Calvin and J. Wright, IEEE Energy Conv. Congress & Expo. ECCE, 2178–2184 (2013)
12. S. Gerber and R.-J. Wang, IEEE Int. Conf. on Ind. Tech. ICIT, 319–324 (2013)
13. M. Fukuoka, K. Nakamura and O. Ichinokura, Int. Conf. Electr. Machines & Sys. (2012)
14. L. Shah, A. Cruden, B. Williams, Int. Conf. on Power Electron. & Drive Sys. PEDS, 989–993 (2007)
15. J. Bird, Proceedings of IEEE Energy Conversion Congress and Exposition (ECCE), 1422 (2014)
16. N. Viola, S. Corpino, M. Fioriti, F. Stesina, *Functional Analysis in Systems Engineering: Methodology and Applications,* (InTech, 2012)
17. M. Atherton, P. Jiang, D. Harrison, A. Malizia, Res. Eng. Des. 29(2), 245–262 (2018)
18. J. Ben Salem, M.N. Lakhoua, L. Amraoui, Int. Conf. on Electrical Sciences and Technologies in Maghreb (CISTEM), IEEE Xplore (2014)
19. T. Pinna, D.N. Dongiovanni, F. Iannone, Fusion Eng. Des. 109–111 (Part A), 795–800 (2016)
20. I. Michalakoudis, P.R. Childs, M. Aurisicchio, N. Pollpeter, N. Sambell, ASME 2014, V011T14A011 (2014).

Experimental study on noise reduction and performance enhancement for internal combustion engines

Ioan Radu Şugar[1] and *Mihai* Banica[1*]

[1]Technical University of Cluj-Napoca, Faculty of Engineering, Victor Babes 62A, 430083, Baia Mare, Romania

Abstract. As the number of cars increases and large cities become more and more crowded, noise reduction becomes more and more important. The decrease of the fuel consumption and the increase of power to the same cylindrical capacity are always current topics. This paper's aim is to bring a contribution to solving these problems. The proposed solution consists in the use of ceramic materials in the design of the combustion chamber.

1 Introduction

Although internal combustion engines were invented in the middle of the 19th century, they went on for more than half a century before imposing themselves in front of electric motors to equip onto cars. This is largely due to Henry Ford's innovative ideas put into practice in 1913 [1, 3, 5].

With global warming and the depletion of fossil fuel reserves after almost a hundred years, the debate between electrical and internal combustion cars has been resumed. On the other hand, on a larger scale, biofuel is required in fuelling internal combustion engines. The cars equipped with electric motors seem ideal at first sight as they are more friendly to the environment. According to some researchers in the field, manufacturing of electric batteries represents approx. 80% of that made by an Otto engine car. Also, recycling the batteries is still an unsolved problem due to their lack of standardization. Internal combustion engines will have a relatively long-life equipping hybrid car [2, 7, 8].

Internal combustion engines will still represent the solution for the propulsion of trucks, military equipment, ships etc [4, 5]. Improving the performance indices of internal combustion engines remains an ongoing research topic to increase the performance of internal combustion engines, the use of new materials in general and the use of ceramic materials may be a solution [3, 9].

"Advanced ceramics" represents a group of structural and functional materials, with superior characteristics to the classical variants. They are designed for special fields of use, where they offer a higher performance compared to the metallic or polymeric materials. Advanced ceramics intervene in applications where the latter classes of products reach their limits [6, 9].

* Corresponding author: Mihai.Banica@cunbm.utcluj.ro

The new class of "advanced ceramic" materials, which entered the market in the 20th century includes much more pure material systems, with specially processed compounds. They were developed mainly for structural and electronic applications. Advanced ceramics are distinguished by their high chemical purity and high values of use characteristics [6, 9].

In the last 25 years, new structural materials, such as ceramics, polymers and composites, have caused revolutionary changes in the field of materials engineering. With advanced ceramics and composites, the concepts of materials and structures taken together have led to a new concept of integrated design. Each projected material consolidates the discrete and functional parts into one, multifunctional structure, which leads to the highest efficiency of material use and the lowest costs [6, 9].

The innovative technological ceramics extend their use in the engineering of applications that exploit their mechanical properties. All applications for cutting-edge technologies require components that have high mechanical and wear resistance, along with high toughness, high breaking resistance through ballistic impact. They add tenacity, chemical inertia and ability to work at high temperatures. In some cases, they may have special functions: electrical, magnetic, optical or chemical-biological [6, 9].

2 Equipment and methods

In the early 1980s, the Japanese researchers from TOYOTA MOTORS made a 100% ceramic engine. The idea was abandoned due to its modest reliability of only 250 hours of operation.

Currently, composite materials are used for building engines: they contain a metal part and a ceramic part. The metal part provides mechanical strength, and the ceramic one achieves the thermal barrier in the most thermally demanded areas. The ceramic part was made by sintering.

Ceramic coatings have been known to implement these directions: isolation chamber, isolating the exhaust path, isolating hot components. Our research focused on isolating the piston head with Al_2O_3. Since the main mechanical stress is given by the gas pressure, the technical solution chosen was adhesive bonding of the ceramic crown on the piston surface.

Fig. 1. Insulated piston with ceramic crown: 1-ceramic crown

The model of the experimental stand is presented in the following figure:

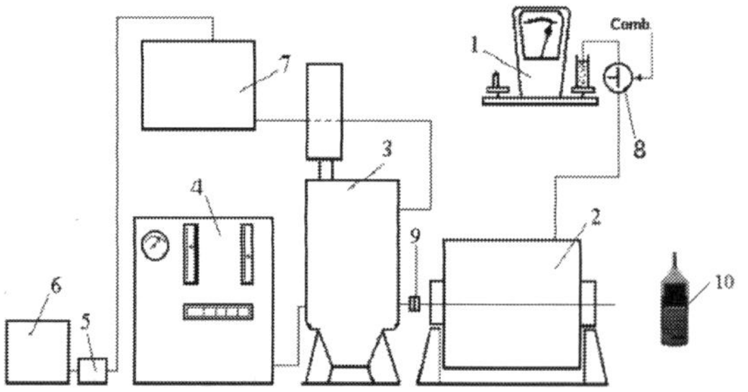

1-mass measurement of fuel consumption; 2-engine 810-99; 3-hydraulic brake with electromotor and analogue apparatus for determining the motor moment; 4-control block with digital display of coolant temperature, oil temperature and pressure and crankshaft speed; 5-electric pump; 6 and 7-coolant basin; 8-way three-way valve; 9-cardan coupling; 10-Digital Sound Level Meter

Fig. 2. Model of the experimental stand.

For the experimental part, an 810-99 engine was used with the following characteristics: SIE motor type; 8.5:1 compression ratio; Cylinder diameter 73mm; Piston stroke 77mm; Total cylinder 1289cm^3; Maximum power 39.72kW at 5250rpm; Maximum torque 95Nm at 3.000rpm; Number of times 4; Positioning of cylinders: Line.

The characteristic parameters and admissible measurement errors are presented in Table 1 and general conditions for carrying out the experiment are summarized in Table 2.

Table 1. Characteristic parameters and admissible measurement errors.

Parameter	*Symbol*	*Unit of measurement*	*Admissible measurement errors*
Actual motor momentum	M_e	N·m	±1%
Brake indication	F	N	±1%
Arm length from the weighing device	l	m	-
Drive shaft speed	n	rpm	±0,5%
Hourly fuel consumption	C, C_h	l/h, kg/h	±1%
Hourly air consumption	C_a	m^3/h, kg/h	±2%
Air temperature allowed	T_{aer}, T_o	K	±2K
Coolant temperature	T_{apa}, T_r	K	±2K
Oil temperature at the measuring site	T_u	K	±2K
Temperature of exhaust gases	T_{ge}	K	±5K
Fuel temperature	T_{comb}	K	±2K
Atmospheric pressure (barometric)	p_{atm}, (p_b)	bar, (KPa)	±0,1KPa
Oil pressure at the indicated location	p_u	KPa	±10KPa
Pressure or depression in the exhaust system	$p_{sist\ ev.}$	bar, KPa	±0,1KPa
Pressure drop on the air filter	p_{filtru}	KPa	±0,05KPa
Depression at the admission gallery	p_{adm}	bar, KPa	±0,05KPa
Depression of control of the advance	p_{av}	KPa	±1%
Ignition advance angle	β_s	RAC	±1%
Time	t	s	±0,1s

Table 2. General conditions.

Engine type	Standard / Modified	
	Engine: *810-99* + Carburettor: *32IRM-A*	
Purpose of the test	Determining the effective power	
Test conditions	Air temperature: *28°C*	Atmospheric pressure: *740mmHg*
	Coolant temperature: *83÷92°C*	Oil temperature: *93÷117°C*
Fuel and lubricant	Petrol: *PREMIUM II (STAS 176-80)*	Oil: *M20W40 (STAS 9171-80)*

3 Results and discussions

The values of measured parameters (*Braking force (F)*, *Fuel mass consumed (m)* and *Sound pressure level (SPL)*) are presented in the Table 3.

3.1 Calculated parameters

Effective power (P_e):

$$P_e = \frac{F \cdot n}{13600} \quad [kW] \tag{1}$$

Actual specific consumption (C_e):

$$C_e = 3{,}6\frac{m}{\tau \cdot P_e} \left[\frac{kg}{h}\right] \quad (2)$$

Actual efficiency (η_e):

$$\eta_e = \frac{3600}{C_e \cdot Q_i} = \frac{3600}{C_e \cdot 43890} \quad (3)$$

where: Q_i is the calorific value of the fuel used.

The values of calculated parameters (*Effective power (P_e)*, *Actual specific consumption (C_e)* and *Actual efficiency (η_e)*) are presented in the Table 4.

Table 3. The values of the measured parameters.

rpm	*Time of trial (τ)*	*Standard engine*			*Modified engine*		
		Braking force (F)	*Fuel mass consumed (m)*	*Sound pressure level (SPL)*	*Braking force (F)*	*Fuel mass consumed (m)*	*Sound pressure level (SPL)*
	[s]	*[N]*	*[g]*	*[dB]*	*[N]*	*[g]*	*[dB]*
1500	32	115,6	40,35	61,0	118,2	39,28	58,0
2000	30	121,0	45,38	62,0	125,3	44,53	60,0
2250	24	125,5	41,25	63,0	128,3	39,62	61,0
2625	25	130,3	52,22	64,5	133,8	50,75	62,5
3000	22	132,4	54,26	67,0	135,2	52,67	64,0
3375	19	128,1	50,33	67,5	132,0	48,75	65,5
3750	18	125,3	50,79	68,0	128,3	48,82	66,5
4125	16	121,0	47,63	69,0	124,3	45,74	67,0
4500	15	115,3	47,37	71,0	117,7	45,11	68,0
4875	15	109,0	49,82	72,5	110,0	47,64	68,5
5250	14	101,3	48,21	74,0	102,9	46,34	71,5
5500	13	91,2	44,22	75,0	94,9	43,66	73,0

Table 4. The values of the calculated parameters.

rpm	*Standard engine*			*Modified engine*		
	Effective power (P_e)	*Actual specific consumption (C_e)*	*Actual efficiency (η_e)*	*Effective power (P_e)*	*Actual specific consumption (C_e)*	*Actual efficiency (η_e)*
	[kW]	*[kg/kWh]*	*[%]*	*[kW]*	*[kg/kWh]*	*[%]*
1500	12,75	0,356	23,04	13,04	0,339	24,20
2000	17,79	0,306	26,80	18,43	0,290	28,28
2250	20,76	0,298	27,52	21,23	0,280	29,29
2625	25,15	0,299	27,43	25,83	0,283	28,98
3000	29,21	0,304	26,98	29,82	0,289	28,38
3375	31,79	0,300	27,34	32,76	0,282	29,09
3750	34,55	0,294	27,90	35,38	0,276	29,72
4125	36,70	0,292	28,09	37,70	0,273	30,05
4500	38,15	0,298	27,52	38,94	0,278	29,50
4875	39,07	0,306	26,80	39,43	0,290	28,28
5250	39,10	0,317	25,87	39,72	0,300	27,34
5500	36,88	0,332	24,71	38,38	0,315	26,04

3.2 Statistical analysis

Comparative analysis of functional parameter values of the two engines obtained in the experiment is presented below:

Table 5. Analysis of functional parameter values.

rpm	Standard engine P_e [kW]	Modified engine P_e [kW]	*Relative increase (P_e) [%]*	Standard engine Ce [kg/kWh]	Modified engine Ce [kg/kWh]	*Relative decrease (C_e) [%]*	Standard engine η_e* [%]	Modified engine η_e* [%]	*Relative increase (η_e)* [%]*	Standard engine SPL [dB]	Modified engine SPL [dB]	*Relative decrease (SPL) [%]*
1500	12,75	13,04	*2,25*	0,356	0,339	*4,78*	23,04	24,20	*5,01*	61,0	58,0	*4,92*
2000	17,79	18,43	*3,55*	0,306	0,290	*5,23*	26,80	28,28	*5,52*	62,0	60,0	*3,23*
2250	20,76	21,23	*2,23*	0,298	0,280	*6,04*	27,52	29,29	*6,43*	63,0	61,0	*3,17*
2625	25,15	25,83	*2,69*	0,299	0,283	*5,35*	27,43	28,98	*5,65*	64,5	62,5	*3,10*
3000	29,21	29,82	*2,11*	0,304	0,289	*4,93*	26,98	28,38	*5,19*	67,0	64,0	*4,48*
3375	31,79	32,76	*3,04*	0,300	0,282	*6,00*	27,34	29,09	*6,38*	67,5	65,5	*2,96*
3750	34,55	35,38	*2,39*	0,294	0,276	*6,12*	27,90	29,72	*6,52*	68,0	66,5	*2,21*
4125	36,70	37,70	*2,73*	0,292	0,273	*6,51*	28,09	30,05	*6,96*	69,0	67,0	*2,90*
4500	38,15	38,94	*2,08*	0,298	0,278	*6,71*	27,52	29,50	*7,19*	71,0	68,0	*4,23*
4875	39,07	39,43	*0,92*	0,306	0,290	*5,23*	26,80	28,28	*5,52*	72,5	68,5	*5,52*
5250	39,10	39,72	*1,58*	0,317	0,300	*5,36*	25,87	27,34	*5,67*	74,0	71,5	*3,38*
5500	36,88	38,38	*4,06*	0,332	0,315	*5,12*	24,71	26,04	*5,40*	75,0	73,0	*2,67*
Min			**0,92**			**4,78**			**5,01**			**2,21**
Max			**4,06**			**6,71**			**7,19**			**5,52**
Average	**30,16**	**30,89**		**0,31**	**0,29**		**26,67**	**28,26**		**67,88**	**65,46**	
Average increase		***2,41%***	***Average decrease***		***5,59%***	***Average increase***		***5,98%***	***Average decrease***		***3,56%***	
**** Informal values***												

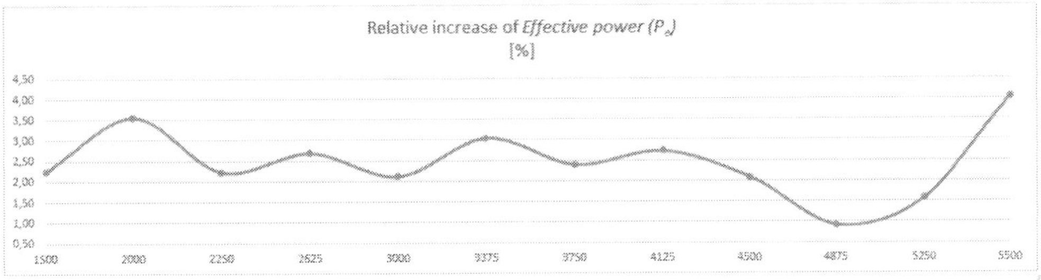

Fig. 3. Relative increase of *Effective power (Pe)*.

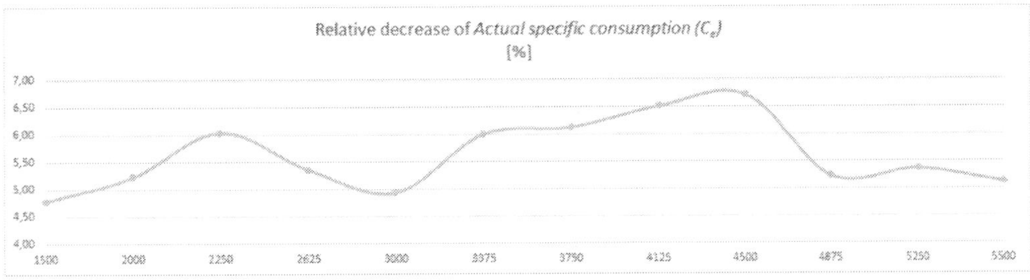

Fig. 4. Relative decrease of *Actual specific consumption (Ce)*.

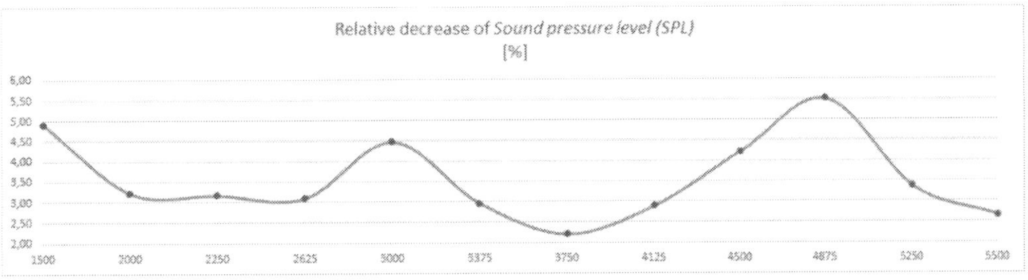

Fig. 5. Relative decrease of *Sound pressure level (SPL)*.

Results of analyse *Modified engine vs. Standard engine*:

i. Regarding the *Effective power* (P_e), there is an increase between 0,92÷4,06%, showing an increasing in the *average value* of 2,41%;

ii. *Actual specific consumption* (C_e), shows a decrease between 4,78÷6,71%; the *average value* of this decrease is 5,59%;

iii. The decrease of *Sound pressure level (SPL)* is between 2,21% and 5,52%, showing a decreasing in the *average value* of 3,56%.

4 Conclusion

Using the ceramic materials in the construction of the combustion chamber of spark-ignition engines seems ideal in terms of mechanical strength and functional benefits. Their main shortcoming is the fragility to shocks and vibrations which led to the successful use of composite materials. By using ceramic materials, the temperature in the firing chamber increases by 150÷200K which favours the firing process. In addition to increasing the actual power, the actual engine torque and the reduction of the actual specific fuel consumption, the noise and, not to be neglected, the chemical pollution are reduced. The replacement of metallic materials with ceramic ones leads to a decrease in the motor mass due to the lower density of ceramic materials. The next step is to redesign the cooling system by using a smaller radiator and implicitly by decreasing the volume of the coolant.

Future research will focus on redesigning the cooling system by using a smaller radiator (implicitly decreasing the volume of the coolant) and improving the current ceramic crown bonding solution on the piston surface.

References

1. N. Bataga, N. Burnete, I. Barabas, A. Cazila, N. Filip, A. Naghiu, F. Dan, R. Szasz, D. Karamusantas, *Motoare cu ardere interna. Combustibili. Lubrifianti. Materiale speciale pentru autovehicule. Economicitate. Poluare/ Internal combustion engines. Fuels. Lubricants. Special materials for vehicles. Economy. Pollution* (U.T.Press, Cluj-Napoca, 2000)
2. G. Evangelos, Energies, **10(11)**, 1882, DOI: 10.3390/en10111882 (2017)
3. A. Chiru, H.R. Anca, C. Cofaru, R. Kuchar, A. Soica, N. Ispas, *Materiale compozite. Vol.I/ Composite materials. Vol.I* (E.U.T., Brasov, 1999)
4. N. Burnete, N. Bataga, D. Karamusantas, *Constructia si calculul motoarelor cu ardere interna/ Construction and calculation of internal combustion engines* (Todesco, Cluj-Napoca, 2001)
5. N. Burnete et al., *Motoare diesel si biocombustibili pentru transportul urban/ Diesel engines and biofuels for urban transport* (Mediamira, Cluj-Napoca, 2008)
6. D. Radu, R. State, Romanian Journal of Materials 41 **3**, 269 (2011)
7. I. Barabas, I.A. Todorut, DOI: 10.5772/25370 (2011)
8. N. Negurescu, C. Pana, M.G. Popa, *Motoare cu aprindere prin scanteie-Procese/ Spark ignition engines-Processes* (Matrixrom, Bucuresti 2009)
9. M.L. Scutaru, A. Chiru, S. Vlase, C. Cofaru, H. Teodorescu-Draghicescu, *Materiale plastice si compozite in ingineria autovehiculelor/ Plastics and composites in automotive engineering* (Matrixrom, Bucuresti, 2013)

https://doi.org/10.1051/matecconf/202031801015

Impact of process parameters on dimensional accuracy of PolyJet 3D printed parts using grey Taguchi method

Kyriaki-Evangelia Aslani[1,3], *Apostolos* Korlos[2], *John D* Kechagias[1,*] and *Konstantinos* Salonitis[4]

[1]General Department, University of Thessaly, Larissa, Greece
[2]Industrial Engineering and Management Department, International Hellenic University, Thessaloniki, Greece
[3]Department of Mechanical Engineering, University of West Attica, Athens, Greece
[4]Manufacturing Department, Cranfield University, UK

Abstract. In this study, the dimensional accuracy of parts fabricated with PolyJet 3D Printing Direct process is investigated. An L_4 orthogonal array was utilized as the design of experiments, while the process parameters examined are layer thickness, build style and scale. A simple prototype was proposed and specified external and internal dimensions were measured using a digital vernier calliper. Grey-Taguchi method was applied for optimizing all dimensional measurements. The effect of each parameter on dimensional accuracy has been identified using ANOM (Analysis of Means), while ANOVA (Analysis of Variances) has been performed to determine each parameter's dominance. Additionally, the results of this study were compared with the findings of a previous optimization study in which the usual Taguchi method was used. It was concluded that 16 μm of layer thickness, glossy style and 50% scale provide the optimum dimensional results, while scale is the most important factor.

1 Introduction

PolyJet 3D printing technique is one of the most popular additive manufacturing (AM) methods. AM processes produce physical models from 3D model data, by depositing material layer upon layer. They can manufacture complex shapes, having applications in a number of fields, such as automobile, aerospace and medical [1]. For the case of PolyJet 3D process, photopolymer resin layers are selectively jetted onto a build-tray via inkjet printing. The jetted photopolymer droplets are simultaneously cured with ultraviolet lamps that are mounted onto the print carriage [2-3].

Dimensional accuracy is a main quality indicator in manufacturing, examined by many researchers around the world [4]. Kent et al. studied the dimensional accuracy and surface finish of parts manufactured with PolyJet technology using three alternative materials [5].

* Corresponding author: jkechag@uth.gr

They found differences in dimensional accuracy and surface profile depending on the orientation of the feature relative to the print head travel direction. Kim et al. [6] compared a number of AM technologies such as stereo lithography (SL), fused deposition modelling (FDM), PolyJet and selective laser sintering (SLS) in terms of mechanical properties, dimensional accuracy, surface roughness, speed and material cost. It was found that SL style gives the best dimensional accuracy. Kechagias et al. [7] studied the effect of the layer thickness, build style and model scale on the surface roughness of parts produced using the PolyJet technology. The classic Taguchi method was used for the surface roughness optimization. In a similar manner, Aslani et al. [8] investigated the effect of the above mentioned process parameters on the surface roughness of PolyJet manufactured parts using Grey Taguchi method. Kechagias et al. [9] examined the effect of the same process parameters on the dimensional accuracy of PolyJet manufactured parts using the classic Taguchi method.

In this paper, the optimization of the dimensional accuracy of parts printed with PolyJet technology is studied. Grey Taguchi method is used along with statistical analysis. An L_4 orthogonal array is utilized as design of experiments, while the process parameters examined are layer thickness, build style and model scale. The results are compared with the findings from the classic Taguchi method examination (see [9]).

2 Experimental Procedure

2.1 Specimen design and measurements

In the present study, a 3D test part was created with three holes on its sides and its base as it can been seen in Fig. 1. Part's nominal dimensions are presented in the same figure in millimetres (mm). The 3D model was designed with the use of Solidworks software and it was extracted in STL format. The 3D printer which was used was Stratasys Objet Eden 250 (see Fig. 2), while the printing material was Objet Fullcure 720 RGD. Dimensional measurements were taken with the use of a digital vernier calliper of 0.01 mm accuracy (see Table 1). Next, dimensional deviation was calculated, which shows the difference between the nominal and the measured dimensional values and represents the dimensional accuracy of the parts.

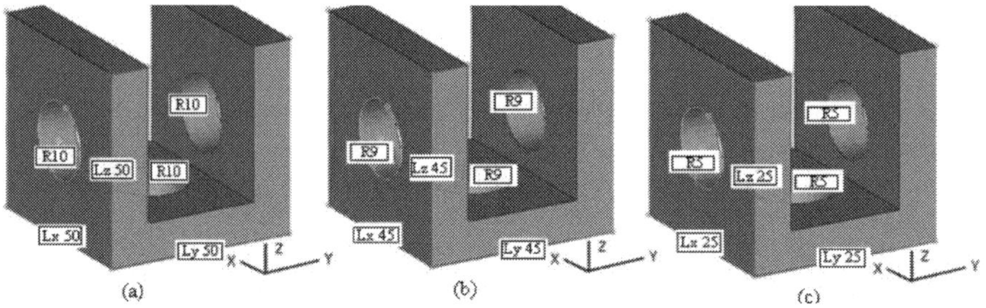

Fig. 1. CAD models [9]: (a) 100% digital part, (b) 90% scaled part, and (c) 50% scaled part.

Table 1. Dimensional accuracy measurements.

Exp no.	**Linear**			**Diametric**		
	DLx [mm]	*DLy [mm]*	*DLz [mm]*	*DDx [mm]*	*DDy [mm]*	*DDz [mm]*
1	0.100	-0.050	-0.050	-0.173	-0.050	-0.145
2	0.060	-0.050	-0.020	-0.226	-0.200	-0.095
3	0.330	0.180	0.060	-0.410	-0.390	-0.150
4	0.110	0.130	-0.060	-0.250	-0.200	-0.175

Fig. 2. Stratasys Objet Eden 250 3D printer [9].

2.2 Design of experiments

In general, the dimensional accuracy of the PolyJet technology depends on the selection of the process parameters. In this study, the process parameters which are considered are: layer thickness, build style and scale. Two levels were selected for each process parameter (see Table 2). In order to investigate the effect of the layer thickness, the build style and the scale, Taguchi's L_4 (2^3) orthogonal array was used. The L_4 orthogonal array that is utilized here is presented in Table 3.

Table 2. Selected process parameters and their levels.

Process Parameters	Levels	
	1	*2*
Layer Thickness [μm]	16	30
Build Style [-]	Mate	Glossy
Scale [%]	50	90

Table 3. Taguchi L_4 orthogonal array.

Exp no.	Layer Thickness [μm]	Build Style [-]	Scale [%]
1	16	Mate	50
2	16	Glossy	90
3	30	Mate	90
4	30	Glossy	50

3 Results

Taguchi's experimental method was developed for single response optimization, but in the case of two or more responses optimization, this method is unsuitable [10-12]. Hence, Grey Taguchi approach can be used for creating a single response from different performance features [13]. In this examination, grey relational analysis is utilized to identify the optimal level combination of all dimensional accuracy measurements.

In the grey relational analysis, all data are first normalized to a range of 0–1. Next, the grey relational coefficient is calculated with the use of the normalized values. The grey relational grade is computed finally, by averaging all grey relational coefficient results. The optimization of the multiple responses is done by optimizing the grey relational grade. The normalized dimensional accuracy values which are the smaller-the-better can be written as:

$$x_i(k) = \frac{\max y_i(k) - y_i(k)}{\max y_i(k) - \min y_i(k)} \quad (1)$$

where $x_i(k)$ is the normalized grey relational value and $y_i(k)$ is the k_{th} characteristic of the i_{th} experiment sequence. The min and the max indicate the minimum and the maximum $y_i(k)$ values. Table 4 tabulates the normalized deviation values of all dimensional accuracy experiments.

Table 4. Normalized deviation values $x_i(k)$.

Exp no.	Linear			Diametric		
	DLx [mm]	*DLy [mm]*	*DLz [mm]*	*DDx [mm]*	*DDy [mm]*	*DDz [mm]*
1	0.8519	1.0000	0.9167	0.0000	0.0000	0.8519
2	1.0000	1.0000	0.6667	0.2236	0.4412	1.0000
3	0.0000	0.0000	0.0000	1.0000	1.0000	0.0000
4	0.8148	0.2174	1.0000	0.3249	0.4412	0.8148

The grey relational coefficient ξ can be computed as:

$$\xi_i(k) = \frac{\Delta_{min} + \zeta\Delta_{max}}{\Delta_i(k) + \zeta\Delta_{max}} \quad (2)$$

where $\Delta_i(k) = |x_0(k) - x_i(k)|$, $x_0(k)$ is the ideal sequence and ζ is distinguishing coefficient $(0\sim1)$. The ideal sequence value is considered as 1 in the research, while the distinguishing coefficient value is considered as 0.5. Again the min and max indicators specify the lowest and the highest $\Delta_i(k)$ values. The grey relational coefficient results for dimensional accuracy are presented in Table 5.

Table 5. Grey relational coefficient $\xi_i(k)$ results.

Exp no.	Linear			Diametric		
	DLx [mm]	*DLy [mm]*	*DLz [mm]*	*DDx [mm]*	*DDy [mm]*	*DDz [mm]*
1	0.7714	1.0000	0.8571	0.3333	0.3333	0.7714
2	1.0000	1.0000	0.6000	0.3917	0.4722	1.0000
3	0.3333	0.3333	0.3333	1.0000	1.0000	0.3333
4	0.7297	0.3898	1.0000	0.4255	0.4722	0.7297

Finally, the grey relational coefficient is evaluated as:

$$\alpha_i = \sum_{i=1}^{n} \xi_i(k) \quad (3)$$

The calculation results for the grey relational grade are showed in Table 6.

Table 6. Grey relational grade results.

Exp no.	Layer Thickness [µm]	Build Style [-]	Scale [%]	Grey Relational Grade [-]
1	16	Mate	50	0.644444
2	16	Glossy	90	0.632882
3	30	Mate	90	0.602564
4	30	Glossy	50	0.669546

In this paper, Minitab 17 Statistical Software was utilized for the statistical analysis. In Table 7, the response table for mean grey relational grade values calculated from the Analysis of Means (ANOM) is presented. It was found that scale is the most influential process parameter followed by build style and layer thickness. As it is shown in the plot of means (Fig. 3), the optimal process parameter levels are: a) Layer thickness: 16 μm, b) Build style: Glossy, c) Scale: 50%.

Table 7. Response table for mean grey relational grade values.

Level	Layer Thickness [μm]	Build Style [-]	Scale [%]
1	0.6387	0.6512	0.6570
2	0.6361	0.6235	0.6177
Delta	0.0026	0.0277	0.0393
Rank	3	2	1

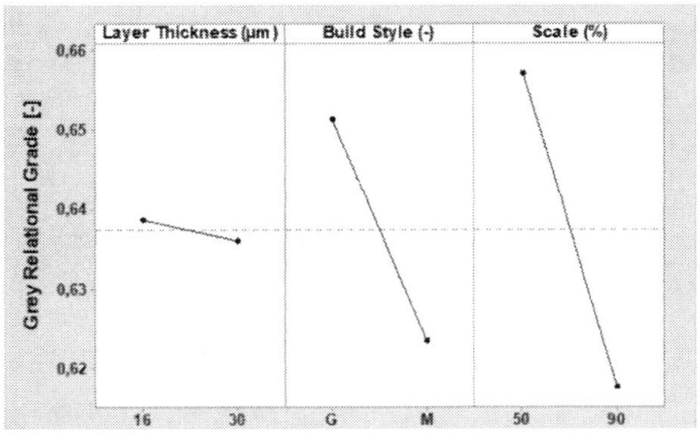

Fig. 3. Plot of means for grey relational grade results.

In this investigation, Analysis of Variance (ANOVA) is utilized to identify the process parameters that exhibit significant effect on the dimensional accuracy. For this reason contribution rate was computed. In general, high contribution rate values mean significance. The ANOVA for grey relational grade of all dimensional accuracy measurements is tabulated in Table 8. It was discovered that scale is the most significant parameter (Contribution = 66.54%), followed by build style (Contribution = 33.16%). Layer thickness was found to be unimportant (Contribution = 0.3%). It should be noted that scale was realized as the most significant parameter both in the ANOM and the ANOVA.

Table 8. ANOVA table for grey relational grade values.

Source	DF	Adj SS	Adj MS	Contribution Rate [%]
Layer Thickness [μm]	1	0.000007	0.000007	0.30
Build Style [-]	1	0.000768	0.000768	33.16
Scale [%]	1	0.001542	0.001542	66.54
Error	0	-	-	-
Total	3	0.002317	-	-

As it can been seen from [9], Grey Taguchi method results are not compatible to a certain extent with the results from the classic Taguchi method. In the case of Grey Taguchi method, scale was found to be the most important parameter, followed by build style and layer thickness which can be considered unimportant. In the case of the classic Taguchi method, different factors are important for every direction (layer thickness for the linear X and Y directions, scale for the linear Z direction, layer thickness for diametric X, Y and Z directions). Scale was found to be almost as important as layer thickness for the diametric directions. Different parameter levels optimize the dimensional accuracy of every direction in the classic Taguchi method (16 μm layer thickness, glossy build style and 50% scale optimize linear X, Y and Z directions, 30 μm layer thickness, mate build style and 90% scale optimize diametric X and Y directions, 30 μm layer thickness, mate build style and 50% scale optimize diametric Z direction). In the case of the Grey Taguchi method, 16 μm layer thickness, glossy build style and 50% scale optimize all dimensional accuracy responses.

4 Conclusions

Grey Taguchi optimization methodology was applied for the optimization of dimensional accuracy of parts produced by the PolyJet technology. An experiment Taguchi array having the orthogonality property was used, while the process parameters studied are layer height (LH), build style (BS) and scale factor (SF) of the build model, having each one two levels. The dimensional deviation from part's nominal dimensions was used as the dimensional accuracy response. The 16 μm LH value,the glossy BS and 50% SF optimize all dimensional accuracy responses. Scale was found to be the dominant process parameter, followed by build style and layer height. Results from the Grey Taguchi method are not compatible to a certain level compared to the classic Taguchi method.

References

1. K-E. Aslani, K. Kitsakis, J.D. Kechagias, N.M. Vaxevanidis, D.E. Manolakos, SN Appl. Sci. **2**, 1016, https://doi.org/10.1007/s42452-020-2823-z, (2020).
2. J.D. Kechagias, S. Maropoulos, *Proceedings of the International Conference on Industrial Engineering*, 150-153 (2015).
3. M.W. Barclift, C.B. Williams, *International Solid Freeform Fabrication Symposium*, 876-890 (2012).
4. K-E. Aslani, D. Haidas, J.D. Kechagias, P. Kyratsis, K. Salonitis, J. Manuf. Mater. Process., **4(2)**, 47, https://doi.org/10.3390/jmmp4020047, (2020).
5. N.J. Kent, L. Jolivet, P. O'Neill, D. Brabazon, Adv. Mater. Process. Tec. **3**, 3 (2017)
6. G.D. Kim, Y.T. Oh, *Proceedings of the Institution of Mechanical Engineers*, 201-215 (2008).
7. J. Kechagias, V. Iakovakis, E. Giorgo, P. Stavropoulos, A. Koutsomichalis, N.M. Vaxevanidis, *Proceedings of the 1st International Conference on Engineering and Applied Sciences Optimization,* 2877-2888 (2014).
8. K-E. Aslani, F. Vakouftsi, J.D. Kechagias, N. E. Mastorakis, *Proceedings of the 3rd International Conference on Control, Artificial Intelligence, Robotics & Optimization,* 213-218, doi: 10.1109/ICCAIRO47923.2019.00041, (2019).
9. J. Kechagias, P. Stavropoulos, A. Koutsomichalis, I. Ntintakis, N. Vaxevanidis, *Proceedings of the International Conference on Industrial Engineering*, 61-65 (2014)
10. C. L. Lin, Mater. Manuf. Process. **19**, 2 (2007).
11. A. K. Sood, R.K. Ohdar, S.S. Mahapatra, Mater. Design **30**, 10 (2009)
12. J.D. Kechagias, K-E Aslani, N.A. Fountas, N.M. Vaxevanidis, D.E. Manolakos, Measurement **151**, DOI:10.1016/j.measurement.2019.107213, (2020).
13. P.A. Sylajakumari, R. Ramakrishnasamy, G. Palaniappan, Materials **11**, 1743 (2018).

Tribological Characterization of Some Elastomers Used at Progressive Cavity and Piston Pumps

Mohamed Iyad Al Naboulsi[1] , *Niculae Napoleon* Antonescu[1,*] , *Alin* Dinita[1] , and *Marius* Morosanu[1]

[1]University Petroleum-Gas of Ploiesti, Department of Mechanical Engineering, Blvd. Bucuresti, no.39, 100680, Ploiesti, Romania

Abstract. The progressive cavity pump (PCP) is a positive displacement pump, consisting of a polished helical-shaped rod (rotor) turning inside a helical elastomer (stator). PCP has many advantages, but the pump durability is manly limited by elastomer behavior. At piston pumps (PP) used for drilling mud piston has an elastomer sleeve that also limit the durability. Standards like ISO 15136.1 & 2 for pumps developed by manufactures and users' committees provides requirements for design, quality design verification etc., but do not define specifically the elastomer for the stator or the metal used for the rotor. Each PCP and PP manufacturer used specific materials at pump construction. The aim of this study was to evaluate the tribological behavior of some elastomers such polybutadiene rubber (BR), polybutadiene acrylonitrile rubber (NBR), polybutadiene acrylonitrile carboxylate (XNBR) and polyamide (PA 6) in couples with hard chromium coated steel, nitride steel and cast iron. Were determined friction coefficients and wear on 2 types of friction couples (plane to plane and shoe to plane) on two tribometers and some mechanical proprieties (Young's modulus, ultimate tensile strength, elongation, hardness).

1 Introduction

In petroleum industry piston pumps (PP) and progressive cavity pumps (PCP) are largely used. The pumped fluid contains formation water, solid particles, dissolved gases such as O_2, CO_2, H_2S, salts and chlorides all increasing fluid aggressivity, [1,2,3,4]. In acidization operation with PP are inserted in well acids to stimulate the formation deposits and to increase the well production. Drilling mud contains a large amount of solid particles. As presented working pumps fluids are aggressive, and in pumps active elements are subjected to wear. To diminish abrasive wear one of active pump element are made of non-metallic materials such as elastomers. PCP is a positive displacement pump, consisting of a polished helical-shaped rod (rotor) turning inside a helical elastomer (stator). PCP has many advantages, but the pump durability is mainly limited by elastomer behavior. At PP used

* Corresponding author: nnantonescu@upg-ploiesti.ro

for drilling mud piston has an elastomer sleeve that also limit the durability. Standards like ISO 15136.1 & 2 for pumps developed by manufactures and users' committees provides requirements for design, quality design verification etc., but do not define specifically the elastomer for the stator or the metal used for the rotor. Each PCP and PP manufacturer used specific materials at pump construction, [4].

The aim of this study was to evaluate the tribological behavior of some elastomers such polybutadiene rubber (BR), polybutadiene acrylonitrile rubber (NBR), polybutadiene acrylonitrile carboxylate (XNBR) and polyamide (PA 6) in couples with hard chromium coated steel, nitride steel and cast iron. Were determined friction coefficients and wear on 2 types of friction couples (plane to plane and shoe to plane) on two tribometers and some mechanical proprieties.

2 Experiments

2.1 Tested materials characterisation

To perform wear tests were selected 3 metallic materials widely used at PP cylinder. Chemical composition of materials was established by spectral optical emission with Foundry Master Pro device. In Table 1 it is shown the results obtained.

Table 1. Chemical composition of tested metallic samples.

Chemical elements of sample 1, [% masic]									
C	*Si*	*Mn*	*P*	*S*	*Cr*	*Mo*	*Ni*	*Al*	*Cu*
0.315	0.1845	0.7698	0.0355	0.0025	1.008	0.1547	0.1489	0.0479	0.229
Chemical elements of sample 2, % masic									
C	*Si*	*Mn*	*P*	*S*	*Cr*	*Mo*	*Ni*	*Al*	*Cu*
0.268	0.2498	0.7459	0.0219	0.0104	0.948	0.2049	0.7198	0.0348	0.289
Chemical elements of sample 3, % masic									
C	*Si*	*Mn*	*P*	*S*	*Cr*	*Cu*			
3.289	2.6589	0.6214	0.0948	0.1048	0.1089	0.2189			

To analyse the metallographic structure of the 3 metallic samples, specimens were prepared, and the surfaces were attacked with 5% Nital solution. The images obtained with Olympus BX 60M are in Figure 1.

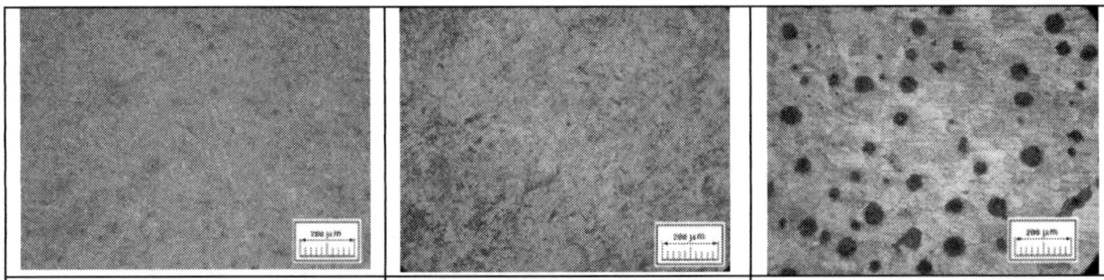

a) Sample 1, X 200	b) Sample 2, X 200	c) Sample 3, X 200

Fig.1. Metallographic structure of the metallic samples.

From Figure 1 the samples 1 and 2 are hypoeutectoid steels and has a martensite and chromium and nickel carbide and sample 3 it is a cast nodular iron in a ferrite-pearlite matrix. The chemical composition presented in Table 1 show that sample 1 it is 34CrMo4, EN 10083-3:2006, sample 2 it is 25CrMo4, EN 10083-3:2006, and sample 3 it is EN-GJS-HB 230, EN 1563:2018. Sample 1 was gas nitride obtaining a hardness of 540 HV1, the sample 2 was hard chromium plated obtaining a hardness of 910 HV1. In Figure 2, the thickness of hard chromium plated stratum (0.1 mm) are presented.

Fig.2. Hard chromium plated stratum thickness.

Elastomers type and main proprieties are presented in Table 2. Were tested polybutadiene rubber (BR), polybutadiene acrylonitrile rubber (NBR), polybutadiene acrylonitrile carboxylate (XNBR) and polyamide (PA 6) with commercial name Ertalon 6 SA.

Table 2. Mechanical characteristics of tested elastomers.

Mechanical characteristic	Elastomer type			
	NBR	XNBR	BR	PA 6
Young's modules, [MPa]	3.5	4.5	4.5	3.25
Ultimate strength, min. [MPa]	20	15	15	76
Elongation at break, min. [%]	350	300	300	50
Hardness, [Shore]	80±3	80±3	70±5	73

There are many pumps manufacturers interested in recommendations for elastomers used in pumps construction. In the present there is no standard to limit the use of elastomers. The ISO sub-committee tasked with reviewing ISO 15136-1 including specialist and pumps manufacturer representatives such as PCM, Weatherford, Netzsch, operators as Exxon, Total, Shell etc. do not impose some specific elastomers, [4]. Thus, the pump manufacturers must perform tests in order to choose the best elastomer material for specific application.

2.2 Wear tests

2.2.1 Wear tests on CSM pin -on-disk microtribometer

Disk samples with Ø 30 mm and thickness of 10mm were prepared of polybutadiene rubber (BR), polybutadiene acrylonitrile rubber (NBR), polybutadiene acrylonitrile carboxylate (XNBR) and polyamide (PA 6).

Static partner was with cubic shape with dimensions of 3.97X3.97X3.97 mm of 34CrMo4, 34CrMo4 nitrided and EN-GJS-HB 230. Due to the small dimensions of cubic samples, the hard chromium plated could not be obtained.

Tests were conducted in dry conditions in air RH=57%, at temperature of 20°C, normal load of 2N (0.12689 Pa), sliding speed of 0.2 m/s at a radius of 10 mm, friction length of 50 m, [5]. In Figure 3 it is shown the image of Instrument X tribometer screen caption for couple EN-GJS-HB 230 with elastomer BR.

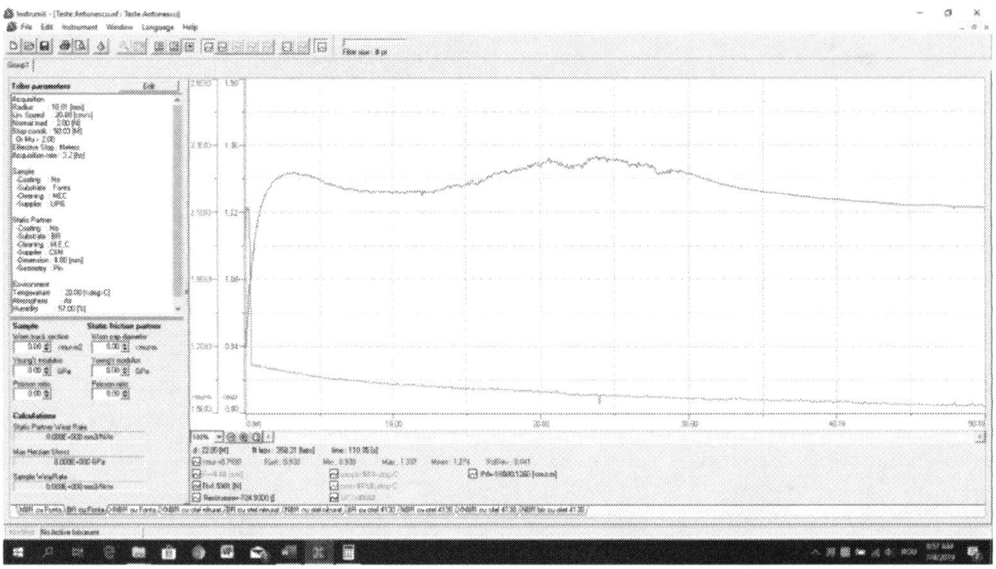

Fig.3. Image of Instrument X tribometer screen caption for couple EN-GJS with elastomer BR.

In Figure 4 we present the friction coefficient variation for steel 34CrMo4 with different elastomers.

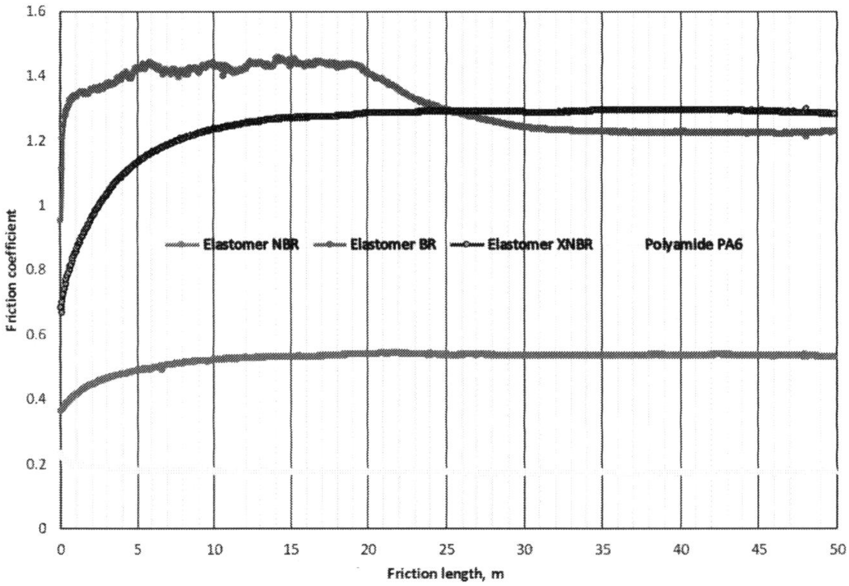

Fig.4. Friction coefficient vs. friction length for couples 34CrMo4- elastomers.

We could observe from fig.4 that the smallest values of friction coefficients were obtained for polyamide PA6 in couple with steel 34CrMo4. Similar results were obtained for 34CrMo4 nitrided and EN-GJS-HB 230. In Table 3 are presented the results obtained for tested materials couples.

Table 3. Friction coefficients values for tested materials couples.

Materials couples	Hertzian pressure [Pa]	Maximum value of friction coefficient [-]	Average value of friction coefficient [-]	Friction coefficient stabilized at 50m [-]
Rubber NBR with EN-GJS-HB 230	0.12689	0.812	0.796	0.812
Rubber BR with EN-GJS-HB 230		1.337	1.274	1.2313
Rubber XNBR with EN-GJS-HB 230		1.006	0.975	0.9768
Rubber NBR with 34CrMo4 nitrided		0.978	0.850	0.828
Rubber BR with 34CrMo4 nitrided		1.269	1.141	1.086
Rubber XNBR with 34CrMo4 nitrided		0.878	0.849	0.878
Rubber NBR with 34CrMo4		0.541	0.521	0.530
Rubber BR with 34CrMo4		1.455	1.315	1.225
Rubber XNBR with 34CrMo4		1.296	1.242	1.281
Polyamide PA6 with 34CrMo4 nitrided		0.258	0.151	0.143
Polyamide PA6 with EN-GJS-HB 230		0.250	0.182	0.175
Polyamide PA6 with 34CrMo4		0.231	0.173	0.171

The metallic materials type does not have an important influence on friction coefficients as we could see from table 3.

2.2.2 Wear tests on Amsler A135 type tribometer

On Amsler tribometer were used cylinder- shoe couples. Cylinder samples with Ø 30 mm were made of 25CrMo4 hard chromium plated, 34CrMo4 nitrided and EN-GJS-HB 230. Shoe samples were made of polybutadiene rubber (BR), polybutadiene acrylonitrile rubber (NBR), polybutadiene acrylonitrile carboxylate (XNBR) and polyamide (PA 6). Tests were conducted in dry conditions (air RH=57% at 20°C ambient temperature), at a sliding speed of 0.314 m/s and a normal load of 500N for 30 minutes. Were established friction coefficients and gravimetric wear.

In Figure 5 we present the friction coefficients for couple cylinder of EN-GJS-HB 230-shoe of rubber BR, NBR, XNBR and polyamide PA6.

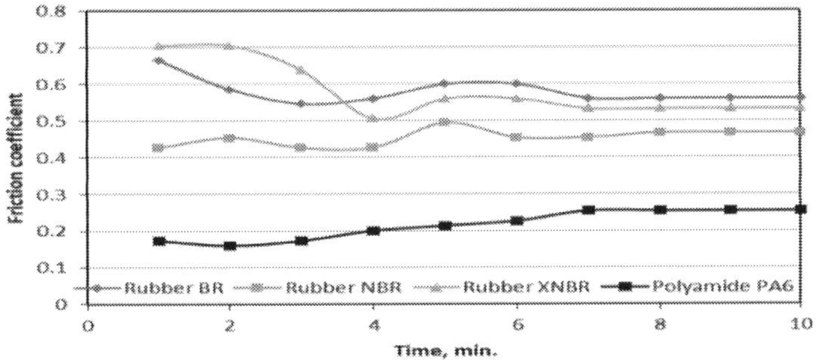

Fig.5. Friction coefficients for couples' cylinder of EN-GJS-HB 230- shoe of rubber BR, NBR, XNBR and polyamide PA6.

Similar results were obtained for all tested couples. In figure 6 are the wear curves for tested elastomers. Gravimetric wear was obtained with a analytical balance type Mettler H35 with 10^{-4}grams precision.

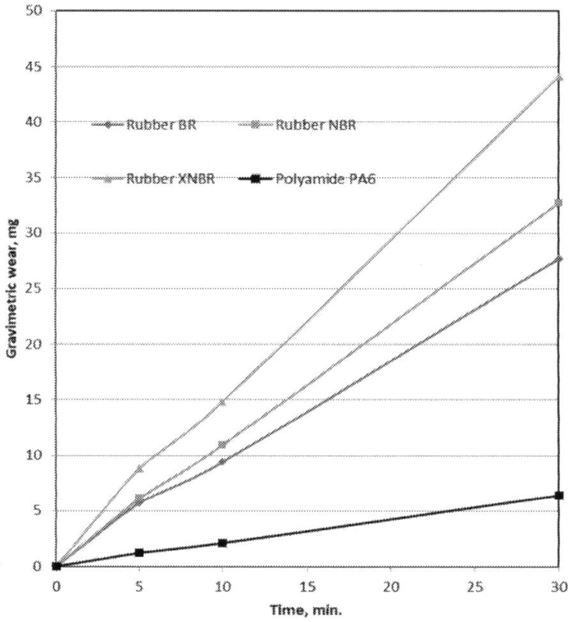

Fig.6. Wear curves for tested elastomers.

Due to the high hardness difference between metallic materials and elastomers were not observed wear at tested metallic materials.

3 CONCLUSIONS

Were analysed 3 largely used metallic materials at pumps construction. Wear tests performed on plane-to-plane couples on CSM tribometer and on cylinder-shoe on Amsler show similar values of friction coefficients for the same materials couples. Differences are caused by temperature influence and the rigidity of elastomer materials. The best wear behaviour was obtained for polyamide PA6. Metallic material has no significative influence on friction coefficients and wear at tested materials couples. This study will be continued with wear tests in the presence of working medium with solid particles and corrosive behaviour.

References

1. H. Wang, S. Wang, and X. Lv, Mechanika, **22**, 4, 308(2016)
2. R.G. Ripeanu, V. Ispas, D. Ispas, FME Transactions, **43**, 3, 192 (2015)
3. B. Guo, X. Liu, X. Tan, *Petroleum Production Engineering* (GPP, Elsevier 2017)
4. K.L. Kishiuchi, *Recommended Practice Elastomer Classification and Characterization* (C-FER Technologies 2004)
5. I. Tudor, A. Popescu, R.G. Ripeanu, A.C. Drumeanu, V. Braic, M. Balaceanu, A. Vladescu, M. Braic, J. Balk. Trib. Assoc., **15**, 2, 156 (2009)

Experimental Researches to Establish the Optimum Hardbanding Technology and Materials of the Heavy Weight Drill Pipe

Mihaela Madalina Caltaru[1], *Razvan George* Ripeanu[1,*], *Marius* Badicioiu[1], and *Dragos Gabriel* Zisopol[1]

[1] University Petroleum-Gas of Ploiesti, Department of Mechanical Engineering, Blvd. Bucuresti, no.39, 100680, Ploiesti, Romania.

Abstract. The heavy weight drill pipe, one of the most expensive components of the drill string, is exposed, beside fatigue and corrosion, to an intensive wear process as a result of the friction with the inner wall of the casing or the borehole wall, that lead to a drastically decrease of it durability. In order to improve the durability, the worldwide specialists apply the hardbanding technology to increase the heavy weight drill pipe wear resistance. The major problem raised by the hardbanding process is the selection of the most suitable wear resistant alloy and the optimum hardbanding technology. The present research work investigate the possibility of hardbanding the heavy weight drill pipes by using the gas metal arc welding process, taking into considerations three different wear resistant materials with the trade name ARNCO 100XT, ARNCO 300XT and FLUXOFIL M58.

1 Introduction

In the petroleum industry, the drill string is the drilling assembly that is used to rotate the bit, to conduct the drilling fluid to the bit, to realize the weight on bit and control the drill bit, [1]. Wear, corrosion and fatigue are recognized as a major cause of drill string failures. In practice, the oil industry attempts to guard against the likelihood of these costly failures by improving the surface characteristics of the drill string components, by using different technologies, [1, 2].

The heavy weight drill pipes, the most expensive elements of the drill string, are inserted between the drill collars and drill pipes in order to ensure that the drill pipe string is kept in tension and realize a flexible transition between the stiffer drill collars and limber drill pipe, [1]. The heavy weight drill pipe has the same outside diameter as standard drill pipe but a much reduced internal diameter, has an extralong tool joint (pin/box) and an integral centre wear pad (midsection/centre upset) which acts as a stabilizer, increasing the overall stiffness of the drill string (fig.1), [1, 2]. Due to friction with the inner surface of the casing (that prevent the borehole from caving in and to support the weak or fractured formations) or the open hole wall, the durability of the heavy weight drill pipes decreased,

* Corresponding author: rrapeanu@upg-ploiesti.ro

and to improve it, the worldwide specialists apply hardbanded wear resistant materials onto the surface subjected to wear, respectively onto the tool joint pin/box of the heavy weight drill pipe and onto the central zone of the pipe (fig. 1), [1, 2].

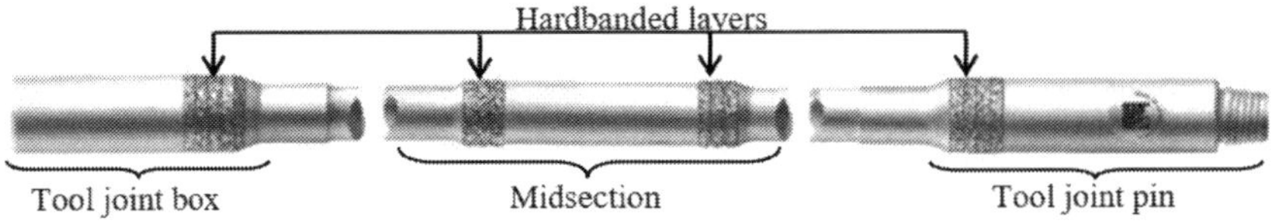

Fig.1. The hardbanding areas of the heavy weight drill pipe.

The deposit hardness is very important in the drilling process because a value higher than 57-62 HRC will cause an excessive wear of the casing, [3, 4, 5]. The present paper bring forward the results obtained in the case of three different wear resistant wires, FLUXOFIL M58, ARNCO 100XT and ARNCO 300XT, used for heavy weighty drill pipes' hardbanding.

2 Experimental procedures

2.1 Materials

The experimentally researches were performed on specimens manufactured by steel AISI 4145H (46MoMnCr10), the reference material utilized in heavy weight drill pipes construction. The chemical composition and the mechanical characteristics of the AISI 4145H steel, according to the manufacturer, are presented in table 1, [3].

Table 1. Chemical characteristics of AISI 4145H alloy steel.

C, %	Mn, %	Si, %	P, %	S, %	Cr, %	Mo, %	Al, %	Ni, %	Cu, %	Hardness, HB	Tensile strength R_m, N/mm^2	Elongation A, %	Striction Z, %
0.43	0.97	0.31	0.022	0.012	0.99	0.21	0.028	0.11	0.14	336	1124.7	14.75	51.0

The specimens used had 1500 mm length, considered to be sufficient to reproduce the heating and cooling conditions of the hardbanded area under real conditions, with different outer diameters, that corresponds to the surfaces subjected to hardbanding, respectively ϕ 133.3 mm (corresponding to the outer diameter of the tool joint pin/box) and ϕ 114.3 mm (corresponding to the midsection), with an inner diameter of ϕ 63.5 mm. A recess grove was machined onto the specimens (fig. 2) to be filled flush with hardbanding material in order to avoid the wear of casing inner diameter, [3]

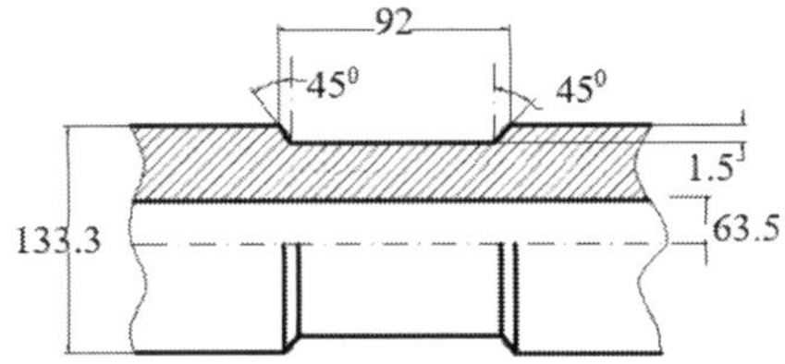

Fig.2. Recess grove of the specimens.

In the experimental works were used tree different wear resistant flux core tubular wires of 1.6 mm diameter, named FLUXOFIL M58, ARNCO 100XT and ARNCO 300XT, having the characteristics, according to manufacturer, presented in table 2, [3]

Table 2. The deposit characteristics.

Type of wire	Chemical composition of the deposit											Hardness
	C, %	*Mn*, %	*Si*, %	*P*, %	*S*, %	*Cr*, %	*Mo*, %	*Nb*, %	*Ni*, %	*B*, %	*Fe*, %	*HRC*
ARNCO 100XT	0.7-0.75	1 max	0.5-07 max	0.030 max	0.030 max	7.0-8.0	0.15-0.20	-	-	-	rest	48-52
ARNCO 300XT	0.75-0.80	0.90-1.2	0.55-0.75	0.030 max	0.030 max	-	-	5.3-5.7	2.0-2.5	3.8 -4.2	rest	min 60
FLUXOFIL M58	0.45–0.85	1.5-2.1	0.50-0.85	0.020	0.020	5.0-7.0	0.35-0.85	-	-	-	rest	57-62

The metal active gas welding process (GMAW) was used for hardbanding.

2.2 Hardbanding equipment

The hardbanding equipment, used, allowed, from the cinematic point of view, the following movements: continuous rotation of the specimen, axial feed of the welding torch and oscillate the welding torch (fig. 3), [3, 4].

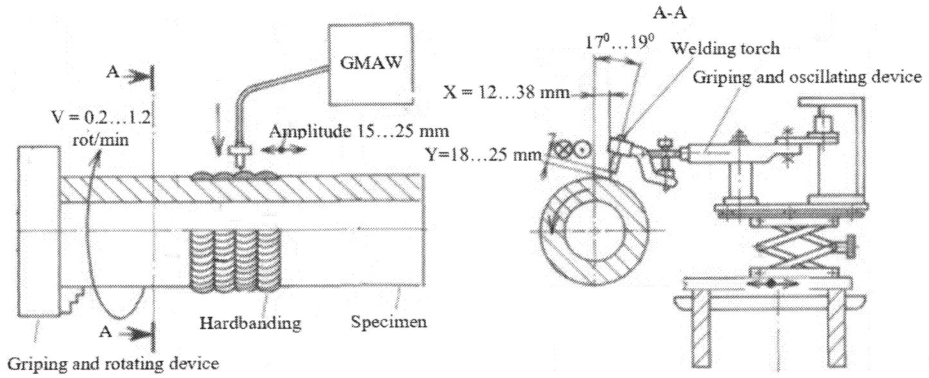

Fig. 3. The hardbanding equipment scheme.

2.3 Hardbanding research methodology

In order to establish the optimum hardbanding technology, based on the own experience, on the wear resistant wires manufacturers' recommendations and on the technical information presented in the specialized literature by the specialists, the methodology presented below has been applied, [3].

A) Specimen preparation - After turning the specimen by using a lathe machine, in order to prepare the recess grove, a visual inspection of the specimen's surface was performed, to ensure it is clean and without rust, dirt, grease, oil etc.

B) Preheating the specimen - Before the hardbanding process, the specimen was pre-heated at 100 degree Celsius, in order to not cause undesirable cracking in the parent metal).

C) Hardbanding the specimen - The number of the hardbanded layers and the order of their applications were different, depending on the width of the hardbanded area (fig. 4).

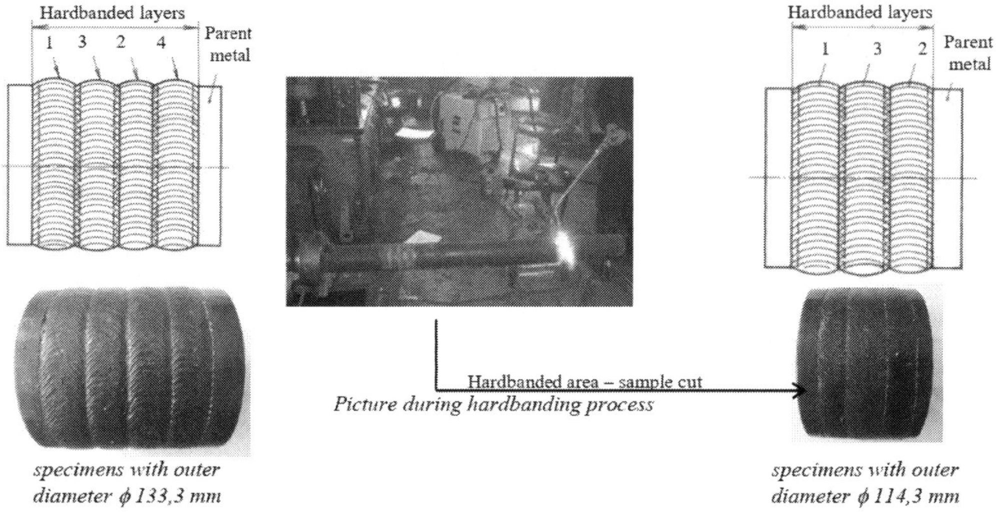

Fig.4. Applications of the hardbanded layers.

The recess grove of the specimen was filled flush with hardbanding material by using the technological parameters mentioned in table 3.

Table 3. Hardbanding technological parameters.

Technological parameter	**Type of wear resistant wire /adopted values**		
	FLUXOFIL M58	*ARNCO 100XT*	*ARNCO 300XT*
Polarity	reverse	reverse	reverse
Amperage I_s, A	250…270	250…270	250…270
Voltage U_a, V	23…25	23…25	23…25
Shielding gas	80% Ar + 20% CO_2	100 % CO_2	100 % CO_2
Shielding gas flow rate Q_g, l/min	13.5	14	14
Filler metal speed v_{sm}, m/min	4.5	4.5	4.5
Y distance (from figure 3), mm	20	20	20
X distance (from figure 3), mm	20	20	20
Rotating speed of the specimen, rot/min - for specimens with ϕ 133,3 mm - for specimens with ϕ 114,3 mm	 0.57 0.66	 0.57 0.66	 0.57 0.66
Welding speed v_s, m/h	14.5	14.5	14.5
Oscillating the welding torch: - oscillating speed, oscillation/min - oscillation amplitude, mm	 60 25	 60 25	 60 25
Preheat temperature T_{pr}, ^{0}C	100	100	100
Maximum interpass temperature, ^{0}C	350	350	350

D) Controlled cooling of the hardbanded specimen - After finishing the hardbanding process, the hardbanded specimens were slowly cooled by using a thermal isolated blanket with a thickness of 100 mm, to avoid the cracks in the hardbanded layers.

E) Quality Control - After cooling of the specimens, the quality of the hardbanded area was verified by: macroscopic analyses, metallographic analyses by optical microscopy, Rockwell hardness and micro hardness measurements.

2.4 Experimental results and discussion

The ***macroscopic analyses*** consisted in investigation by naked eye of the hardbanded layers had indicated that the profiles of the hardbanded layers are flat (deviation from this kind of profile can cause premature wear of the casing, when the profile is convex, or premature wear of the heavy weight drill pipe, when the profile is concave, [4]). The surface of the hardbanded layers with ARNCO 100XT and FLUXOFIL M58 doesn't contain visible cracks, but in the case of ARNCO 300 XT the cracks were observed. The adhesion between the substrate (parent metal) and the hardbanded layers, also between two adjacent layers is good (without any discontinuities or voids).

The ***metallographic analyses by optical microscopy*** were performed on metallographic samples cut from the transition zone parent metal-hardbanded layer, also from the overlap zone between two adjacent hardbanded layers. In figure 5 are presented the metallographic structures obtained for the hardbanding. The analysis shows a good adhesion between the hardbanding and the parent metal, and without any cracks for layers hardbanded with ARNCO 100XT and FLUXOFIL M58. In the case of layers hardbanded with ARNCO 300XT, the high percent of niobium leads to the separation of primary niobium carbide, and the high percent of boron leads to the separation of borides, and the separation of these chemical compounds (crystals of carbides and borides) places this material in the field of cast irons. The transition area from the hardbanded layer to the parent metal is more obvious, in this case, which also explain the appearance of cracks (fig. 5).

FLUXOFIL M58 (Attack: Nital)

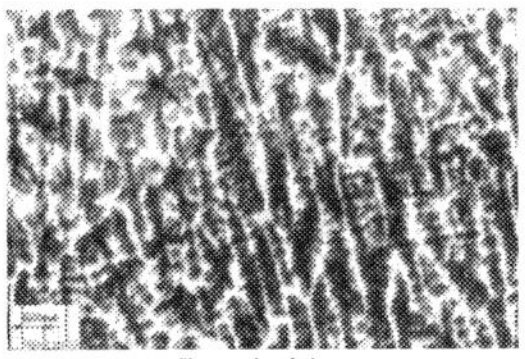

Hardbanded layers: dendritic casting structure

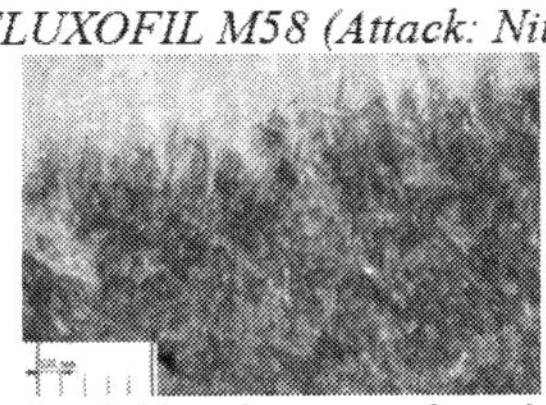

Hardbanded layer/parent metal transition area: martensitic-bainitic structure in the heat affected zone

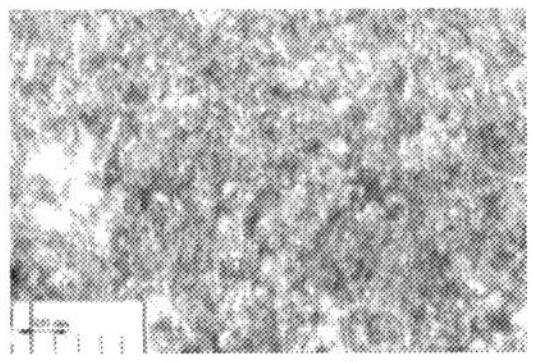

Parent metal: sorbitic -bainitic structure

ARNCO 100XT (Attack: Nital)

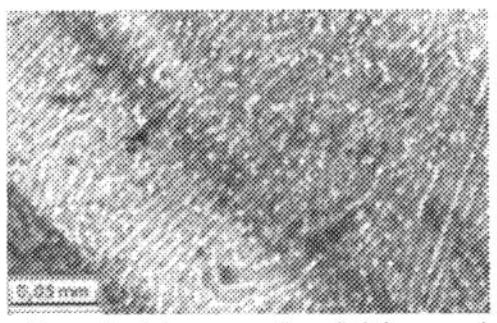

Hardbanded layers: dendritic casting structure with carbides separation

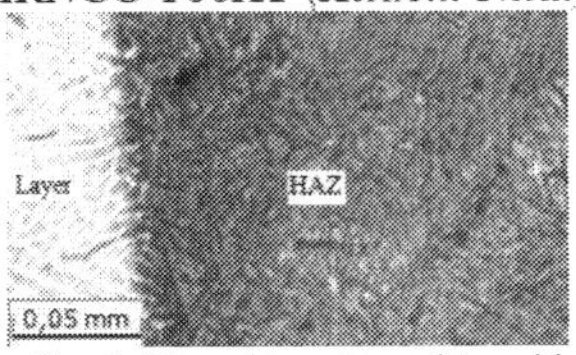

Hardbanded layer/parent metal transition area: martensitic-bainitic structure in the heat affected zone

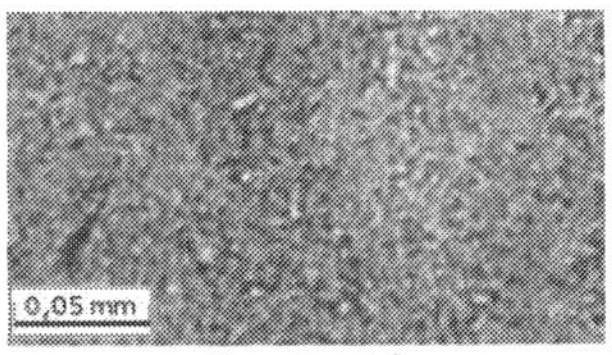

Parent metal: sorbitic -bainitic structure

ARNCO 300XT (Attack: Nital)

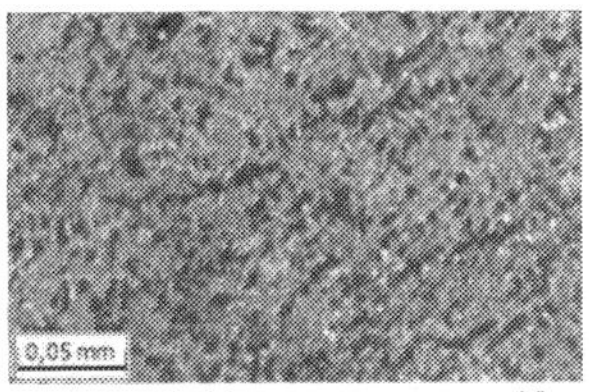

Hardbanded layers: Structure with separation of primary niobium carbides and borides (crystals)

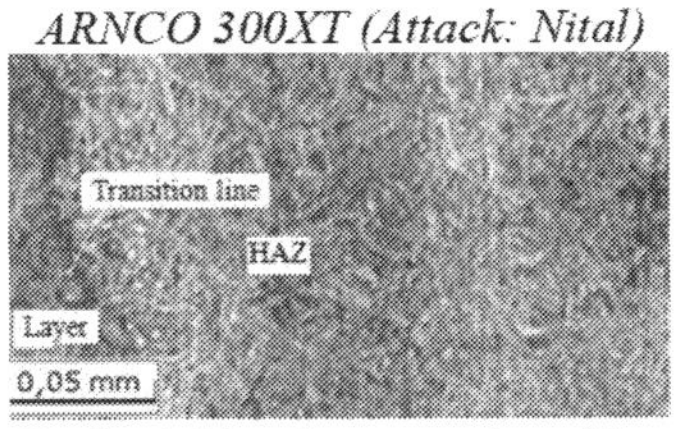

Hardbanded layer/parent metal transition area

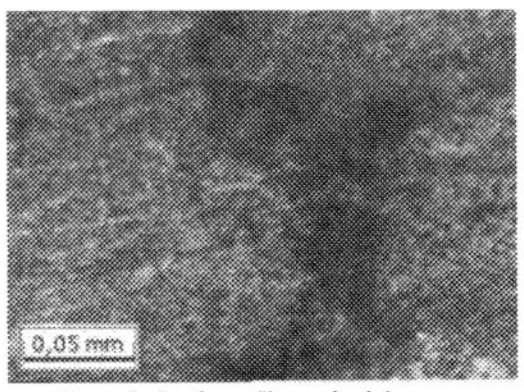

Crack in hardbanded layer

Fig. 5. Metallographic structures for the hardbanding.

The ***Rockwell hardness*** measurements were performed on metallographic samples, and the average values of HRC hardness are presented in table 4.

Table 4. Rockwell hardness measurements.

Type of wire	Rockwell hardness, [HRC]		
	Hardbanded layer	**Heat affected zone**	**Parent metal**
FLUXOFIL M58	55	50	36
ARNCO 100XT	58	52	36
ARNCO 300XT	62	55	37

The ***micro-hardness measurements*** were performed on metallographic samples with a load of 100 g. The measurements performed shown high values of hardness in the hardbanded layers and a maintaining initial hardness in the parent metal (fig. 6).

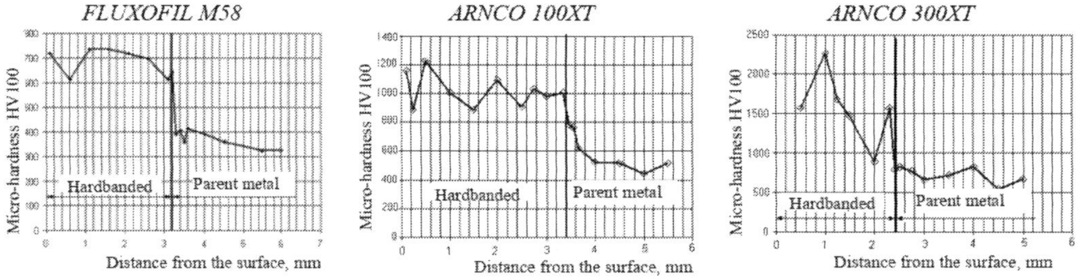

Fig.6. Micro-hardness of the hardbanding.

3 Conclusions

The hardbanding equipment and the hardbanding technology developed in the frame of this research work allow to perform, in good conditions, the hardbanding with wear resistant alloys of the heavy weight drill pipes, by using the gas metal arc welding - GMAW process. The comparative researches performed with three different wear resistant materials, have shown that ARNCO 100XT wire, successfully used worldwide for hardbanding of the tool joints and the drill collars, can be replaced with the FLUXOFIL M58 wire, available in our country, allowing the delivery of heavy weigh drill pipes hardbanded, competitive worldwide from the point of view of the quality of the hardbanded layers and, at the same time, at a lower price. The experimental results obtained do not recommend the wire ARNCO 300XT to be use for hardbanding of the heavy weigh drill pipes because the hardbanded layers cracks.

References

1. V. Ulmanu, *Drillpipe & BHA Design and Components*, Well Academy Program, Petrom-OMV, p. 71, (2009).
2. V. Ulmanu, *Petroleum Equipment Manufacturing and Repairing Technology* (in Romanian), ILEX Publ. House, (2002).
3. V. Ulmanu, M. Minescu, Gh. Zecheru, Gh. Drăghici, D. G. Zisopol, M. Călţaru, M. Bădicioiu, C.N. Trifan, *Manufacture the heavy weight drill pipe competitive worldwide by hardbanding with wear resistant alloys* (Research project no. 2053/2004 - in Romanian), Petroleum-Gas University, (2006).
4. Arnco Technology Trust Limited, *Arnco hardbanding specification for the preparation, application and inspection of Arnco hardbanding products (100XT, 150 XT, 300XT and 350XT)*, Version 2.1. June 28th (2012).
5. K. Zhang, Z. Wang and D. Wang, *Research Progress of the Drill String Hardbanding Materials*, DOI: http://dx.doi.org/10.5772/intechopen.90013, (2019).

Research, study, design and development of an artificial ear splint model by using a 3D printer

Athanasios Argyropoulos [1,*], *Pantelis* N. Botsaris [1]

[1]Democritus University of Thrace, Dep. Of Production Engineering and Management, Lab. of Mechanical Design, Greece

Abstract. The aim of this research study centers around the design and development, using a three-dimensional model of an ear-splint. This study results from the need of a non-invasive method, as the current techniques used in otoplasty have succeeded in reducing the risk of complications, but without drastically reducing the rate of postoperative deformity. Indeed, stick out ears present a deformity that occurs in 5% of the population. Factors such as psychosocial impact and intimidation lay the foundations for the need for such a method. This paper essays to create, through 3D printing, a case customized into the ear of any person of all ages. The splint was made on Zortrax M200 3D printer model, equipped with the V3 Hotend head, and the material used for printing was 16 grams of Z-FLEX. It is noted that the creation of the case is innovative and original. After putting the ear-splint into test, by applying it on a human ear, two conclusions emerge. First, it is very light, weight, just 16 grams, and second, it offers comfort and convenience to the user. In order to get results for its practicality, a clinical rehabilitation study is recommended.

1 Introduction

Among the most common genetic abnormalities that can occur in the head areas are the stick-out protruding ears or prominent ears. The stick-out protruding ears are a condition that affects both sexes equally. The most common problems are the deformation anomaly of concha and the unformed antihelical fold od of the ear. The diagnosis of prominent ears is always subjective. Although various measurements have been made for the dimensions of the ears, the terms normal ears and dumbo ears are unclear. The incidence of stick out protruding ears is about 5% of the world's population, as it is the most common ear deformity. Otoplasty is one of the most common pediatric otolaryngological procedures. The signs of this deformity are mainly aesthetic and the corrective procedures are performed by a wide range of doctors, either in a private clinic, or by plastic surgeons or otolaryngologists and pediatricians. Surgical correction of the protruding ears is usually performed at the ages of 4 to 5 years, when the ear completes its development. There is no specific timetable for surgical or non-surgical implementation, however, the ultimate goal is to prevent children from being bullied at school. [1, 2, 3] Today, the correction is covered by a non-invasive method through the product "Earbuddies" as it seems in the figure 1. [4] It should be noted that this method is based on the technique that a splint placed in the ear canal is fastened with tape. The pressure of the splint suspends the spring of the cartilage of the ear, while correcting its natural curves. When the cartilage hardens over time, a permanent and controlled shape is created. Both methods of otoplasty, invasive and non-invasive, present a number of problems that the invention is going to remedy. Specifically, there are both aesthetic and functional problems after completing the treatment. These include pain during sleep, increased sunburn, bleeding problems, and discomfort when patients wear glasses, a hat or hearing aids. The most basic disadvantage of all these methods, whether invasive or non-invasive, is the fact that after some time the ears come back into their initial place. [1, 2] Therefore, there is a clear need for a non-invasive method that can meet the needs of the patients. Namely, the presence of a product that does not require the intervention of an otolaryngologist or a plastic surgeon, but the role of the physician is only advisory. The creation of the case that fits into the ear will fill exactly this gap that exists today on the world market. The case can be used by people of all ages, since one or more cases are created for each human individually. All the necessary steps that were followed during experimental procedure for the study, design and construction of the artificial ear splint are illustrated bellow.

* Corresponding author: franki.thanasis@gmail.com

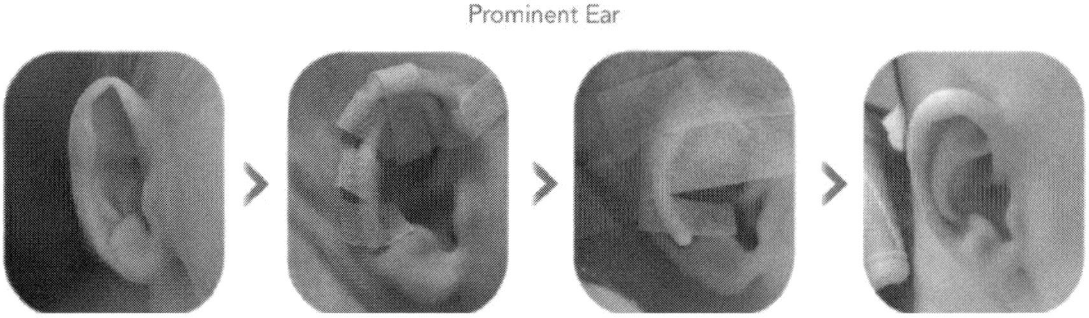

Fig. 1. Illustration of the product Earbuddies and how it corrects the deformity of the stick out protruding ears. [4]

2 Methods

2.1 Design of the artificial ear splint model

During the design of the artificial ear splint model a novel method was created. This model aims to restore stick-out ears back to their normal position, according to human anatomical features, through forces exerted on the helix and antihelix of the ear. The natural complexities of the ear surfaces presented the main challenge towards the success of this project. For this reason, a specific methodology had to be followed, involving photogrammetry technique, design and reverse engineering principles. Photogrammetry has emerged as a leading approach for photorealistic digital replication and 3D scanning of real-world objects, particularly in areas of cinematic visual effects and interactive entertainment. While the technique generally relies on simple photography methods, the foundational practices for the field of human photogrammetry remain relatively undocumented. Human subjects are significantly more complex than still life, both in terms of photogrammetric capture, and in digital reproduction. [5] Initially we had to found the sample of the human ear and one way to scan it. This was achieved by using the photogrammetry technique. First, we had to find the appropriate photographic instrument. A mobile phone with a 48 megapixel camera and multi-taking mode, was the medium of producing a range of photos of the same object from different angles. The purpose of many shots was to illustrate the final model with more details and eliminate the "noise" from it. This was followed by the selection of the space, as the lighting is the main factor for the taking of a photo. First, one hundred photos were taken outdoors with natural light and then one hundred photos were taken inside by using the camera's flash. The next step was to import all the photos into the Agisoft (Metashape) program. Agisoft is a software that performs photogrammetric processing of digital images of the ear and produced its three-dimensional digital model. The process was completed with one problem, the lack of scale from the digital model. [6] This problem was solved in the Meshlab program, an open source system that provides a set of tools for editing and defining project points. More specifically, 3 points were selected between the helix and the head, the actual distances were measured and then, by selecting the same 3 points within the program, the scale of the design was adjusted.

For display, data exchange, and construction, the 3D digital model was divided into a three-dimensional model consisting of 120 different vertices. However, the 3D representation can be modified and after numerical simulation it was no longer corresponds to the original model. Then required a method of reverse engineering in order to recreate a three dimensional model. [7, 8] The solution was provided by using Geomagic, a software that provides accurate three dimensional digital models and assemblies of CAD type natural objects. This software helped to modify the model file from .obj to x_t file. The .obj file format is a simple data format that represents the geometry of two-dimensional and three-dimensional designs, while the .x_t file format indicates a Parasolid file that makes it possible to process model surfaces in a CAD program. The whole process is schematically described in Fig. 2.

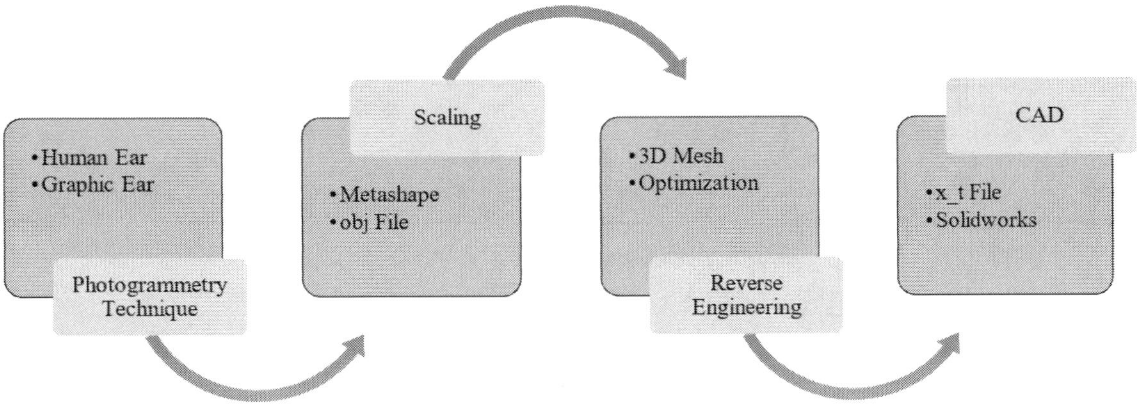

Fig. 2. Conversion process from Human ear to CAD model ear. The human ear is converted to graphic ear (obj file) through photogrammetry technique and after some modifications to 3D mesh, making tests and optimizations, the CAD model ear (x_t file) is created.

2.2 Production of the artificial ear splint model

As for the design part, together with the 3D printing process, they are the two most difficult and most important parts of the artificial ear splint production process. In particular, the design was carried out in the Solidworks program and first step needed to be done, was to cut the outer material around the ear and then remove the inner material, leaving only specific surfaces. The shape that emerged from the procedure was also the final shape of the splint, with the only problem being the lack of scale, because the splint had to be applied to the ear. Therefore, after various tests and measurements, we ended up with an increase in thickness by 1,005 mm. [9]

Next, the most important step in the preparation of the ear splint was the 3D printing method, filament deposition material (FDM). The M200 model of the Zortrax equipped with the V3 Hotend head was the 3D printer in which all the experiments were conducted. First of all, in the software provided by this company, Z-Suite, the necessary commands had to be set and it had to be decided which materials will be used. The experiments were performed with Z-ABS, while the basic material of the ear splint is Z-FLEX. At the same time, the first successful 3D printing attempt with Z-FLEX material on a M200 3D printer model was made, according to the company's literature, because the head was changed from V2 to V3 in order to print the Z-FLEX material.

Figure 3 shows the illustration of the design procedure for the artificial ear splint model that have been described before. In particular, it seems the whole methodology, how from a simple human ear, it can be fabricated, 3D printed an object that can offers a solution to everyday problems, such as the deformity of stick out protruding ears. Especially, the last picture shows the back view of a human who wears the ear splint.

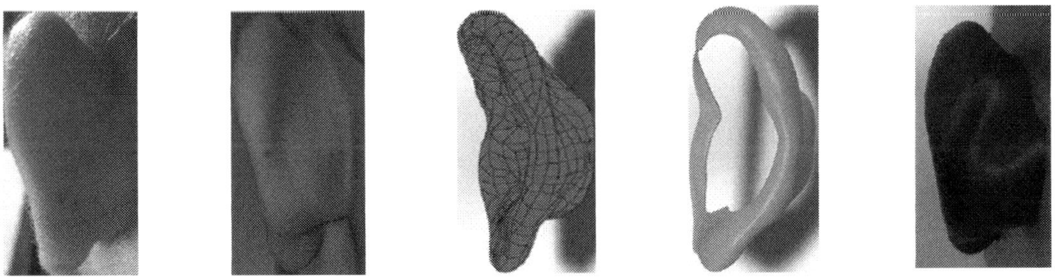

a) b) c) d) e)

Fig. 3. Illustration of the design procedure for the artificial ear splint model. The human ear (a) is used to give the digital model of the ear. (b) This shape is used as a basis for the computer aided design (CAD) model (c), which gives the final form (d), after some modifications, the artificial ear splint model is worn by a human (e). [6, 9]

3 Experiments

3.1 Experimental procedure

As for the experimental process, it took only 16 grams of flex material, setting the appropriate settings on the program's platform, and after 4 hours and 43 minutes the ear splint was ready for use. Its dimensions according to the model of the human ear used are 79.36 mm long, 34.01 mm wide and 44.05 mm high just as it is applied to the ear. The commands set in the program are shown at table 1.

Table 1. 3D printing settings for materials flex and abs.

3D Printing settings	Flex	Abs
Nozzle Diameter	0.4 mm	0.4 mm
Layer Thickness	0.09 mm	0.09 mm
Extrusion Temperature	230 °C	230 °C
Platform	60 °C	50 °C
Infill Density	80 %	90 %

3.2 Abs material

The abs material is a flexible and economical material that perfectly suits the needs of users who are dealing with 3D printing for the first time and have one of the M200 or M200 Plus models of Zortrax. Its low cost allows for many effective and different tests across the available range of dimensions for the thickness of each layer. This material is characterized on the one hand by resistance to high temperatures, resistance to basic operating tests and on the other hand, by a high level of shrinkage. The problem with this material was that it did not meet the basic requirement of the splint, to give elasticity and therefore it was used exclusively for the study and correction the problems of the final ear splint design. Therefore, there was a need to find an elastomeric material that is human-friendly and provides the splint with the required elasticity of an object that fits right on a human ear. Figure 4 shows the ear splint model which was successfully fabricated with abs material on a 3D printer. [10]

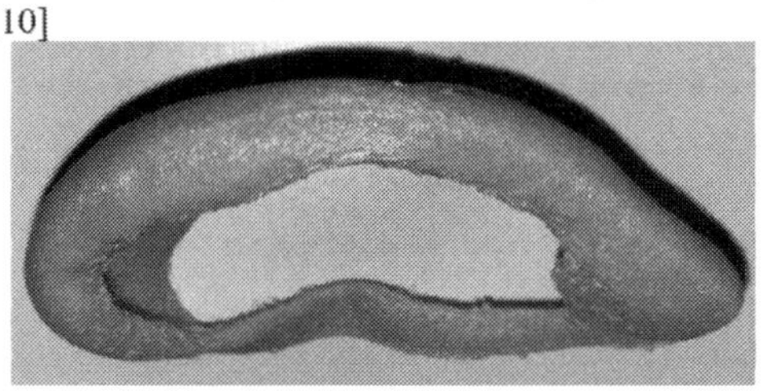

Fig. 4. Illustration of the artificial ear splint model, after successfully 3D printed with abs material.

3.3 Flex material

The flex material was the one who solved the problem. An elastomeric and thermoplastic material of the polyester category with a print speed of 45 mm per second (mm / s). The thread has a large adhesion between the layers and a tolerance of dimensions less than +/- 0.02 mm (mm). Also, flex is elastic, can bend without breaking and is resistant to various chemicals such as gasoline, ethyl alcohol, butane and carbon monoxide. The only shade of this filament is black. Flex is ideal for 3D printing models that are intended to be folded frequently, such as tires and elastic hinges. Its elasticity, texture and the fact that it is human-friendly are the reasons that it was chosen as the basic material for printing the artificial ear splint. Figure 5 shows the ear splint model which was successfully fabricated with flex material on a 3D printer. [11]

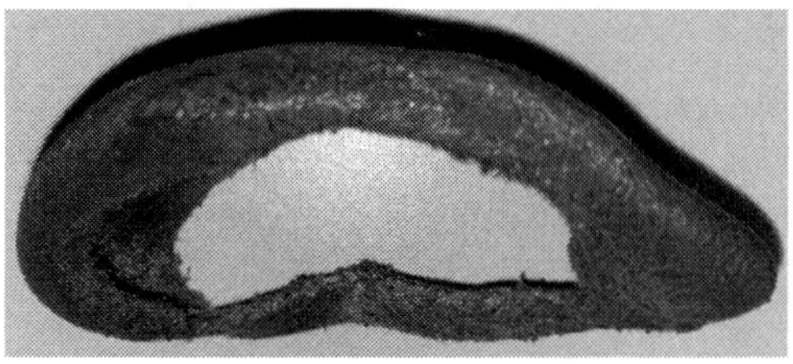

Fig. 5. Illustration of the artificial ear splint model, after successfully 3D printed with flex material.

4 Results

We presented a non-invasive method of restoring the stick out protruding ears by using an artificial ear splint model. The way it was designed and manufactured allows it to be applied to different types of people with prominent ears, regardless of age and geometry of the ear, since one or more cases are created for each human individually, using reverse engineering and 3D printing techniques. Applying the case to the human ear, any patient could gain the substantial benefits of its usage. On the one hand, it is very lightweight, since it weighs only a few grams, and on the other hand, it provides comfort and convenience to the wearer, while the other already existing methods are causing pain and other functional problems. Even during sleep when the head is pressed against the pillow no pain or discomfort has been observed. The use of an elastomeric and thermoplastic material, namely flex, has contributed to the aforementioned results. Having been used to fabricate the ear splint, this material has been proven to give it the flexibility it requires/is necessary for such an object being placed on the head. The tray model was printed on a 3D printer, specifically the Zortrax 3D printer M200 model. The V3 head was added in order to print the Z-FLEX material. The invention can be manufactured on any 3D printer model that allows. It is worth mentioning that there was no abrasion, itching or pain at any point during the ear splint application. Meanwhile, it should be emphasized that it is possible to use either myopic glasses or sunglasses over the case.

As for the manufacturing process, there are some modifications that could be made. For example, a different way of approaching the sample, which includes scanning from a modern 3D scan with the ability to capture the object at an angle of 180 degrees. In this way the human factor is removed from the process and the possibility of an error occurring, which in the case of being studied, contributed to the taking of the photographs of the ear model. Also, there could be used arms that hold the camera in place and provide the ability to take photos from all possible angles and perspectives.

5 Discussion

The creation of the splint was carried out at an initial level in the context of the preparation of an undergraduate dissertation, and a laboratory product was prepared. However, many future improvements that can be made in the near future have been discussed and observed, evolving the final product. Initially, it is helpful to conduct a series of experiments with the flex filament as a basic material to prove that it is human-friendly, although in the literature does not mention the opposite. In addition, a future proposal for Zortrax is the preparation of the filament from transparent material (Natural Transparency), as the only available shade for the flex filament at the moment, creates an aesthetically strange result.

Next, it should be noted that without measuring the tension exerted by the ear splint, since no specialized measuring instruments were available, it was observed that force applied on the helix and antihelical fold of the ear. Therefore, there is a need for further research and study of the method with the assistance of specialized scientists, such as otolaryngologists, invasive surgeons, etc. This research can be carried out in the context of a clinical rehabilitation study such as the term in medical science, in order to obtain experimental results on the practicality of the artificial splint. While after the completion of the clinical study and the taking of a sample, the modeling of the object is proposed, with the aim of its possible subsequent industrial production.

This project is supported by the "Athena Research and Innovation Center- Institute of Language and Speech Processing (ILSP)".

References

1. A. L. Schneider, D. M. Sidle, Department of Otolaryngology-Head and Neck Surgery **26**, 19-29 (2018)
2. A. Gantous, A. J. Tasman, J. C. Neves, Facial Plast Surg Clin North Am **26**, 181-192 (2018)
3. L. Kajosaari, J. Pennanen, T. Klockars, Int J Pediatr Otorhinolaryngol **100**, 52-56 (2017)
4. www.earbuddies.com
5. T. Boe, C. Carter, International Journal of Computer Graphics & Animation **10**, (2020)
6. www.agisoft.com
7. R. Bénière, G. Gesquiere, F. Breton, W. Puech, G. Subsol, Computer-Aided Design **45**, 1382-1393 (2013)
8. S. Gauthier, W. Puech, R. Bénière, G. Subsol, LIRMM Laboratory **93**, 67-83 (2017)
9. www.Solidworks.com, S.N: 9020 0085 4501 9225 RTC2 22H7, (2019)

10. www.zortrax.com/filaments/z-abs/
11. https://zortrax.com/filaments/z-flex/

Design and Analysis of a 5-Axis Gantry CNC Machine Tool

Esra Yuksel[1], *Emre* Özlü[1*], *Ahmet* Oral[2], *Fulya* Tosun[2], *Osman Fatih* İğrek[2], *Erhan* Budak[1]

[1]Manufacturing Research Lab, Sabanci University, Tuzla, İstanbul, Turkey
[2]İgrek Makina Döküm Inc.,OSB A.O. Sönmez Bulv. No:10 16140 Bursa, Turkey

Abstract. In this study, design and analysis of a gantry-type 5-axis CNC machine tool is presented with experimental results on a manufactured prototype. Critical points in the design of a large-scaled and heavy-duty machine tool is discussed. Moreover, FE analysis results is also presented with detailed discussion. The measurement results on structural dynamics is shown together with the FE results. Furthermore, the final performance of the machine tool is demonstrated thorough position and velocity measurements of the axes.

1 Introduction

The fast and competitive development of aerospace, automotive and die/mould industries drive machine tool companies to offer better solutions for heavy-duty machining applications. Gantry-type machine tools are one of the most suitable structures with expandable machining span, significant structural rigidity properties and motional precision and accuracy promises [1]. A competitive heavy-duty 5-axis gantry milling machine can process workpieces that are ranged from 16 to 20 tones, with 8 to 10 μm axial precision at X and Y of machine coordinates, while positional accuracy ranges from 10 to 15 μm for these axes. Generally, fork head spindle structures are employed for heavy-duty 5-axis gantry-type milling machines with 95 to 120 degrees rotation capacity at A-axis, while C-axis can rotate 360 degrees. Axial and positional accuracy/precision of Z column can change between 3 and 5 μm.

In this study, the subjected 5-axis milling machine has X-axis (gantry) with length of 6700 mm. Similarly, Y-axis has a length of 3500 mm Thus, there must be three main considerations as well as the other considerations mentioned in [2]: i) production technique of the machine tool components and material selection, ii) structural reliability and stiffness, iii) thermal behaviour and robustness.

Generally, welded or casted components are employed for the construction of gantry-type milling machines. The usage of welded components and joints-especially during the construction of machine base and main frames- is an emerging trend due to lower manufacturing costs compared to casted complex machine tool parts [3,4]. Moreover, welding techniques allow usage of carbon-fiber based composites in machine tool

* Corresponding author : emre.ozlu@sabanciuniv.edu

structures which can be structurally reliable as casted components [5,6]. However, anisotropic properties of these composites cannot maintain the required stiffness levels for long structural components and spindle structures of machine tools [5]. In addition, Gas Metal Arc Welding (GMAW), Manual Metal Arc Welding (MMAW) and Laser Arc-Hybrid Welding (LAHW) are employed for welding of thick metal components [7], and welded materials with these techniques reaches their melting points which can cause distortion, porosity and weak material properties [8]. As a result, Y-axis strokes between 2500 and 3000mm are accepted as standard lengths for large-scaled 5-axis gantry-type machine tools which are produced with welding techniques. Thus, to be able to extend the stroke of X and Y axes, casted components with superior damping properties are required to maintain the desired stiffness properties [9]. Another way to forecast and maintain the required stiffness values is employing FEM and structural optimization techniques [10], such as topology, shape or crossbeam optimizations [1,10,11]. Modelling of mechanical interfaces in FE analyses is critical during the simulation and structural optimization for even regular-sized machine tools [13]. Thus, massive component sizes and weights make contacts more critical for large-scaled machine tools. Moreover, thermal compensation techniques are not very effective for large-scaled machine tools, and thermal design/optimization requires advanced techniques for thermal distortion detection and compensation even for a medium gantry-type 5-axis CNC machine [10]. Therefore, thermal distortion preventing designs are more effective on large-scaled machine tools.

Regarding the above-mentioned facts, this study demonstrates casting method-benefited design strategies for a 5-axis gantry-type CNC machine tool. The effectiveness of the proposed strategies is evaluated with FE analyses and experiments on a prototype machine tool manufactured.

2 Design of the Machine Tool and Control System

The casting-method benefited design strategies are as follows: The main frames are designed as casted mono-block parts, which makes this design unique compared to other CNC structures in the machine tool industry. The mono-block main bodies will imitate a single-volume structure when geometric and dynamic behaviours are considered. The superiority of the casted materials over the welded ones is their ability of damping. Especially, damping ability of Gray iron is at least threefold of steel structures [14]. Frames of X and Y axes and machine base is made of GG25, and GG50 is employed to build Z-axis. The base part of the milling machine is also casted as a mono-bloc piece which weighs 24 tons. Therefore, the first mode of the structure is expected to demonstrate a mode shape of a single-volume structure. The targeted features of the machine tool are shown in Fig.1a. The X-axis of the machine (gantry) which is longer than the usual standards, thus, the usage of casted components may not enough to maintain the required stiffness. Therefore, the internal structure of the gantry is supported with special ribbing elements. Additionally, pretensions are directed to the gantry surface during the casting process. Moreover, rails of the linear guides are casted with these mono-block components to avoid structural deformations on X and Y axes of the machine. The pretensions and casted railways are indicated in Fig.1b. Additionally, a rack and pinion mechanism based linear motion system is employed instead of classical drive screw systems. This type of a linear motion system exposes lesser efficiency losses. Moreover, the structural configuration of the rack and pinion mechanism on the gantry provides an additional support for the Z column as shown in Fig.1c.As mentioned earlier, thermal distortion preventing designs are more effective compared to thermal compensation techniques during design stage of large-scaled machine tools. Thus, as a cooling precaution, there are pipelines for water circulation under the slideways of X and Y axes, to keep the temperature change in the minimum. These pipes are inserted during the casting process. The pipeline locations for main frame cooling can be seen on the manufactured prototype in Fig.1d.

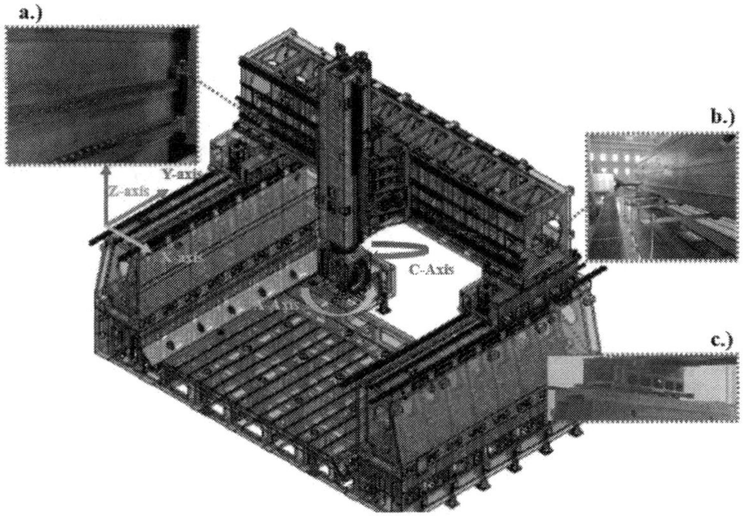

Fig.1: CAD data of the 5-axis CNC with general machine coordinates, **a.)** Target features **b.)** The pretensions on the gantry surface due to casting, and casted rails of linear guideways, **c)** The rack and pinion system used for linear motion system, **d)** The pipeline locations for the cooling of main frame.

3 FE Analyses

The CAD model of the assembly is simplified and composed of 377 parts in total. 14,117,354 pyramid mesh elements are employed with 2,538,140 nodes after Jacobian and skew ratio checks. The pure penalty algorithm (prone to assign higher contact stiffness values [13]) is employed with 0.2 relaxation tolerance on the contact interfaces.

3.1 Static Analysis

The gravitational force is applied at the- Z-axis direction of the machine tool coordinates. The structure is fixed from the pads under the machine base. The FE model is positioned at the least stiff position to be able to detect the maximum deflection amount. The total deflection of the CNC structure is shown in Fig.2i.a. which is 0.164 mm. Fig.2i.a. indicates that spindle-gantry connection is the weakest point of the structure. However, the total deformation of the gantry structure is predicted as 0.109 mm while these deformations are detected as 0.002 mm for X-axis. Deflection values of 0.03 mm and 0.09 mm are obtained for Y and Z axes, respectively. The directional deformations are illustrated in Fig.2b, c and d. The maximum deformation is found as 0.002 mm on the X-axis, and it proves that the mono-block casted frames behave like a single volume. 0.07 mm and 0.149 mm deformation are detected on the Y and Z axes, respectively.

3.2 Dynamic Analysis

Subsequently, a modal analysis is employed to find the natural frequencies of the structure. The results indicate a natural frequency range between 20 and 70 Hz, which is smaller compared to regular-sized machine tools. However, it is an expected result when the large sizes of the CNC structure are considered.

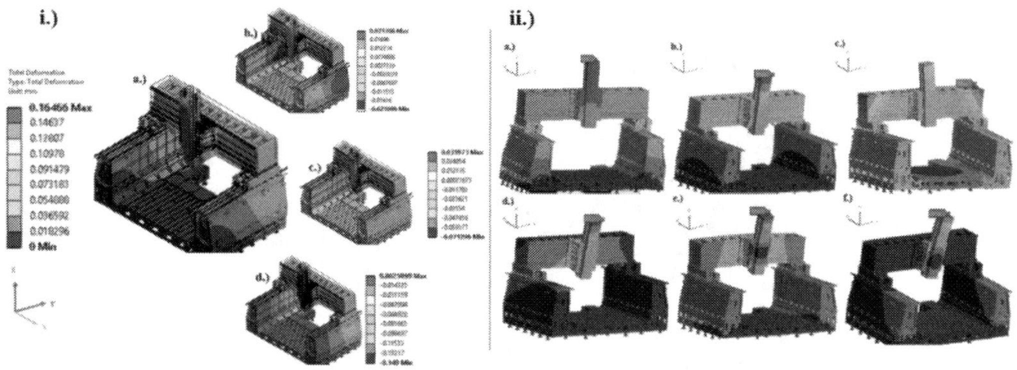

Fig.2: ***i.)*** **a)** The total static deformation of the CNC under gravitational force at the worst position. The directional static deformation results **b)** X-axis, **c)** Y-axis, **d)** Z-axis. ***ii.)*** The mode shapes of the FE analysis. **a)** 1st Mode (21.8 Hz), **b)** 2nd Mode (23.9 Hz), **c)** 3rd Mode (41.8 Hz), **d)** 4th Mode (43.8 Hz), **e)** 5th Mode (64.2 Hz), **f)** 6th Mode (69.3 Hz) shapes.

The mode shapes are indicated in Fig.2 ii.) a, b, c, d, e and f. The 1st mode shape is 21.8 Hz and makes a swinging motion along X-axis. The 2nd mode shape is 23.9 Hz, and it demonstrates a leaning motion along Y-axis. These two adjacent modes prove the casted mono-block structures behave like a single-volume structure. The 3rd mode is around 41.8 Hz and shows torsional features which is undesired for machine tool design. The 4th mode shape is 43.8 Hz which caused from the gantry structure. The 5th and 6th modes are detected as 64.1 Hz and 69.3 Hz, respectively. These mode shapes are related with the Z column and gantry structures. The natural frequencies predicted by FE is expected to be lower after the 4th mode due to contact parameters, since the related modes are related with linear guides and connection elements, however, the used contact algorithm is prone to assign higher contact stiffness values [13].

4 Experimental Investigation

4.1 Modal Analysis

FRF measurements from 25 points on the body columns are conducted by impact hammer testing. The results are presented in Table 1 together with FE results. Table 1 indicates that the natural frequencies calculated by FE analyses are close to the measured data of tests. As a result, the body components of the machine tool are produced at the predicted dynamic rigidity values. Impact Hammer tests were also conducted on the gantry which correlates the motion between Y-axis and the Z column. In these measurements 20 FRF data is collected in total. The natural frequencies and dynamic parameters obtained from the modal tests are tabulated in Table 2 together with FE analyses. As can be seen in Table 2, the 1st Bending Mode was calculated with a lower error rate compared to the 2nd Torsional Mode by employing FEM.

The hammer tests were performed on the Z column which carries the spindle. All measurements were repeated in the XX and YY directions results in total of 36 FRF data. Impact hammer test results can be seen in Table 3. The spindle has close natural frequencies to those of the machine tool structure at the X and Y directions. However, the most flexible mode of the spindle in the Y direction is 2 times weaker than the most flexible mode in the X direction.

Table 1: The Comparison of the results obtained from the FE and impact hammer tests for the body columns (PCB 626B03 accelerometer, PCB 086C05 oversized hammer)

Modes	*Natural Frequencies obtained by the FEM (Hz)*	*The results obtained by the Impact Hammer Test*			
		Natural Frequency (Hz)	*Modal Stiffness (N/m)*	*Damping Ratio (%)*	*Receptance of FRF (μm/N)*
1st Bending	21.8	18.4	3.6×10^{8}	2.1	0.022
1st Torsional	41.8	48.8	4.8×10^{9}	0.9	0.011

Table 2: The Comparison of the results obtained from the FE and impact hammer tests for the gantry. (PCB 626B03 accelerometer PCB 086C05 oversized hammer)

Modes	*Natural Frequencies obtained by the FEM (Hz)*	*The results obtained by the Impact Hammer Test*			
		Natural Frequency (Hz)	*Modal Stiffness (N/m)*	*Damping Ratio (%)*	*Receptance of FRF (μm/N)*
1st Bending	43.8	50.1	2.46×10^{10}	0.15	0.028
1st Torsional	64.2	84.6	1.33×10^{10}	1.30	0.005

Table 3: The top 3 most flexible modes of the Spindle and Z column measured by Impact Hammer tests. (PCB 353B33 accelerometer, PCB 086C05 hammer)

Modes		*The results obtained by the Impact Hammer Tests*			
		Natural Frequency (Hz)	*Modal Stiffness (N/m)*	*Damping Ratio (%)*	*Receptance of FRF (μm/N)*
XX direction	*1st Bending*	26.9	2.63×10^{7}	4.92	0.392
	2nd Bending	89.5	1.31×10^{8}	3.92	0.115
	3rd Bending	171.9	1.50×10^{8}	3.42	0.122
YY direction	*1st Bending*	23.7	2.44×10^{7}	2.81	0.823
	2nd Bending	106.5	8.58×10^{7}	4.89	0.048
	3rd Bending	190.9	4.98×10^{8}	3.31	0.037

4.2 Motion Performances of the Axes

The physical capability tests were carried out to measure the velocity, acceleration and jerk behaviours of the X, Y, Z, A and C axes of the machine tool by laser sensor. The results can be seen in Figure 3.a, 3.b and 3.c for linear axes. From Figure 3.a, it can be concluded that the feed rates given to the linear axes can continuously go up to 1500 mm/min axial velocity, and the axes can reach 1600 mm/min axial velocity as the maximum for the feed rates that are higher than 1500 mm/min. This is caused by the 5-mm measurement distances. A competitive regular-sized CNC with a processing capacity of - 500 mm x 500 mm x 500 mm- volume can go up to 3500 mm/min axial velocity at a 5-mm distance. Therefore, the measured velocities in Figure 3.a can be considered as satisfactory when the large sizes of the machine tool are considered. The acceleration and jerk behaviours of the linear axes are indicated in Figures 3.b and 3.c, respectively. The measured maximum of accelerations and jerks are accepted as normal values compared to other competitive large - scaled machine tools. The values obtained for the rotary axes can be seen in Figures 3.d, 3.e and 3.f. As can be seen from Figure 3.e, C-axis can reach 2000 mm/min feed rate properly, while A-axis could not catch this value.

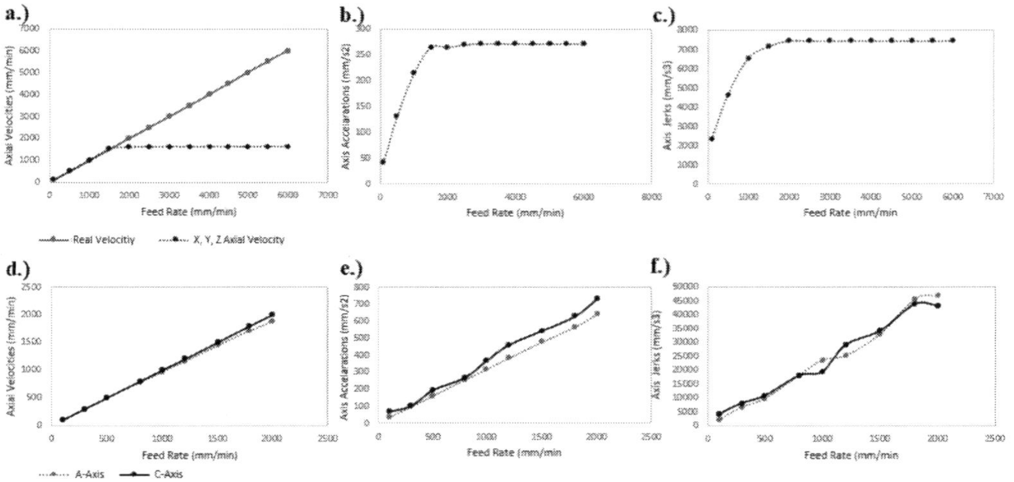

Fig.3: a.) Linear Axial Velocities, **b.)** Linear Axial Accelerations, **c)** Linear Axial Jerk profiles of X, Y, Z axes for a 5-mm translation; **e.)** Rotational Axial Velocities, **d.)** Rotational Axial Accelerations, **f)** Rotational Axial Jerk profiles of A and C axes for 3° rotation.

5 Conclusions

Design and analysis methods are presented for a gantry-type machine tool with large sizes. The significance of casted mono-block components is emphasized to enable the structural rigidity during the manufacturing of these massive structures. Casting-method benefited design strategies are proposed to maintain the stiffness values to build a large-sized gantry-type CNC machine tool. The effectiveness of the proposed methods was evaluated by FE analyses and experiments special to machine tools on the manufactured prototype. The results indicate competitive performance by using the proposed casting-method benefited design strategies.

References

1. S. Liu, Y. Li, Y. Liao, Z. Guo, Str. and Mult. Opt, 50, 2, 297-311. (2014).
2. X. Yang, K. Cheng, Procedia Manufacturing, **11**, 1454-1462, (2017).
3. P. Boral, T. Nieszporek, R. Gołębski, *MATEC Web of Conf.*, **157**, (EDP Sci, 2018).
4. M. Guillo, L. Dubourg, Robotics and CIM, **39**, 22-31, (2016).
5. J. Do Suh, H.S. Kim, J.M. Kim, Comp. Sci. Tech., **64**, 10-11, 1523-1530, (2004).
6. E.F. Kushnir, M.R. Patel, T.M. Sheehan, ASME-PUBS.-PVP, **432**, 133-146, (2001).
7. G. Sproesser, Y.J. Chang, A. Pittner, M. Finkbeiner, M. Rethmeier, *Sustainable Manufacturing*, 71-84, (Springer, Cham., 2017).
8. X. Lu, X. Lin, M. Chiumenti, M. Cervera, Y. Hu, X. Ji, W. Huang, Additive Manufacturing, **26**, 166-179, (2019).
9. R. Schaller, Journal of Alloys and Compounds, **355**, 1-2, 131-135, (2003).
10. Y. Altintas, C. Brecher, M. Weck, S. Witt, CIRP annals, **54,** 2, 115-138, (2005).
11. L. Zhang, L. Ma, D. Wu, Y. Zhou, The Int. J. of Adv. Manuf. Tech., 104,1-4, 245-260, (2019).
12. W.U. Tao, L.G. CAI,. DEStech Tr. on Eng. and Tech. Res., (amee), (2018).
13. E. Yuksel, A.S. Erturk, E. Budak, E. J. of Manuf. Sci. and Eng., **142**, 8, (2020).
14. T. Murakami, T. Inoue, H. Shimura, M. Nakano, S. Sasaki, Mat.Sci.and Eng, A, **432**, 1-2, 113-119. (2006).

Design of a spiral bevel gear acc. to ISO 23509:2006 standards

Ioannis Tsiafis[1], *Polychronia* Mamouri[1,*], and *Kyriakos* Kyriakidis[1]

[1] Laboratory of Machine Tools and Manufacturing Engineer, Aristotle University of Thessaloniki, Greece

Abstract. When the transmitting of power and torque between parts of a machine requires high efficiency, the most appropriate type of gears are the bevel gears. The two main forms of bevel gears are straight and spiral bevel gears. Manufacturing of bevel gears has been implemented using dedicated bevel gear cutting machines. These machines make the production of this type of gears quite costly. This is the main reason that forced many manufacturers start using CNC milling machines for cutting bevel gears. A combination of a 5-axis milling center machine and a modern computer aided manufacturing (CAM) system that supports a vast number of complicated machining procedures, offers the opportunity of machining high quality bevel gears. This paper concerns the presentation of a program written in MATLAB that calculates the basic designing features of a bevel gear pair, as they described by ISO standard. It is also presented the calculated tooth profiling that is used in order to be designed a 3-D model of the gears in a CAD software.

1 Introduction

Gear is one of the few mechanical devices that not only can satisfactorily transmit power between rotating shafts, but can also do so with uniform motion and reliability. Spiral bevel gear is a type of gear that is required to turn the corner from a horizontal engine to the vertical rotor shaft. These gears carry large loads and operate at high rotational speeds. For many decades, information on bevel gear geometry has been developed and published by the gear machine manufacturers. It is clear that the specific formulas for their respective geometries were developed for the mechanical generation methods of their particular machines and tools. In many cases, these formulas could not be used in general for all bevel gear types. This situation changed with the introduction of universal, multiaxis CNC machines, which in principle are able to produce nearly all types of gearing.[1],[2] ,[3] ,[4]
The goal of this paper is to present program that calculates all the geometrical features of a spiral bevel gear. The algorithm of this program is developed in a programming platform, in MATLAB.

2 Method

A complete information on design of bevel gearing is provided by the *American Gear Manufactures Association* (ANSI/AGMA standards) and the *International Organization for Standardization* (ISO 23509:2006 standard). Loading, speed, accuracy requirements and special operating conditions influence the design. In this paper the algorithm for the calculation of geometrical features of a spiral bevel gear pair is developed according to ISO 23509:2006 standard.
A design system is selected to be used in order to produce a spiral bevel gear set. Users should determine the cutting method prior to proceeding.

As it is known bevel gears have tapered elements because they are generated and operate, in theory, on the surface of a sphere, Fig. 2. Pitch diameters of mating bevel gears belong to frusta of cones, Fig. 1. [1],[6]

*Corresponding author: mamourip@meng.auth.gr

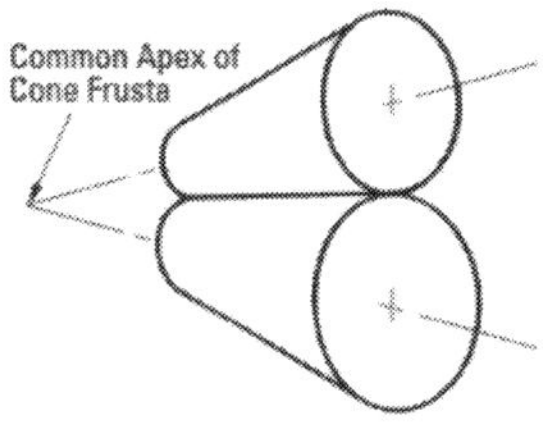

Fig. 1. Pitch cone frusta

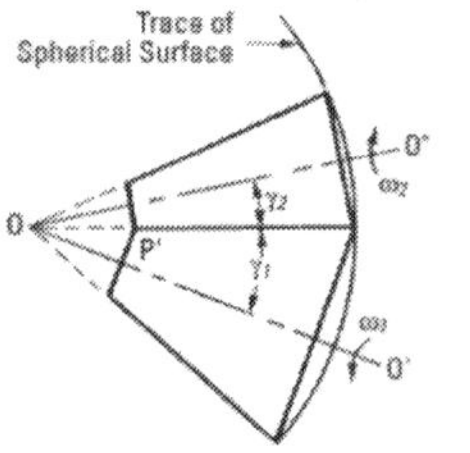

Fig. 2. Pitch cones and the development sphere

Before proceeding design of the geometrical features of each gear of the pair, the pitch cone parameters of the pair should be calculated. One of the recommended procedures of this international standard that is used to begin the calculation of the pitch cone parameters is given in the Table 1. [3]

Table 1. Initial data for the calculation of pitch cone parameters [3]

Symbol	Description	Method 0	Method 1	Method 2	Method 3
Σ	shaft angle	X	X	X	X
a	hypoid offset	0,0	X	X	X
$z_{1,2}$	number of teeth	X	X	X	X
d_{m2}	mean pitch diameter of wheel	—	—	X	—
d_{e2}	outer pitch diameter of wheel	X	X	—	X
b_2	wheel face width	X	X	X	X
β_{m1}	mean spiral angle of pinion	—	X	—	—
β_{m2}	mean spiral angle of wheel	X	—	X	X
r_{c0}	cutter radius	X	X	X	X
z_0	number of blade groups (only face hobbing)	X	X	X	X

The required data of selected method is imported to the algorithm.
Afterwards, one of the two common production methods of spiral bevel gears should be chosen.
In gear industry the production methods are face-milling and face-hobbing, both of them have similar cutting movements. The two different manufacturing methods are shown in Fig. 3 and Fig. 4. [9],[10],[11]

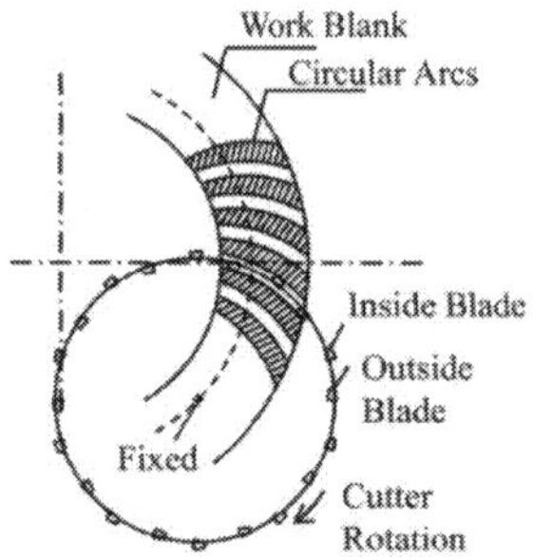

Fig. 3. Face-milling method

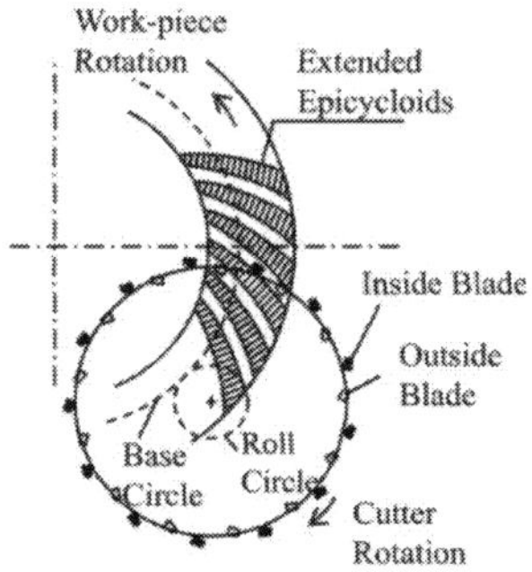

Fig. 4. Face-hobbing method

Continuing, having already been calculated the pitch cone parameters, a set of additional data is now necessary for the calculation of gear dimensions. Table 2. [3]

Table 2. Additional data for gear dimensions [12]

Data type A		Data type B	
Symbol	**Description**	**Symbol**	**Description**
α_{dD}	Nominal design pressure angle, drive side[a]	α_{dD}	Nominal design pressure angle, drive side[a]
α_{dC}	Nominal design pressure angle, coast side[a]	α_{dC}	Nominal design pressure angle, coast side[a]
$f_{\alpha lim}$	Influence factor of limit pressure angle[a]	$f_{\alpha lim}$	Influence factor of limit pressure angle[a]
x_{hm1}	Profile shift coefficient[b]	c_{ham}	Mean addendum factor[b] of wheel
k_{hap}	Basic crown gear addendum factor[b]	k_d	Depth factor [b]
k_{hfp}	Basic crown gear dedendum factor	k_c	Clearance factor [b]
x_{smn}	Thickness modification coefficient [b]	k_t W_{m2}	Thickness factor[b] or Mean slot width of the wheel
	For data types A and B:		
j_{mn}, j_{mt2}, j_{en}, j_{et2}	Backlash (one of four)		
θ_{a2}	Addendum angle, wheel		
θ_{f2}	Dedendum angle, wheel		

When all required data has been imported to the program, the result of the algorithm gives the geometrical features of the tooth profile, Fig. 5.

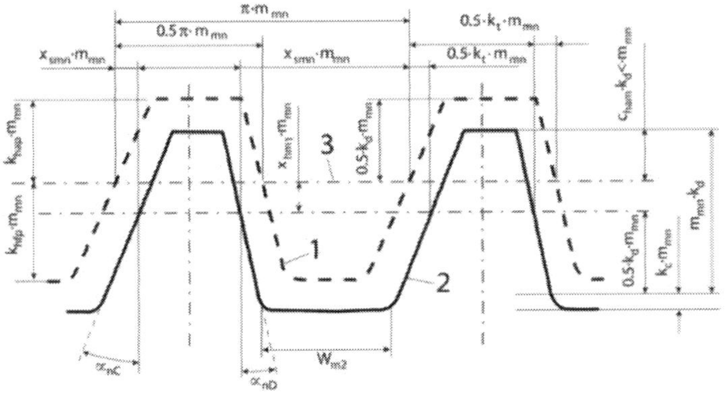

Fig. 5. Basic rack tooth profile of wheel [12]

The calculated points determine a tooth profile of the generated gear in a 3-D space. Fig.6.

 https://doi.org/10.1051/matecconf/202031801020

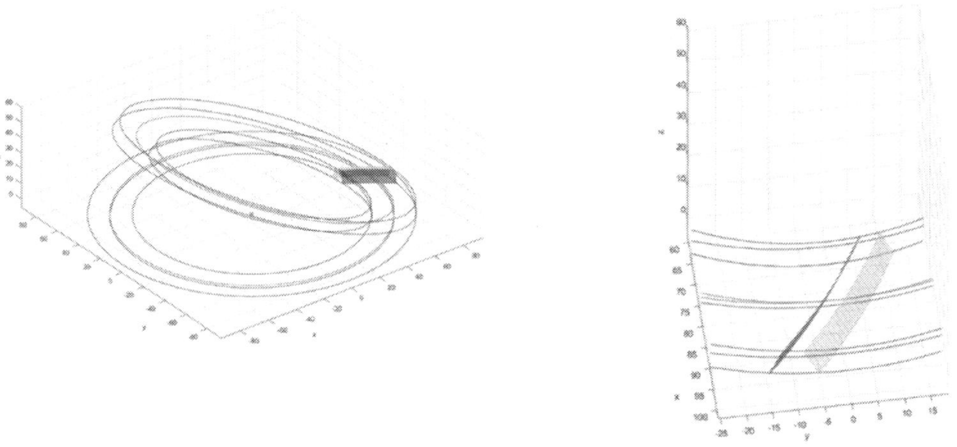

Fig. 6. A tooth profile generating by the program

Commonly, spiral bevel gears are cut with a generating machine that uses a series of cutting blades mounted on a circular tool holder. Instead of a basic rack, a virtual basic crown gear is used, which has straight tooth profile. The generated gear can be flat or take a conical shape. [12]

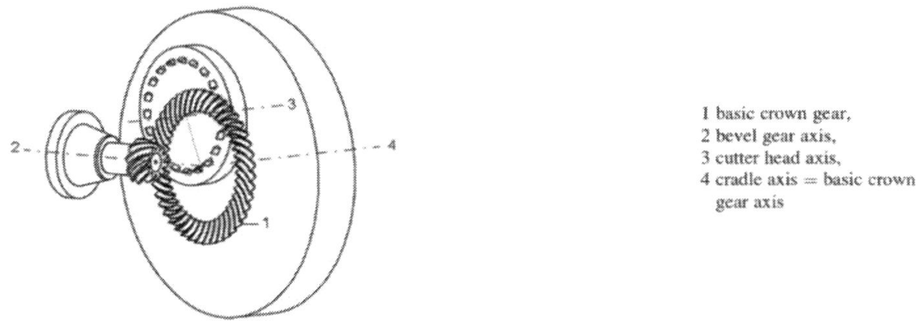

Fig. 7. Virtual crown gear in a bevel gear generator [12]

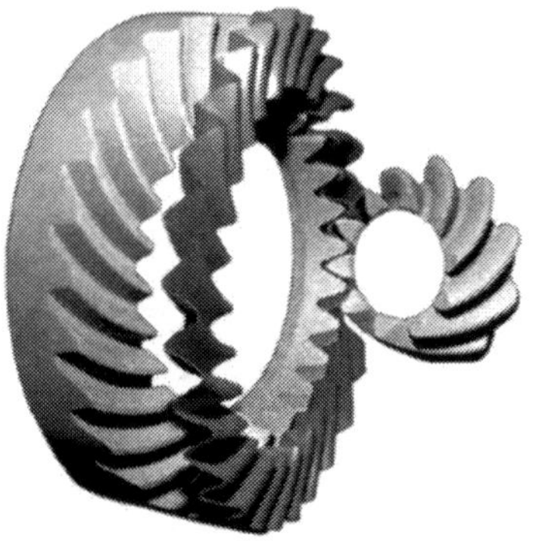

Fig. 8. Bevel gear pair with associated virtual crown gear [12]

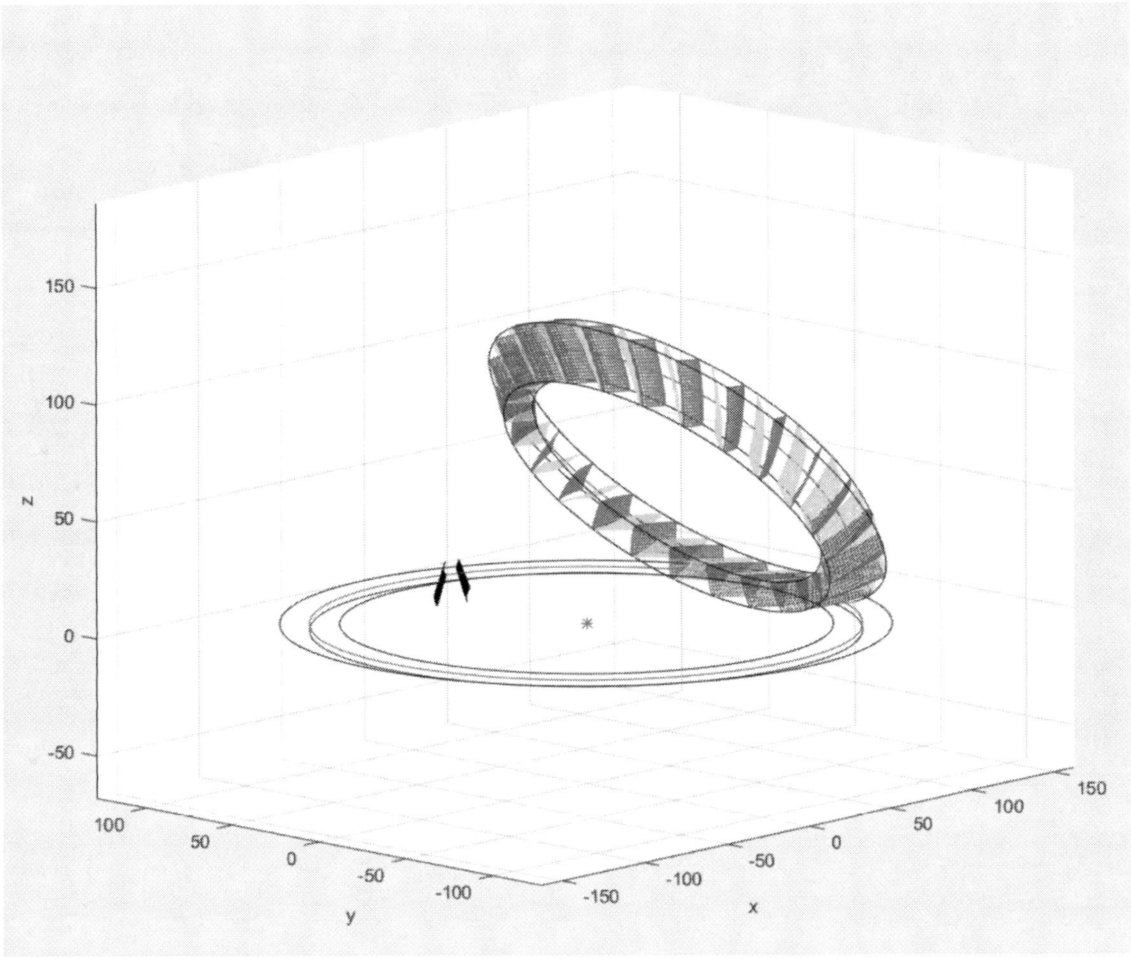

Fig. 9. Calculated generating gear by the program in Matlab

3 Conclusions

The effort to create a program that calculates the basic geometrical features of a tooth profile and generates a virtual gear has come up to a successful procedure.

The spiral bevel gears design remains complex since tooth geometry is affected by the manufacturing process. There are different cutter systems to choose from, depending on the method of producing gears and the supplier of the cutter system. Using the calculated geometrical features of a spiral bevel by the above program, a 3-D CAD model can be created. A 5-axis milling machine center can be used for the machining of the spiral bevel gear. The increased availability of 5-axis milling machine center and the continuous development of modern CAM systems that today support a vast number of complicated machining procedures, offer the opportunity of machining high quality bevel gears avoiding the restrictions that govern the traditional bevel gear manufacturing community.

References

1. Dr. Stadtfeld H J, *The Basics of Spiral Bevel Gears*, 31-38 (Gear Technology, 2001)
2. Radzevich S P, *Dudley's Handbook of Practical Gear Design and Manufacturing* , (CRC Press ,2012)
3. *BS ISO Standard 23509:2006* British Standards
4. *Design Manual for Bevel Gears ANSI/AGMA – 2005 – D03* American National Standard 102
5. Dr. Hermann J Stadtfeld, *Mathematics for Bevel & Hypoid Gears*, 50-56 (Gear Technology , 2015)
6. Radzevich S P , *AGMA/ANSI/ISO Standards on Bevel and Hypoid Gears,* 45-56 (2017)
7. *Elements of Metric Gear Technology* SDP/SI www.sdp-si.com
8. *Introduction to bevel gears and related cutting methods* www.tokugawa-gears.com
9. Dr. Stadtfeld H. J., *The Basics of Gear Theory Part 2,* 52-58 (Gear Technology, 2015)
10. T.J. 'Buzz' Maiuri, *Spiral Bevel and Hypoid Gear Cutting Technology Update,* 28-39,(Gear Technology, 2007)
11. Shih Y P, LAI K L, Sun Z H and YAN X L, *Congress* DOI Number: 10.6567/IFToMM.14TH.WC.OS6.03 (*The 14th IFToMM World,* Taipei Taiwan, 2015)
12. Ian Klingelnberg, *Bevel Gear Fundamentals and Applications*,40-121 (Germany, 2016)

Skills Requirements for the 4th Industrial Revolution: The Additive Manufacturing case

Panagiotis Stavropoulos[1,*], *Andreas K.* Lianos[1], *Harry* Bikas[1], *Dimitris* Mourtzis[1]

[1]Laboratory for Manufacturing Systems & Automation, Department of Mechanical Engineering and Aeronautics, University of Patras, Patras 26504, Greece

Abstract. This work analyses the required expertise knowledge for the European workforce under the implementation spectrum of the technologies from Industry4.0. The advancement of the conventional manufacturing technologies with complementary monitoring and control systems combined with the rapid growth of unconventional manufacturing technologies, calls for the equivalent advancement in the workforce's expertise. The Industry's 4.0 skills are mapped and categorized based on the knowledge requirements derived from the major technologies involved. The competences' categorization is what further determines the Professional Profiles and skills requirements for the Industry4.0. As Additive Manufacturing is one of the most significant manufacturing technologies implemented from Industry4.0 a case study for the required AM skills is performed. The outcome of this work indicates that the AM Professional Profile is a multi-dimensional quantity with multiple competence units that require validation and further evaluation in order to meet the skills requirements imposed by the industry.

Terminology

Digital Complementary Technologies	All secondary technologies that: are aiding an existing and integral technology of the overall product development, the part manufacturing processes or are introduced as a separate stage in the production in order to increase the overall performance. The complementary technologies can be subtracted without effecting the production's ability to be completed. The "digital" complementary technologies are mainly related to computational means for automation, AI and cloud-based technologies.
Competence/ Competency	Competence: refers to a specific function and its effective accomplishment. It is based on the outcome-based approach that screens the individual skills and knowledge as the required foundation. Competency: Refers to the sum of behaviors and traits that are required to perform a function These two terms both appear in the literature and are used interchangeably to describe traits under the scheme of the Professional Profiles. It can be derived that the term competence is used for specific technical skills whereas the term competency for general in nature soft skills. In this paper the term competence is exclusively used.
Competence Model	The result of a competence identification which is a description of competences for an identifiable group and its key characteristics [1].
Competence Unit	Competence units are the smallest modules that shape the model. They are extracted from different types of analysis including occupational analysis, value chain analysis, functional analysis, and proficiency level analysis. Often, only one of the analysis methods is dominant considering the nature of the industry [2].
Competence Standards	They document the expected performance for the competence units and are used for qualification procedures at both the early stages of the Professional Profiles definition and at the later training or the workforce.

*Corresponding author: pstavr@lms.mech.upatras.gr

Competence Level	Is a quantified metric that screens the educational level of the Professional Profile. The existing classification has six levels of competence levels form graduate diploma to generic certificate [2].

1 Introduction: Industry 4.0 and Professional Competences

Along the lines of Industry 4.0 paradigm for the manufacturing industries, a series of competence units, requirements and new skills have been introduced. The Professional Profiles of the dominant industries are being redefined as the mechanisms of how the progress and tasks are being released is radically changing compared to the status quo of no more than 20 years ago [3,4]. This initial approach referred to all-rounded abilities a worker/engineer should master, and included all technical, leadership and social skills.

The above professional competences status and requirements have been extensively described and reviewed from a soft-skills prism for Industry 4.0 [5]. That is, the current state of the manufacturing industry, considering the uptake of advanced manufacturing technologies, requires a further definition of the Professional Profiles in terms of technical/hard skills from technician certificates to graduate diplomas [6,7]. This created a need for establishing training and educational mechanism. This applies for both products and services as the mentality of how they are perceived, designed, made and finally validated and monitored is changing due to the innovative outcomes of Industry 4.0. The decomposition of a Professional profile to its characteristic subsets is shown in Fig. 1.

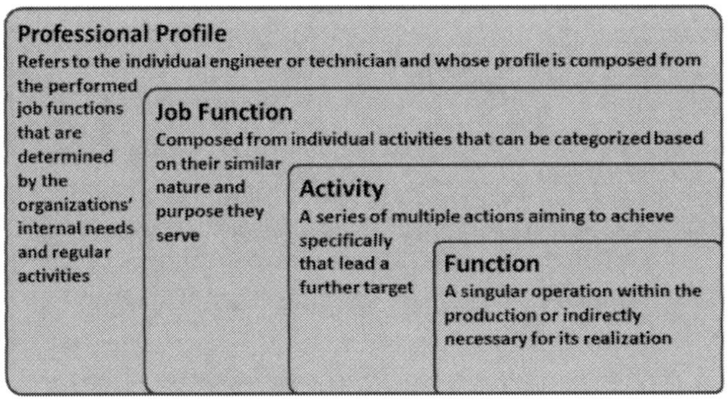

Fig. 1. The Professional Profile's sub elements

2 The Shift of the Professional Profiles

Although the nature of the manufacturing and its processes remain the same at their core, the involvement of digital complementary technologies targets these exact stages in order to effectively optimize their performance and reduce costs [8]. The above advancements are driven from four main factors that are sorted based on the frame in Fig. 2. The introduction of a new technology has two implications; it creates an additional production step in the production chain and it also increases the demand for expertise linked to this technology. This work's main thematic axis is focused on the Advanced Manufacturing technology of Additive Manufacturing. That is, the framework and the technological drivers of Industry 4.0 that are used to establish the main aspects that are causing Professional Profiles' shift can be extrapolated to the rest of the technologies.

The current skills' gap is traced back to the misalignment of the AM industry with the knowledge and training providers that is further increased with the introduction or advancement of new complementary technologies and materials [9]. For instance, a machine operator for a DED AM process, apart from the generic expertise for robot and AM head operation, is now required to operate in-line monitoring and control equipment which have increased the Competence Units for their Competence Level. The high growth rate of the AM industry cultivates an equally high demand for skilled AM workers and engineers which combined with the limited in number and capacity training and education institutions specialized in AM, results in the current insufficient status of the AM workforce and expertise.

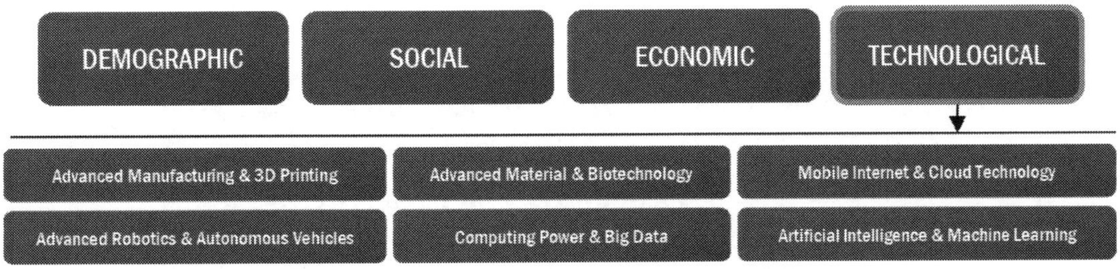

Fig. 2. Categorized Drivers of Industry 4.0 that effect the Professional Profiles

3 Establishing the Professional Profiles for Additive Manufacturing

The method presented below, to establish the AM Professional Profiles, is derived from the process-driven modelling approach [10] and the Organization Strategy and Objectives Model [11]. The combined methods are to best serve the AM current state and nature for the AM sectorial skills. The method can be separated to two stages:

1. Initiation from the objectives of the end user and the required operations that are gathered form specific activities. In the case of AM, the objectives are to realize product development and part manufacturing.
2. These activities are delegated to the people within the organization and further grouped to form a job function and finally a Professional Profile.

Determining the Professional Profiles for AM and the blocks of knowledge, expertise and skills for these Competences Units will define the required educational and training content for the upcoming AM workforce. This will additionally set the requirements for the Competence Standards and qualifications of the AM engineers and technicians.

The different stages of the overall AM production impose activities and operations that must be carried on by the equivalent experts. This creates specific roles within the whole product development and production chain as shown in Fig. 3. The definition of the above functions and activities for the Professional Profiles is a task being undertaken by major AM European players to further expand the use of AM in production applications.

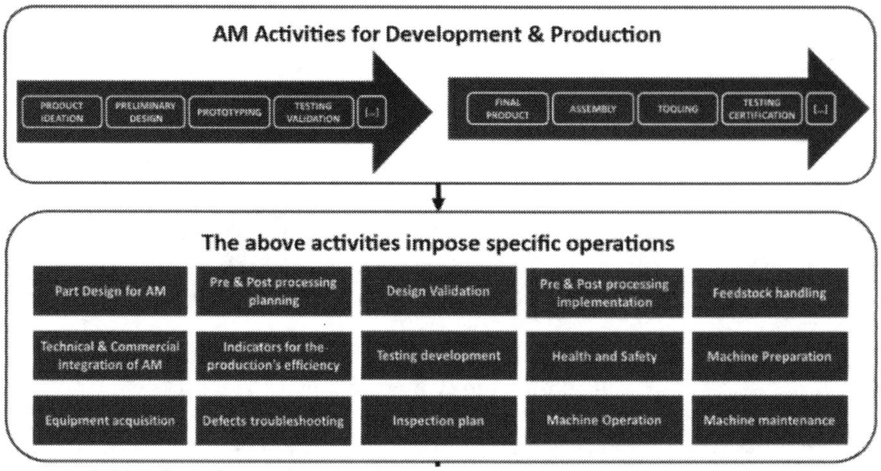

Fig. 3. AM Activities from the product development to the final manufacturing of the product.

The Professional Profiles and the specific functions each person perform within the AM design and production stages have been defined as shown in Fig. 4.

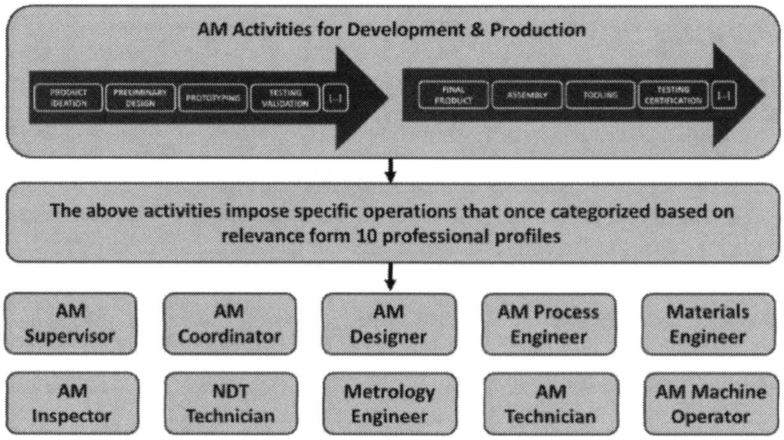

Fig. 4. The generation of the AM Professional Profiles

The Professional Profiles' descriptions are presented in Table 1:

Table 1. AM Professional Profiles' descriptions

PROFILE	DESCRIPTION
1. AM SUPERVISOR	Supervises the AM production on shop floor ensuring quality and HSE procedures.
2. AM COORDINATOR	Evaluates manufacturing suitability for end-use customers' requests defining which AM and post processes are fit for the request. Coordinate the work within the AM team.
3. AM DESIGNER	Designs optimized parts for function and cost, ensuring that the buildability constrains are secured and validating that parts can be produced cost-effectively and efficiently; Validates specific AM process design projects by verifying requirements with production engineer.
4. AM PROCESS ENGINEER	Creates the AM manufacturing process for the efficient production planning of additively manufactured parts including post-processing. They also carry the responsibility of the final design validation and tasks coordination between the operators according to the work planning.
5. MATERIALS ENGINEER	Specifies the optimal material to meet the functional and manufacturing requirements of the part and implement material handling processes for the entire material life cycle. They also provide insight for part failure troubleshooting that is traced back to material properties.

6. AM INSPECTOR	Carries out the inspection of the parts after their manufacturing and during the post -processing stages to the specified requirements and quality assessment.
7. NDT TECHNICIAN	Carries out the safe and reliable non-destructive testing of complex geometry additively manufactured parts.
8. METROLOGY ENGINEER	Specifies the optimal measurement method to meet the functional and manufacturing requirements of the part.
9. AM TECHNICIAN	Carries out the safe and reliable non-destructive testing of complex geometry additively manufactured parts.
10. AM MACHINE OPERATOR	Professionals that perform reliable production based on standardized procedures including post-processing, fitting and setting up AM printers, maintenance and machine repairs.

This is to identify the current knowledge gaps in terms of technical competences for AM technologies under the umbrella of Industry 4.0. The initiative of the training and educational European institutions aims to determining the content of their future programs which makes this process of establishing the AM Professional Profiles highly critical.

4 Conclusion and Discussion

The current and near future of AM workforce faces a shortcoming of specialized people having the expertise to perform AM engineering and AM machine operation. That is, the Industry 4.0 future AM workforce framework is to be based on the established Professional Profiles and Competence Units for both the processes of educating and certifying in the AM world. This is to be realized via three main channels:

1. Educational institutions. This channel involves the early education of pupils and students and is undertaken by organizations in education (e.g. universities, schools, online course providers). The Teaching Factory initiative was found to the best educational approach, as it effectively creates a communication channel, connecting an industrial AM entity with the undergraduate engineers [12].
2. Training Centers. For certified technicians and high Competency Level diplomas the existing European organizations, legislated for providing training courses, are to establish and validate the content for the AM Professional Profiles. This content will be commercialized by their programs or shared within their networks.
3. Internal company trainings. For the most cases the end-users of AM engineers and technical personnel, are organizations (e.g. R&D departments, part or equipment manufacturers) that utilize an AM technology in one of their productive stages. The current status sees most of these organizations with their own internal educating and training system that is based on their specific needs and AM activities in present time. Once the Professional Profiles are established with the modular fashion of the proposed method, these organizations will have the ability to recruit AM personnel based on the Job Functions they are performing and the Competence Standards of the training and certification centers.

The process described above calls for a close contact between the training and educational institutions with the AM industry. This is due to the fact that the constant shift of needs and operations withing Industry 4.0 calls for the equivalent adaptation of the training and educational context via periodic redefinition: every two years for rapidly changing industries and every 5-10 years for conservative industries. Furthermore, the validation and publication of the skills mapping is to be established by standardization organizations to be aligned with the European Qualifications Framework (EQF).

The continuation of the initial determination of the Professional Profiles with the proposed method is their validation and then their periodic re-evaluation. This requires a two-side party cooperation. The first party is the standardization organization with universal acceptance and legal presence that validates and publishes the Professional Profiles based on the Competence Standard for the Competence Level it provides. The second end of the two-side party is a panel of key-actors in the AM field, that performs the initial evaluate and then the update of the competence units for the Professional Profiles. The current expertise demand appointed by the industrial actors combined with the foresight of the training and educational organizations acts as input for the Job Functions' complementary definition.

The long-term sustainable mechanism to meet the everchanging skills requirements is Education 4.0 where the AM competences are fully developed. This aims to create technical learning outcome for the future engineers in order for them to fully exploit the freeform advantageous nature of Additive Manufacturing [13].

The authors would like to acknowledge the support of the European Union's ERASMUS+ SAM project http://www.skills4am.eu/

References

1. W. J. Rothwell, J. E. Lindholm, Int. J. of Training and Development, **3(2)**, 90–105. doi:10.1111/1468-2419.00069 (1999)
2. H. Bound, M. Lin, Workplace Learning and Assessment (Stage I), Singapore Workforce Skills Qualification (WSQ), Institute for adult learning Singapore. (2011).
3. A. Cerezo-Narváez, M. Otero-Mateo, A. Pastor, Development of professional competences for industry 4.0 project management, (2017).
4. D. Mourtzis, S. Fotia, N. Boli, E. Vlachou, Int. J. of Production Research. 1-14. 10.1080/00207543.2019.1571686. (2019).
5. K. Grzybowska, L. Anna, Key competencies for Industry 4.0. 250-253. 10.26480/icemi.01.2017.250.253. (2017).
6. Deloitte Insights, A Deloitte series on the skills gap and future of work in manufacturing (2018).
7. D. Mourtzis, D. Tsakalos, F. Xanthi, V. Zogopoulos, Optimization of highly automated production line: An advanced engineering educational approach. *Procedia Manufacturing*, **31**, 45–51. doi:10.1016/j.promfg. (2019).
8. P. Stavropoulos, P. Foteinopoulos, A. Papacharalampopoulos, H. Bikas, International Journal of Lightweight Materials and Manufacture. doi:10.1016/j.ijlmm.2018.07.002 (2018).
9. SAM Project, Professional Profiles / Set of Skills′ Roadmaps http://www.skills4am.eu/documents/SAM%20D1.4%20Skills%20Roadmap%20VF.pdf (2019)
10. W. Rothwell, H. Kazanas, Mastering the instructional design process: A systematic approach, San Fransisco, Jossey-Bass (1998).
11. V.S. Chouhan, S. Srivastava, Understanding Competencies and Competency Modeling — A Literature Survey (2014).
12. G. Chryssolouris, D. Mavrikios, D. Mourtzis, Manufacturing Systems: Skills & Competencies for the Future. Procedia CIRP, **7**, 17–24. doi:10.1016/j.procir.2013.05.004 (2013).
13. H. Bikas, A.K. Lianos, P. Stavropoulos, Int. J. Adv. Manuf. Technol. **103**, 3769–3783 (2019). https://doi.org/10.1007/s00170-019-03627-z (2019).

Green Face of Packaging – Sustainability Issues of the Cosmetic Industry Packaging

Jelena Drobac[1,*], *Vesna* Alivojvodic[1], *Predrag* Maksic[1] and *Marina* Stamenovic[1]

[1]Belgrade Polytechnic, Brankova 17, Belgrade, Serbia

Abstract. Packaging is a steadily growing industry that faces many challenges as well. Ever growing consumer expectations, though competition, technological advances don't come even close to the huge issue of sustainability within the industry. Packaging for the beauty industry is dependable on the looks more than any other with massive changes in the recent years since extensive grooming became a part both genders lifestyle today and spreading those standards across the globe. The global cosmetic packaging market was valued at USD 25.9 billion in 2018 and is expected grow 4.8% in the next 5 years flying high on the innovative package designs, increasing demand for cosmetics focused on the middle class and youth as well as e-commerce. The first stepping stone is that more than half of the materials used for the massive cosmetic industry packaging are plastics. Finding alternatives and making a shift to a greener and circular way will be a challenge in the years to come.

1 Introduction

Consumer product companies have evolved from small firms to large global companies. We know these products today as global brands that are mostly sold by a small number of global corporations. Today packaging of beauty products is a steadily growing industry that faces many challenges. Ever growing consumer expectations, though competition, technological advances don't come even close to the huge issue of sustainability within the industry.

Packaging for the beauty industry is dependable on the looks more than any other with massive changes in the recent years since extensive grooming became a part both genders lifestyle today and spreading those standards across the globe. The first stepping stone is that more than half of the materials used for the massive cosmetic industry packaging are plastics. Finding alternatives and making a shift to a greener and circular way will be a challenge in the years to come. In the following paper we will analyze these alternatives and show where greener and circular ways are taking cosmetic industry packaging today.

2 Beauty in context

* Corresponding author: jdrobac@politehnika.edu.rs

2.1 Beauty as a term

Beauty in a particular context can mean "indisputable excellence", something like health. A visited place of beauty that is more or less indisputable is visited by the beauty of man - the outer and inner beauty. Beauty of body and face, beauty of soul. By relocating the term beauty to the world of creation, one cannot escape the fact that beauty in the context of creation comes to the "exaltation of the beauty of art" [1]. Although in Art (external) beauty is an initial and certainly superficial characteristic, it actually comes from the inner beauty of the art work itself. When we talk about art, we consider not just fine arts, but we consider design as well.

In the world of art, and even in design, beauty is an immutable, permanent, self-contained category. It is fixed in forms that emerge from art, it is deep, it is not superficial, it can be hidden, it is sometimes disturbing, indestructible, it is not ephemeral as time and is a fact in nature. Beauty is a lofty term for art and design, even the ideal one strives for. In life, beauty does not have to be obligatory, while in art and design it is an indispensable element.

Beauty cannot be separated from the social and cultural context in which it is defined. Beauty reaches agreement with the social context, which is inconsistent with class, race and other social elements. Looking at the historical context of beauty, it can be portrayed as the possession of the elite or elitist. For example, the field of photography is the segment of art where beauty is most triumphant. The photographic eye or the photographic way of observing the world around us has actually led to the beauty of being appreciated much more than before.

In fact, when one looks at the whole history of beauty, when one looks at the present and assumes the future, beauty can be illustrated as an ideal and as perfection. It identifies with the body of man, with the body of the animal, with spirit or soul, with existence itself, with nature and health.

Beauty is there to admire and observe, which brings us to the beauty industry. It brings us to all those beauty magazines, salons, products, and so on. Beauty is an integral part of the industry of the beautiful, a history of idealization and of solace. As Susan Sontag says: "The beauty that is human, and the beauty that is made (art) - both raise the fantasy of possession." [1]

2.2 The globalization of the beauty industry

In the following text, we will present the development of the history of the beauty industry in the context of globalization, in order to understand the current state of this industry. Like many other consumer products, beauty products had its own development. The companies that created these products have evolved from small firms to large global companies. We know these products today as global brands that are mostly sold by a small number of global corporations.

It is worth mentioning that this industry produces products that are used on the body of humans for beauty purposes. Here we are talking about cosmetic products such as soaps, deodorants, baths, and all other products for children and adults. The greatest development of this industry occurred in the second half of the twentieth century. [2]. The modern beauty industry is actually based on hygiene concerns. Leaning on this, hygiene is identified with social morality and a way of placing it on the social hierarchy.

It all began in the nineteenth century with the technological revolution of soap making. Then came products like toothpaste, and others. Of course, products and sales were accompanied by marketing, especially in the United States. For example, Colgate began its sale in 1873 by selling toothpaste packaged in jars. It was not until 1896 that the toothpaste tube as we know it today was created. A similar happened with shaving products. Gillette, a metal maker, offered a safety razor to the market in 1901, and only later sold shaving creams. The products were made by pharmacists and chemists [2]. One pharmacy designed the Nivea Cream - the first long-lasting moisturizer.

Between World War I and World War II, the beauty industry increased significantly. In 1915, retailing cosmetics in the United States was $ 45 million, and in 1920 it even reached $ 129 million [3]. Then a number of companies developed that produced cosmetics, makeup, skin and hair care products, toilet soaps and so on. These include companies such as Max Factor, Maybelline or Helene Curtis.

Contemporary beauty industry overlapped with the accelerating globalization during the twentieth century. The value of companies and the quick development of this industry have led the entire industry to take on a quasi-ideological role. Certainly, toilet soaps have led the entire process of globalization. For example, in the United States and European countries soap was widely sold.

The end of World War II showed that the United States represented the largest beauty market. The communist world, on the other hand, was lagging behind this market due to the poor purchasing power of the population. In the 1950s, North America accounted for two-thirds of the consumption of makeup and personal care products [3]. With the development of technology in the 1960s and 1970s, the potential for growth and globalization of this industry emerged.

In developing countries, products coming from the Americas and Europe, such as soaps, toothpaste, etc., have been replaced with traditional, mostly craft-made cosmetic products. In the globalization process of the 1970s, companies began to build factories and manufacturing facilities in third world countries. Other companies included Gillette, Palmolive and Unilever [2].

We should be aware of that cosmetic brands have transmitted their beauty ideals to these countries. For example, American companies relied on their views of beauty that were largely based on popular culture in the form of Hollywood movies or fashion. At the end of the century, when globalization was on the rise, marketing pushed beauty products to developing countries. To this day there is a great struggle of companies in the global market. Surely, besides the product, the product packaging has played a huge role and therefore the design of the packaging.

The globalization of the beauty industry, according to the literature, has not produced homogeneity in the concept of beauty [2]. After the fall of the Berlin Wall, heavy globalization was followed by increasing segmentation by gender, ethnicity, age and more. Yet, the ideals of beauty have become widespread around the world. In the historical context of corporate strategy, they have to some extent united the understanding of beauty around the world, and certainly packaging design has played a key role there.

3 Beauty packaging industry today

Packaging for the beauty industry is dependable on the looks more than any other with massive changes in the recent years since extensive grooming became a part both genders lifestyle today and spreading those standards across the globe. Design plays a huge role in the game a big players have a new market they are trying to please – dominantly Muslim Asia with expectations of doubling market.

Back in 2014, the sales were $20 billion and expectance is to reach $45 billion by the end of 2020. Muslim culture requires a lot of adaptations from process to the visuals. Expecting this culture of "halal" quality means that only certain types of oil can be used, and all used materials are always indicated on the packaging. Also, the visuals like material, shape and imagery must not be "*haram*" and prohibited under Shariat law. Thus, labels on packaging showing female body features are strictly prohibited.

Today, packaging is a steadily growing industry with the global cosmetic packaging market valued at USD 25.9 billion in 2018 and is expected grow 4.8% in the next 5 years [4]. It is fast growth owes a lot to the innovative package designs, increasing demand for cosmetics focused on the middle class and youth as well as e-commerce.

3.1 Green beauty – inside and out

Reading about market peaks and numbers sounds delightful but it is not all rosy in the cosmetic packaging industry. Just like the all the other segments of the packaging industry, cosmetics packaging will have to face the challenges of the modern living: packaging for the one-member household, online shopping demands, personalized packaging expectations and smart packaging. Ever-growing consumer expectations, though competition, technological advances don't come even close to the huge issue of sustainability as the biggest obstacle in the industry. It is estimated that 166 billion of packaging units is sold [4] globally, with 61% market share in 2015 being plastic packaging like as bottles, jars, tubes etc. Plastics rules supremely as a dominant material for cosmetics and personal care products and it is not showing signs of slowing down. As the industry grows each year, as bottles remain the primary packaging form with a 28% share of the global market [4].

Finding alternatives and making a shift to a greener and circular way will be a challenge in the years to come. Many consumers, especially in Europe, the cosmetic epicenter of consumption with 77 billion euros annually [4], are opting for green cosmetic products inside and out. At the top of the wish list is nature-friendly packaging. This means that picky European shoppers are choosing more and more vegan products, and the packaging must be made of natural materials with as little resources as possible – in terms of production, transportation and waste.

Over the years, German cosmetic giant DM created its natural cosmetics brand Alverde. Recently they presented foldable boxes that were printed with biodegradable offset ink and made 80% of recycled raw materials. These inks are made using green electricity, originating from renewable sources that require 76% less carbon dioxide and contain no minerals.

In Germany, one of the seven world's leading spray can consumers, DM paired with Unilever in a recycling programme called R'cycle!. They collected spray cans and made almost 800 bicycles from the recycled aluminum for children charities.

The Frosch brand, another German giant, has unveiled 100% high density polyethylene packaging plastic recycling waste. These "Frosch Senses" shower bottles made from PE-HD post-consumer recyclables (PCR) have been on the market since mid-2019.

Many other big global players are running for innovation in recycling like P&G's with its PureCycle, see by some as a possible breakthrough on the global recycling market. Others turn to refill tactics like Estée Lauder's make-up line MAC Cosmetics. It's 'Back-to- M·A·C Programme' encourages consumers to return empty primary packaging to store. In exchange for six returned cosmetics empties shoppers get a free eye shadow, while sub-contractors take care of recycling. The same tactics is used in Austria - every returned packaging unit from the Vichy is marked with a stamp in a recycling passport and for six stamps you get a free Vichy shower gel.

Sustainability frontrunner, UK based Lush Cosmetics recently announced that it will replace reusable aluminum containers with cork containers. Cork is durable, waterproof and 100% biodegradable. They partnered up with authorities in Portugal, number one producer of the cork in the world, to create a sustainable harvesting program while ensuring locals'benefit.

France and Germany make 53% of the huge European cosmetic market [4]. Therefore, the French cosmetic business ZAO Make Up has made a complete range of packaging products made out of bamboo. Bamboo, which is highly renewable and extremely durable, has allowed them to create a sleek and modern design that exudes luxury and nature. All ZAO products are made from natural, vegan materials while the packaging can be refilled, so bamboo packaging can only be purchased once.

With the expansion of e-commerce, the practicality of packaging is becoming more important while the volume, weight and handling during transport are becoming crucial. In this regard, online retailers are increasingly turning to materials that are lightweight and durable while design is becoming increasingly minimalist. Plastics could be in certain cases replaced with glass, if it becomes less heavy and breakable. This opens the door for experiments, nano-materials or completely new technology.

3.2 New materials for beauty industry

Beside bamboo – seaweed, cornstarch, mushroom fibers, avocado nuts are now gained popularity as plastic alternatives. Bioplastics are highly durable and made from corn that can be used to make bottles, films and other containers. New eco cellophane "NatureFlex" in made out of wood pulp, while biodegradable plastic similar to polystyrene is made out of milk.

Seaweed is used to make the first eatable water container. The material called "notpla" by the British start-up Ooho could be sustainable solution to the problem of the ecological footprint of plastic bottles. Similarly to this, USA company Pacific Shaving Co. designed "Single Use Minis Shaving Cream". One dose of shaving cream is packaged in water-soluble film that vanishes when put under water stream. These resolvable pouches are made from high-barrier PET film laminated to LLDPE (linear-low-density polyethylene). This packaging hits many high tones of the contemporary packaging goals:

- it is single use package
- it is light to carry around
- it is zero waste
- it complies to the flight regulation laws

An alarming thought – if we are to assume that all of us use ten plastic bottles of shower gels and shampoo each year, this amounts to 75,000 tons of plastic waste – enough to fill about a thousand Olympic swimming pools [6]. Around 90% of shampoo and gels is water and about 80% of it's plastic packaging ends up in landfilled and water. No wonder that a lot of the efforts are pushed in this direction. One of the initiatives came from New Zealand, from Brianne West, who created Ethique, cosmetic company that produces shampoos and conditioners as solid bars. Her products are natural and come in a plastic free form.

Another design arouses the attention of the social media, academics and companies. Product designer Jonna Breitenhuber created SOAPBOTTLE—biodegradable, zero waste packaging made from soap. Concept of her master project at UDK Berlin is "packaging-free cosmetics by turning the product into its own packaging". Each bottle consists of 80 grams of soap and can hold one liter of liquid, lasting for approximately a month, thanks to its water-insoluble layer coating that slows down the dissolving process [6]. The empty bottle can be transformed into a hand soap or into detergent by grating it down and adding soda and natron. Her design is minimalistic, with strong forms.

In the similar tone, Mi Zhou, *Material Futures* master's degree student from Central Saint Martins in London [7]. She has created bottles and jars made of soap called Soapack by using vegetable oil-based soap that melts away once they are empty. A thin layer of beeswax is used to line the bottles to make them waterproof, preventing the liquid contents from leaking and even the "paper" instructions dissolve in water [7,8]. Differently from Breitenhuber's project, Zhou created a much more sophisticated aesthetics and shapes while pastel gradients are achieved using pigments from minerals, plants, and flowers.

4 Conclusion

The packaging industry has seen a record increase in production volume in recent years, while at the same time facing growing public pressure to reduce the amount of packaging materials used to package products, primarily. The increasing need to change the usual concept of product packaging is linked to the requirements of sustainable development and the environmental impact of packaging materials (especially in the post consumption phase) plastics, but also caused by changes in people's lifestyles as well as contemporary market demands that introduces the concept of personalized packaging, smart packaging and online shopping. The need to find new materials that are acceptable from an environmental point of view, while striving to preserve the market, influences the development of a whole new area of packaging materials for which it is necessary to connect different branches of science, different sectors of production and turn to a new approach in packaging design plastics [9]., which will be able to meet all set requirements.

References

1. S. Sontag, **Vol. 134**, No. 4, MIT Press, 208-213 (2005)
2. G. Jones, *Blonde and Blue-Eyed? Globalizing Beauty*, c.1945-c.1980, The Economic History Review, New Series, **Vol. 61**, No. 1, Wiley, 125-154 (2008)
3. J. Willett (ed.), *The American Beauty Industry Encyclopaedia*, Greenwood Press, Santa Barbara, California (2010)
4. *Growth of the global cosmetics packaging industry* (2017) https://www.interpack.com/en/TIGHTLY_PACKED/SECTORS/COSMETICS_PACKAGING/News/Growth_of_the_global_cosmetics_packaging_industry (visited 2020, April 18)
5. P. Demetrakakes (2017, March 17). Zero-waste shaving cream pods are convenient and travel friendly. Retrieved from https://www.packagingdigest.com/flexible-packaging/zero-waste-shaving-cream-pods-are-convenient-and-travel-friendly-2017-03-17 (visited 2020, April 18)
6. K. Barba-Court (2019, October 14). This ingenious liquid soap packaging by Jonna Breitenhuber is made from soap. Retrieved from https://plainmagazine.com/jonna-breitenhuber-soapbottle/ (visited 2020, April 20)
7. E. Taggart (2019, July 12). These Elegant Bottles Are Made of Soap Instead of Plastic to Create Zero Waste. Retrieved from https://mymodernmet.com/plastic-packaging-alternative-soapack-mi-zhou/ (visited 2020, April 20)
8. M. Waqas, I. Yuncheol, K. WooJeonc, J. of Clean. Prod., **Vol. 245**, Elsevier Ltd, 118827, (2020)
9. B. Mau, J. Leonard, Institute without Boundaries, *Massive change,* Phaidon Press, UK (2004)

Circular economy framework for sustainable product design strategies

Vesna Alivojvodic[1,*], *Jelena* Drobac[1], *Predrag* Maksić[1] and *Marina* Stamenovic[1]

[1]Belgrade Polytechnic, Brankova 17, 11000 Belgrade, Serbia

Abstract. Limited global resources, climate changes, growing population, combined with demand for sustainable products, indicate the need for change of existing production and consumption practices. In order of achieving more regenerative, restorative, and circular economy, design has been recognized as a catalyst for moving away from the traditional linear economy model (take-make-dispose). As shifting focus from quantity to quality, and finding new ways of working in cooperation with the nature, slowly becoming paradigm, it is important for designers committed to sustainability to have in mind environmental, social and economic performance of product design. This paper aims to address position of different product design strategies correlated with circular economy concept, as well as the opportunities for designers' to shift their mindset from the position of *creators* to *solution providers*.

1 Introduction

Creativity in reusing objects for different purposes, marked human history in many ways, but inherent creativity towards innovative reuse of resources has been lost because of cheap products that have become part of our everyday lives. Growing populations and scarcity of raw materials, as well as exponential and adverse increase in global environmental conditions nowadays, indicated the need to change existing production and consumption practices. Wasted resources (materials and energy), products with wasted lifecycles, or wasted capacity, and built-in values (components, materials, and energy) are four recognizable types of waste within a linear economy concept. Despite the fact that modern technologies can provide product design in a way that the value of incorporated materials can be recaptured for use in future products, or for up cycling of materials in a cost-effective manner, many businesses still operate in a linear way (wasting raw materials, and making new products with short lifespan). [1]

On the other hand, growing awareness of need to shift toward more sustainable way of consumption of goods resulted in several different schools of thoughts in last half of century, as an attempt to somehow "design out" waste. This deep rooted idea started formally to develop in 1970s, with the concept of Regenerative Design, and later through the Performance Economy, and Cradle to Cradle concept, then, through Industrial Ecology, and Biomimicry, up to the concept of Circular Economy of nowadays. [2]

*Corresponding author: valivojvodic@politehnika.edu.rs

One of the aims of circular economy is to provide high quality, functional and safe products, which are efficient and affordable, last longer, and are designed for reuse, repair, upcycling and high-quality recycling, with additional opportunity for developing whole new range of sustainable services, product-as-service models, and digital solutions that will bring about a better quality of life. [1] As a catalyst to obtain a shift from a traditional business-as-usual approach, to economic growth, but in sustainable way, special role has been given to design, in order to help to achieve a more regenerative, restorative, and circular economy. To obtain this, in other words, not only to help in shaping our world, but also to determining our impact on it, position of design within circular economy concept need to be considered within all its complicity, where design of sustainable products is only small but important part of the mosaic.

2 Circular economy and design principles

Shift from linear model of production and consumption of goods, has direct implications for the development of efficient and effective take-back systems and the expanding of product and business model design practices, in order to obtain more durable products, facilitate disassembly and refurbishment, and consider product/service shifts, where appropriate. [3]

The Ellen MacArthur Foundation has led the way in articulating the value of shifting global business models to circular economy. The circular economy according to them is based on a few simple principles to eliminate waste:

- Waste does not exist when the biological and technical components of a product are designed by intention to fit within a biological or technical materials cycle, designed for disassembly and refurbishment.
- It is important to build the systems more resilient in the face of external shocks, not only built simply for efficiency (building resilience through diversity).
- Try to rely on energy from renewable sources. [3, 4]

The "waste" can become a valuable resource, whose value can be captured by eliminating of [1,5]

- Wasted resources (through use of materials and energy that cannot be effectively regenerated over time, such as fossil energy and non-recyclable material).
- Wasted capacity (through products and assets that are not fully utilized throughout their useful life).
- Wasted lifecycles (through products reaching end of use prematurely due to poor design or lack of second-use options).
- Wasted embedded value (through components, material, and energy not recovered from waste streams).

With the aim to redefine the value of waste, five business models were developed within a circular economy concept (Figure 1.). Each of them can be taken individually, or combined, and they are not mutually exclusive, but using simultaneously can give the best results. Through these models minimization or even elimination of waste, pollution, and inefficiencies could be obtained. [5]

There are five underlying business models that Accenture has identified in its analysis of more than 120 case studies of companies that are generating resource productivity improvements in innovative ways. [6]

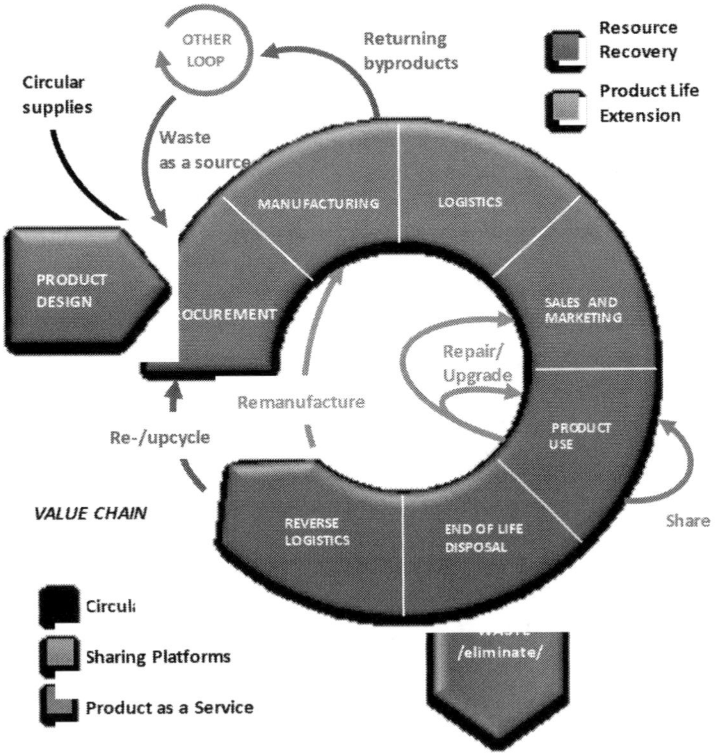

Fig.1. Five circular business models (*by Accenture*) [6]

While three of the models are more focused on production: Circular Supplies, Product Life (Use) Extension, and Resource Recovery, the other two: Sharing Platforms and Product as a Service, target consumption and the relationship between the product and the consumer. [1] Essentially, the models cover the full value chain of circularity.

Circular Supplies focus on the "ingredients" that go into products at the design, sourcing, and manufacturing stages. These inputs, such as renewable resources, aim to eliminate wasted resources (including toxic and single-use materials) and are a stepping stone for all other models. Product Life (Use) Extension focuses on maximizing the use of a product. To do so, companies have to start from the very beginning, from product design and responsible sourcing, in order to avoid wasted lifecycles and to keep products in use for as long as possible. [1, 5, 6]

In addition, Product Use Extension plays an important role in enabling the Product as a Service and Sharing Platforms models. The latter two go further by reinventing "product utility" in entirely new ways (e.g., buying a function or service, like mobility, rather than the product itself, a vehicle). Once a product reaches end of use, the role of Resource Recovery is to return the embedded materials or energy back into the production cycle, thereby "closing the loop" of the product from sourcing to usage and back to sourcing. [1]

Since design sits at the start of a product's lifecycle, with this position it is extremely important to incorporate circular principles.

As Richard Buchanan distinguish four domains of design (design of symbolic and visual communications,design of material objects, design of activities and organized services, and design of complex systems or environments for living, working, playing, and learning). [1]Sustainable design, as well as the design process itself, for each of these domains, as an extension, can incorporate broader strategies (even policies). Design can clearly no longer exist, and successfully practiced, without technology and economics.

Various disciplines, such as engineering and computer science, act jointly, and generate new forms of interaction. In future it will become increasingly important to develop and communicate design knowledge both vertically (within the discipline) and horizontally (between disciplines). [7]

Design strategies need to incorporate circular principles (Table 1.) to allow consumers to extend product usage, as well as to enable recovery of valuable materials (in other words to obtain product circularity). It is important to notice thatmore than 80% of the environmental impact of a product can be determined at the design stage. [3]

Table 1. Circular design principles [6]

Objectives	**Illustrative principles**
Product composition and production - use only circular or more sustainable materials	- Reduce or eliminate the need for materials and packaging - Use recycled/renewable materials and remanufactured components - Reduce stock keeping units and excess inventory waste - Choose alternative, less resource-intensive, and nontoxic materials - Design to eliminate or minimize waste during production processes
Product use - keep in useful life for as long as possible and reduce impact while in use	- Design for durability, modularity, repair/upgradeability, and efficiency while in use - Design to avoid single use and obsolescence - Deploy technologies to extend product use and enable recovery (e.g. for asset monitoring) - Deploy circular business models, designing products for sharing, leasing, and second life
Product recovery - enable materials or components to be easily brought back into a value chain	- Design for end of use disassembly, refurbishment and remanufacture (upcycling) - Choose materials that are recyclable or compostable at end of use

Today the majority of low-to-mid value products are designed for short-term use only (often with built-in obsolescence). Obsolescence is also an issue for high-value products (e.g. consumer electronics, with design features like seamless edges make disassembly particularly difficult). [6] Given the rapid pace of technology changes, obsolescence is likely to remain a considerable challenge in the coming years, highlighting the need for more circular design practices. [1, 6] On the other side, the problem definition for the product design will also come from deeper understanding of the goals the people are trying to accomplish, and the obstacles they experience.

3 Product design strategies

Defining design strategies to prolong useful life of product, for as long as possible and to reduce its impact while in use, starts with the notion that design concerns the whole economic role of the product.

For design of products within circular economy framework the goal is to provide high quality, functional and safe products, which are efficient and affordable, last longer and are designed for reuse, repair, and high-quality recycling. [8] Integrating circularity into product designs, can be a significant shift for teams of engineers, designers, and technicians that have traditionally been focused on cost or performance, to adopt a new way of thinking to incorporate new, principles. [1] Product design within a circular economy concept can be referred as "Circular Product Design". [9] It is about keeping a product as close as possible to its original state over time, for instance through longer use, repair, upgrading, refurbishment or remanufacturing, in other words, extending the useful life of products is not just about recycling of materials. So far, the attention in that domain has focused mainly on closing the material gap, going from new to recycled, and back to (almost) new again.[9] In the case of production, this can lead to a combination of requirements, with respect to clean processing and efficiency, and in the period of product use, there is expectation towards sustainable consuming of energy, and preferably producing some. Ease of disassembling, or maintaining product's value through whole it's life cycle, are also desirable outputs of sustainable products. Also, by mean of better understanding of systems, designers can be able to identify the intervention points that lead to desired outcomes with existing products. To obtain all this currently six strategies have been defined, that can work perfectly well in certain combinations [9]:

1. **Design for attachment and trust**; Explores the way in which users develop a certain bond with the objects they use. Challenge: strategy only partly depend on designer influence.
2. **Design for durability**; – is based on defining optimum product reliability. It is a well-defined technical field. Ideally a product's durability should match its economic and stylistic lifespan.
3. **Design for standardization and compatibility** –is constantly evolving (field of digital technology, for instance).
4. **Design for ease of maintenance and repair**; This is complex field, having in mind that maintenance and repair are currently divided among the original manufacturer, the gap exploiting service provider and users.
5. **Design for adaptability and upgradability**; Implies to incorporate possibilities to change a product.
6. **Design for di- and reassembly**; Easy disassembly is a classic requirement for sustainability. The possibility to reassemble is similar to the previous three strategies, but may include assembly with components of other products to become something different.

If products ought to be easy to repair and disassemble, that starts with design. New business models for repair, reuse, recovery, and product as a service represent the area of opportunity. In 2018, IKEA, the world's biggest producer of affordable design furniture, received more than one million orders for spare parts from customers who were able to perform do-it-yourself repairs, thanks to the company's modular product designs. [10]

It became obvious that what is needed is a systemic shift across the product value line. With increasing Extended Producer Responsibility (EPR) legislation, however, products that need to be designed with circular and disassembly consideration will slowly gaining more attention of designer and innovation departments at leading organizations. Still, for the majority of those items, current marketplace economics do not yet encourage circular designs. [6]

4 Conclusion

Traditional linear economy model of production and consumption of goods could be replaced by circular economy concept. The full implementation of this new concept will result in a shift in perception within the design, because of the need to prolong product lifespan (a cornerstone of the circular economy). Design can become the catalyst for changes by putting focus from quantity to quality, and finding new ways of working in cooperation across different disciplines.

Having in mind that accepting and implementing product design principles within the circular economy concept is actually a difficult task, it is important to note that incorporating design strategies in other areas contributes to the development of sustainable products in the context of environmental, social and economic challenges. In order to further develop creativity within the product design process, greater connection between disciplines is required. Contemporary technologies provide the ability to design products so that the materials from which they are made can be reused of reused materials, slowing down material flow, while at the same time technologies enable material utilization in a cost-effective way. Consideration should be given to new methods of the design process as well as production habits, which would result in improved products and services, better customer relationships and savings of resources and energy. Through design, we can address some of the new solutions to identified challenges, by implementation of contemporary design strategies that include durability, standardization, compatibility, ease of maintenance, product repair, adaptability and upgrade, di- and re-assembly or incentivization of recycled materials to minimize waste, all in order to figure out most acceptable ways of closing materials loops.

References

1. P. Lacy, J. Long, W. Spindler, *The Circular Economy Handbook - Realizing the Circular Advantage,* ISBN 978-1-349-95968-6, (2020)
2. V. Alivojvodić, F. Kokalj, *Upravljanje otpadom i cirkularna ekonomija*, text book, Belgrade Polytechnic, Belgrade, ISBN 978-86-7498-077-4, (2018)
3. Ellen MacArthur Foundation, *Towards a Circular Economy - Economic and Business Rationale for an Accelerated Transition*, Vol.1, (2013)
4. Ellen MacArthur Foundation, *Intelligent Assets*: *Unlocking the Circular Economy Potential*, Report, (2016)
5. P. Lacy, J. Rutqvist, *Waste to wealth: The circular economy advantage, Accenture*, Palgrave Macmillan, ISBN 978-1-137-53070-7, (2015)
6. Accenture Strategy, *Circular Advantage: Innovative Business Models and Technologies to Create Value in a World without Limits to Growth*, Accenture, 12, (2014)
7. B. Bürdek, *Design: History, Theory and Practice of Product Design*, ISBN 9783035603941, (2015)
8. European Commision: A new Circular Economy Action Plan For a cleaner and more competitive Europe, COM (2020) 98 final, (2020)
9. C. Bakker, M. Den Hollander, E. Van Hint, *Products That Last: Product Design for Circular Business Models*, BIS Publishers B.V. ISBN10: 9063695225, (2014)
10. A. Pownall, *IKEA to Begin Renting Furniture as Part of Wider Sustainable Push*, Dezeen, https://www.dezeen.com/2019/02/20/ikea-rental-furniture-circular-economy-design/ /Accessed: March, 2020/, (2019)

A novel printing channel design for multi-material extrusion additive manufacturing

Pinar Urhal[1, *]

[1]University of Manchester, Mechanical, Aerospace and Civil Engineering Department, M13 9PR, United Kingdom

Abstract. Additive manufacturing has a great potential in terms of its capability to produce components with complex geometries and to make multi-material and composite products by combining different materials in a single manufacturing platform. Current trends for the multi-material extrusion additive manufacturing process are categorized by multi-nozzle systems and multi-material inlet systems. In the case of multiple nozzle system, materials are deposited from different nozzles in sequence. On the other hand, in the case of multi-material inlet system, different materials are sent into a mixing tube and deposited as a mixture of materials. In this case, functionally graded parts can be fabricated by changing the volume fraction of two or more materials. Hence, the fabrication of parts with a continuous material supply by varying ratios for the extrusion technologies requires the development of printing heads with suitable printing channels, capable of varying the mixing ratio of different materials. To evaluate the effect of different printing channel designs on the material's flow pattern and the functionally graded material printability, this paper presents a three-dimensional transient computational fluid dynamics (CFD) simulation of the two miscible liquid-liquid system in a printing channel. Different geometries and materials are considered

1 Introduction

Bioprinting describes a novel tissue engineering approach to create cell-laden constructs [1,2]. In this approach, additive manufacturing is used to print a bioink formed by a hydrogel containing cells and growth factors [3]. Ideal hydrogels should present relatively low viscosity, a shear-thinning behaviour and good permeability characteristics [4]. A wide range of hydrogels such as alginate, gelatine, pectin have been reported [5,6].

Initially, bioprinting was used to create cell-laden constructs based on a single material and a single cell type (e.g. fibroblasts, keratinocytes, chondrocytes, stem cells) [7]. Advances in the field of additive manufacturing allowed to create multi-material and multi-cellular constructs using multi-head printing systems [8,9]. However, each printing head usually operates with a single bioink.

* Corresponding author: pinar.urhal@postgrad.manchester.ac.uk

At the University of Manchester (UK), we are investigating novel printing heads (e.g. co-axial and tri-axial printing heads, multi-material inlet screw-assisted printing heads or multi-material inlet pressure-assisted printing heads) to allow the fabrication of more biomimetic constructs, functionally graded constructs and tissue interfaces. The use of printing heads based on multi-material inlets, where the different materials are sent into a mixing tube before printing, allows to change the volume fraction of the different material thus creating functionally graded bioinks.

As this system will be used for the printing of low viscous hydrogel or polymer materials, a finite element based computational fluid dynamics (CFD) model was developed to investigate the flow patterns of some printable miscible materials (Alginate-Gelatine Solutions). This study also investigates the effect of inlet positions on the change of initially determined parameters (velocity and pressure) at the output of the printing channels.

2 Modelling and Simulation

2.1 Printing Heads

In the case of printing heads with multiple material inlets without any additional element (e.g. a screw) to guarantee good mixing conditions before printing, it is important to identify the best configuration (angle between the inlet channels and the mixing channel) to guarantee that. Therefore, three different configurations were considered as shown in Table 1.

Table 1 Geometric characteristics of the printing heads

3D Configurations of Channels	Dimensions	Features
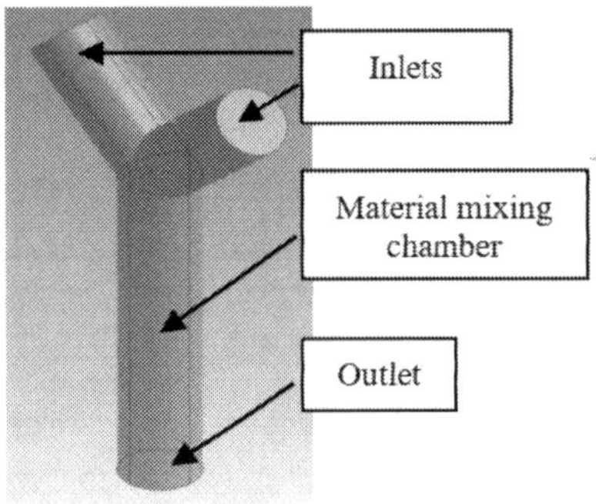	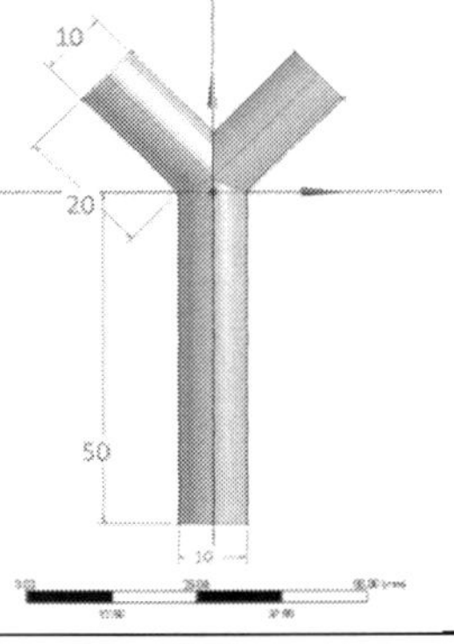	"Y" shape configuration with a 10 mm of internal diameter, and with an angle of 135°
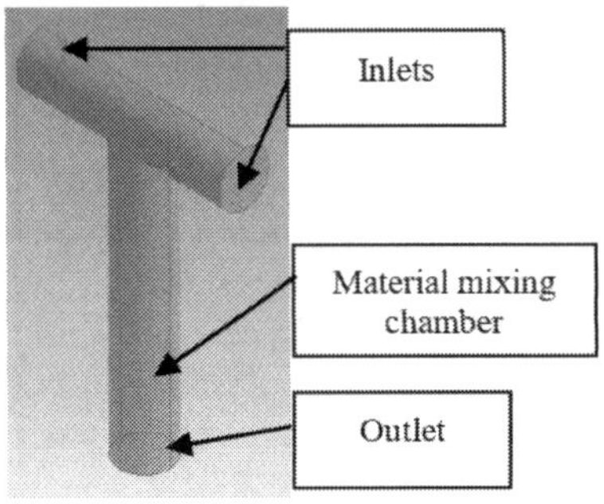	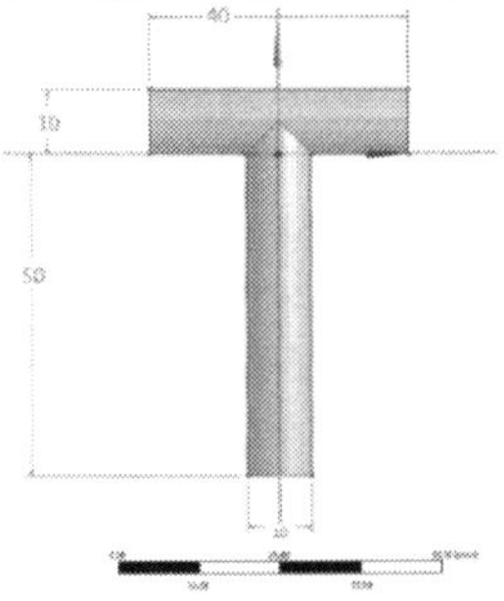	"T" shape configuration with a 10 mm internal diameter with an angle of 90°

Table 1 (cont.) Geometric characteristics of the printing heads

3D Configurations of Channels	Dimensions	Features
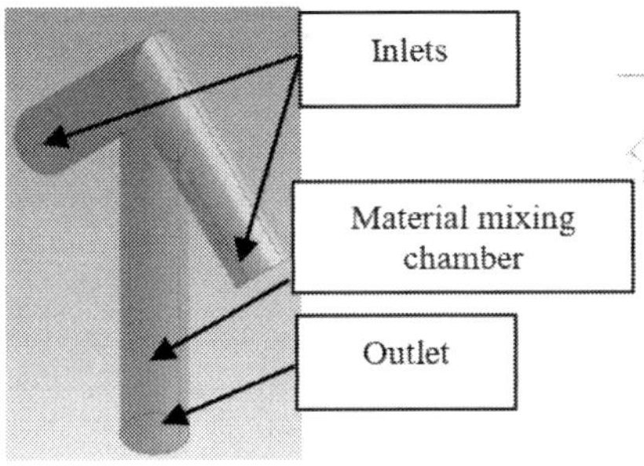	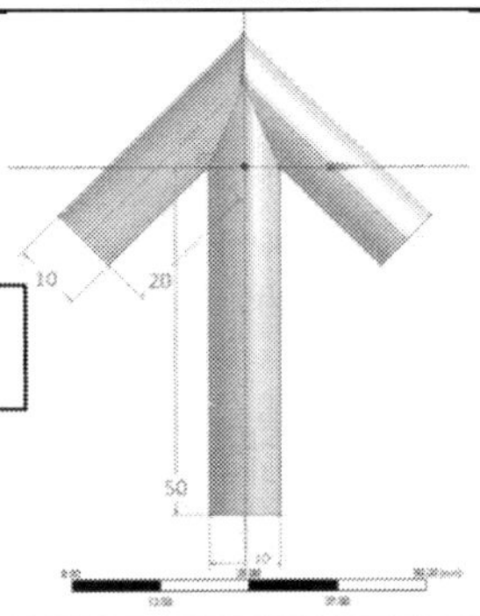	"↑" shape configuration with a 10 mm of internal diameter with an angle of 45°

2.2 Simulation Conditions

Computational Fluids Dynamics (CFD) simulations, were performed using a commercial fluids simulation software (ANSYS, Inc., USA). Printing heads were automatically meshed considering tetrahedral elements and two hydrogels (alginate and gelatine) commonly used for bioprinting applications (see Table 2 for material properties). Based on an extensive literature search the suitable printing concentrations were determined as 2-4% for sodium alginate and 8% for gelatine solutions [10,11]. For solutions containing 3% of alginate the extrusion velocity is assumed to be 50 mm/s, which allows viscous flows with Reynolds numbers ranging between 0.003 and 0.015, far below the turbulent flow transition boundary.

Table 2 Material Properties

	Solution	**Viscosity(kg/m.s)**	**Density(g/cm^3)**
Alginate	3%	0.00925	1,256
Gelatine	8%	0.01	1.3

The numerical analysis was performed considering the Non-Reactive Multi Species Method, using the species transport equations to evaluate the interaction and volume fraction information of the two considered hydrogels. In the species transport model mixture formulation, the components are mixed at the molecular level and both velocity and temperature are the same for all materials. In order to investigate the effect of the printing channel geometry, all the simulation models were performed under the same boundary conditions (Table 3).

Table 3 Boundary Conditions

	Inlet 1	**Inlet 2**	**Outlet**
	Velocity inlet	Velocity inlet	Pressure outlet
Velocity magnitude (m/s)	0.05	0.05	-
Initial gauge pressure (KPa)	103	103	-
Temperature (K)	300	300	300
Initial species mass fraction	Gelatine=0 Alginate=1	Gelatine=1 Alginate=0	Gelatine=0 Alginate=0

3 Results and Discussion

Material fractions and mixing ratios of different materials provide relevant information on the suitability of a specific printing head to print functionally gradient materials. Therefore, the influence of the material inlet positions on the material fraction distributions were evaluated by considering the mass fraction contours across the printing channels and outlet cross-sections of those channels.

The material fractions for the gelatine solution (red colour), pumped from the inlet 1, are presented in Figure 1, showing the distribution of the inflowing materials both along the tube and on the outlet. Figure 1a and 1b show that the configurations corresponding to angles of 135° and 90° respectively have no effect on the material mixing as it is possible to observe that each material is filling the half of the tube and materials can be identified by the colour gradation of each material's mass fraction values. On the other hand, Figure 1c is presenting different colour gradation behaviour in terms of material mixing. Similarly, it can be seen from the outlet cross-sections of the 45° inlet position (Figure 1c), the bright red colour for gelatine solution in Figure 1c and for alginate solution is fading into blurred blue and green colours as the fluids flow out from the mixing chamber, which is a clear indication that printing channel with 45° has a better material mixing performance during the printing process.

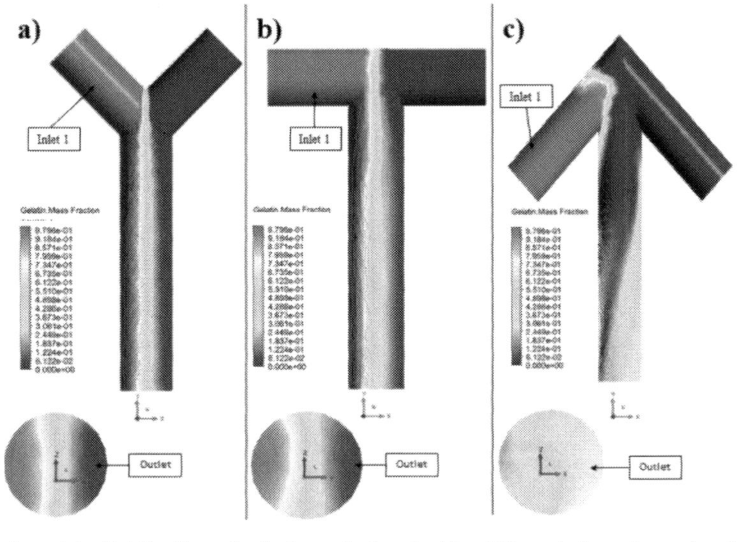

Fig. 1. The volumetric distribution of gelatine solution inside of the printing channels, a) Material inlet angle135°, b) Material inlet angle 90°, c) Material inlet angle 45°

The effect of the material inlet position on the velocity profile of the fluids along the printing channels is shown in Figure 2. The initial inlet velocities for all configurations was 50mm/s. As can be seen from Figure 3, the configurations a and b (Material inlet angle135° and Material inlet angle 90°respectively) show similar velocity profiles, but for the third configuration (Material inlet angle 45°), the velocity is reaching its maximum value (165mm/s) at the beginning of the material mixing chamber and decreasing towards the outlet.

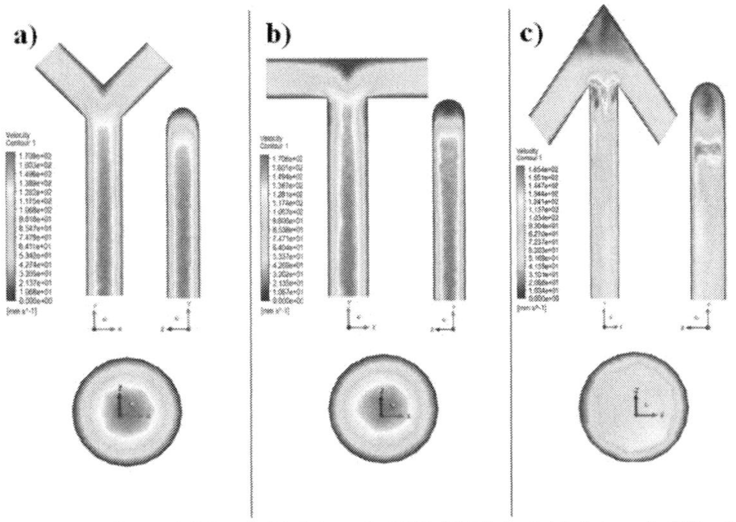

Fig. 2. Velocity contour plots, a) Material inlet angle135°, b) Material inlet angle 90°, c) Material inlet angle 45°

Finally, the pressure distribution plots for each geometry are presented in the Figure 3.

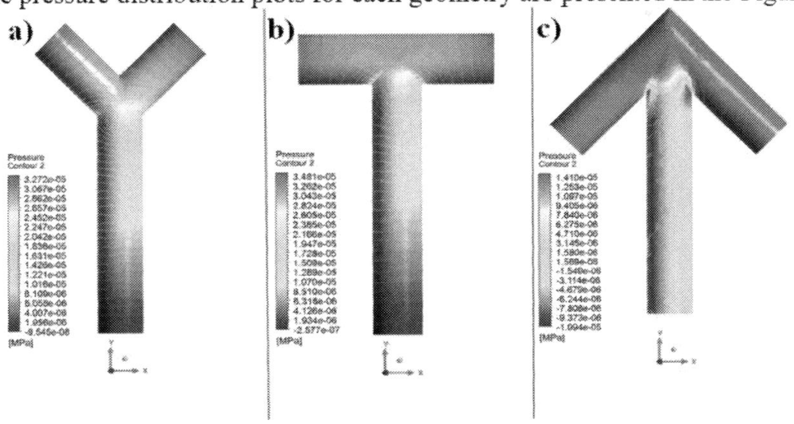

Fig. 3. Pressure distribution plots, a) Material inlet angle135°, b) Material inlet angle 90°, c) Material inlet angle 45°

4 Conclusion

In this study we aimed to investigate the effect of the position and configuration of the inlet of printing channel on the functionally graded material printability. A numerical method was used to determine the flow and mixing behaviours of two miscible biomaterials. Three different printing channels (each with different material inlet angles: 135 °, 90 °, 45 °) were designed. Based on the simulation results, it was possible to observe that the angle of the material inlets has a significant impact on the material mixing rations of materials at the outlet of the printing channel. The highest material mixing performance was observed for the 45 ° material inlet printing channel. As mentioned earlier, although screw-assisted extrusion additive manufacturing systems enable better material mixing, the results of this study indicate that functionally graded materials can be additively manufactured by using air-assisted extrusion AM method if the suitable material inlet angle is chosen. Moreover, screw-assisted systems, contrary to air-assisted extrusion systems, are not suitable to print bioinks (hydrogels containing cells).

References

1. D.-W. Cho, B. S. Kim, J. Jang, G. Gao, W. Han, N. K. Singh, D.-W. Cho, B. S. Kim, J. Jang, G. Gao, W. Han, and N. K. Singh, in *3D Bioprinting* (Springer International Publishing, 2019)
2. S. V Murphy and A. Atala, Nat. Biotechnol. **32**, 773 (2014)
3. D.-W. Cho, B. S. Kim, J. Jang, G. Gao, W. Han, N. K. Singh, D.-W. Cho, B. S. Kim, J. Jang, G. Gao, W. Han, and N. K. Singh, in *3D Bioprinting* (Springer International Publishing, 2019),
4. A. Subash, B. Kandasubramanian -Rushikesh, S. Ambekar, B. Kandasubramanian, and P. Rastogi, (2019)
5. M. Hospodiuk, M. Dey, D. Sosnoski, and I. T. Ozbolat, Biotechnol. Adv. **35**, 217 (2017)
6. J. Li, C. Wu, P. K. Chu, and M. Gelinsky, Mater. Sci. Eng. R Reports **140**, 100543 (2020)
7. F. You, X. Wu, N. Zhu, M. Lei, B. F. Eames, and X. Chen, ACS Biomater. Sci. Eng. **2**, 1200 (2016)
8. M. Toursangsaraki, Comput. Sci. Phys. (2018)
9. C. De Maria, G. Vozzi, and L. Moroni, MRS Bull. **42**, 578 (2017)
10. Y. He, F. Yang, H. Zhao, Q. Gao, B. Xia, and J. Fu, Sci. Rep. **6**, 29977 (2016)
11. M. Di Giuseppe, N. Law, B. Webb, R. A. Macrae, L. J. Liew, T. B. Sercombe, R. J. Dilley, and B. J. Doyle, J. Mech. Behav. Biomed. Mater. **79**, 150 (2018)

Optimization of a novel external fixator for orthopaedic applications

Mohammed S. Alqahtani[1,2*]

[1] Mechanical Engineering Department, College of Engineering, King Saud University, Riyadh, Saudi Arabia
[2] School of Mechanical, Aerospace and Civil Engineering, The University of Manchester, Manchester, UK

Abstract. The use of external fixation devices is a very common method for the treatment of bone fractures. However, these fixators present some limitations in terms of mobility, significant risk of infection, and induce pain and discomfort. Moreover, they are also not fully customized to suit individual patients. To avoid these limitations, this paper presents a novel patient-specific external fixator developed using reverse engineering, finite element analysis and additive manufacturing. The fixator was designed based on a set of computer tomography (CT) scan images of a patient and optimized considering different thickness values and materials. New lightweight designs were produced through a manual process (regular distribution of circular and hexagonal voids) and topology optimization. Different polymeric materials (Polylactic acid (PLA); Acrylonitrile butadiene styrene (ABS) and Polyamide (PA)) were also considered for the fabrication of these designs. It was found that although both PLA and ABS allow to meet the design requirements, and that the best mechanical properties were obtained with fixators made of PLA. Results also showed that the best results in terms of mechanical performance and weight reduction was obtained with topology optimization.

1 Introduction

Locomotion problems and bone disorders are often caused by car accidents, falls, wars and natural disasters. These injuries represent a major health issue as their treatments are complex and expensive affecting the public health system and are usually treated using an external fixation method [1]. External fixation is a biomechanical method widely used for the treatment of bone fractures and soft tissue pathologies, playing a major role in preventing amputation [2,3]. It is often implemented through the use of bone fixing elements (i.e. pins, wires) in case of complex fractures or without these elements in case of simple fractures [4]. Currently, there are several commercially available external fixators that can be used for the treatment of a certain fractures. However, these fixators present several limitations as they correspond to a painful treatment process, presenting also other disadvantages such as skin irritation, discomfort, heavy weight, risk of infection and long

* Corresponding author: Mohammed.alqahtani@manchester.ac.uk

treatment time [5-7]. The design of these devices presents a major challenge and it is considered one of the major causes for device failure. This occurs when the fixators are designed based on a standardized model to suit a wide range of patients' anatomy. Therefore, these orthotic devices need to be customized and personalized taking into account the individual patient's anatomy employing digital and advanced technologies (i.e. computed tomography, CAD modelling and additive manufacturing) in order to allow the design and fabrication of a more efficient and functional orthopaedic device [8,9]. Additive manufacturing (AM) gives designers and engineers the flexibility to design a lightweight but mechanically strong patient-specific device, which precisely fits the anatomy of the patient, increasing patient's comfort as well as device functionality [9-11].

This paper presents preliminary results related to the design of a novel patient-specific device for lower limb injuries to be produced using additive manufacturing.

2 Methodology

The production flow of customized external fixation devices comprises several steps as shown in Figure 1. The process starts with data acquisition using a scanning system (e.g. CT, MRI or laser scanning). The scanned data is post-processed in order to generate a model of the affected part of the patient. This model is then used as a reference to design and 3D model of the external fixator which perfectly fits to the patient. The CAD model is then assessed through finite element analysis to determine its mechanical performance. Finally, suitable models are produced using additive manufacturing.

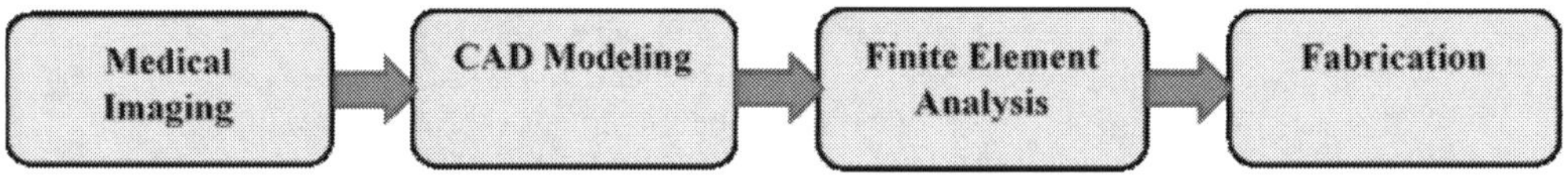

Fig. 1. Workflow process for fabricating personalized orthopaedic devices from patient-specific data.

2.1 Image acquisition and processing

A CT scan of a patient was used to obtain the anatomic details of the lower limb. The scanned data was exported in a common medical file format known as digital imaging and communications in medicine (DICOM). The DICOM file is imported into the three dimensional (3D) Slicer Software (www.slicer.org) to perform the anonymization process, enabling to convert the DICOM format into a nearly raw raster data (NRRD) format, removing sensitive patient's information. This file is then processed using the Embodi 3D cloud service (www.embodi3d.com), converting it into the 3D STL model of the lower limb. Finally, the STL model is imported into the Autodesk Meshmixer Software (Autodesk, Inc., San Rafael, CA, USA) for segmentation of the region of interest, correcting also potential errors. The process of transforming the DICOM file into an STL model of the patient's leg is shown in Figure 2.

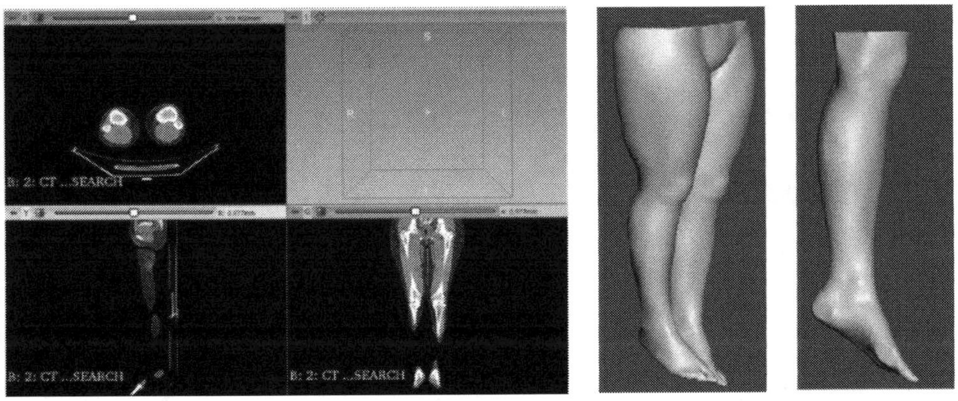

Fig. 2. Post-processing of CT scan data to obtain the STL model of the patient's lower limb.

2.2 Modelling and simulation

The novel external fixator (Figure 3) was modelled based on the STL model of the patient's leg using Solidworks Software (Dassault Systems, Massachusetts, USA). The fixator was designed through the creation of multiple planes along the patient's leg and spline curves to accurately represent the irregular geometry of the limb. These splines were connected together by the loft command to obtain a smooth surface. The solid model was then obtained through the offset of the generated surface and then thicken the offset surface in the outward direction. Finally, the model was divided into three regions (top, middle and bottom) allowing to remove material from the middle region and to keep both the top and bottom regions as solid regions in order to maintain the integrity of the fixator.

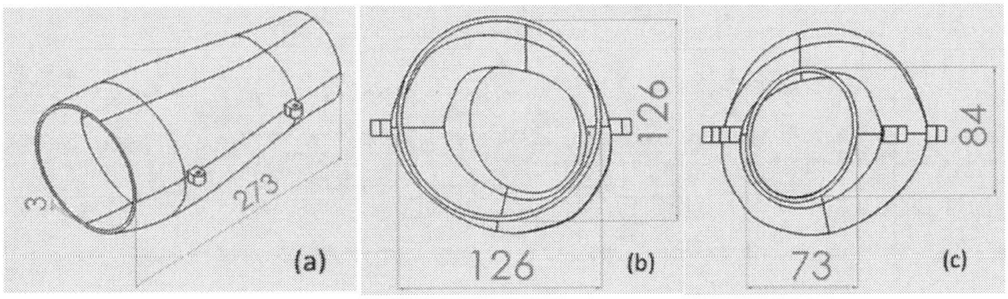

Fig. 3. The dimensions of the fixator (a) isometric view, (b) top view, (c) bottom view.

The solid model of the fixator was optimized in terms of thickness, materials and cut out configurations. Different thickness values (3, 4, 5, 6 mm), commonly used to design orthopaedic devices [12-14], and three materials (PLA, ABS, PA), commonly used for additive manufacturing, were considered. Besides the solid model, different cut out configurations (circular, hexagonal and topological models) (Figure 4) were also designed and evaluated. In the case of circular and hexagonal hollow-patterned models, the material removal was achieved by distributing the holes uniformly on the structure, without considering stress distribution. On the other hand, in the case of topology optimization, the stress distribution is considered, and the material removal is achieved by removing materials in low stress regions, resulting in an initial topology model with rough and sharp element edges. The model is then imported into the CAD software to perform the interpretation step, thereby obtaining a smooth and refined geometry of the topology optimized model. An arbitrary value of 15% of mass reduction was imposed to all designs.

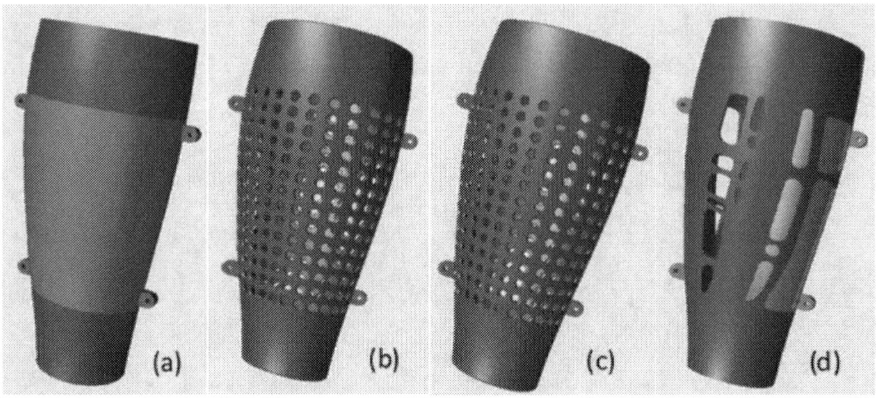

Fig. 4. (a) solid design, (b) circular design, (c) hexagonal design, (d) topology design.

The mechanical behaviour of all designs was investigated using the Ansys Workbench Software (Ansys, Inc., Pennsylvania, USA). A mesh of tetrahedral elements with a 1 mm element size was used. A static compressive load of 700 N that represents the average weight of an adult human (70 kg) was applied on the top face of the model, while the bottom face of the fixator was completely fixed. Three different materials were also considered (Table 1). The purpose of this analysis was to determine the Von Mises stresses and maximum displacements in the fixators according to the applied loading conditions. Each design should meet specific criteria and if one criterion has not met the design will be considered as not a suitable design. These criteria are: i) the maximum stress should be less than the yield strength of the material and, ii) the maximum displacement should be less than the allowable displacement (2 mm) [15].

Table 1. Material properties

Property	Material		
	PLA	ABS	PA
Young's Modulus (GPa)	2.35	1.62	0.58
Poisson Ratio	0.39	0.39	0.35
Yield Strength (MPa)	49.5	39.0	27.8
Density (g/cm^3)	1.24	1.10	1.14

3 Results and discussion

3.1 Thickness selection

As the effect of thickness on the overall performance of the fixator is independent on the material, simulations were performed considering ABS and a solid fixator. Numerical results in terms of maximum displacement (mm), maximum stress (MPa), stiffness (N/mm), and specific stiffness/mass ratio (N/g.mm) are presented in Table 2. The results for maximum displacement and maximum stress are the two main criteria used for the thickness selection. As observed, all models show max stress below the yield strength of the material (σ_{ABS} = 39 MPa) and max displacement below the allowable displacement (2 mm). Therefore, all results obtained were acceptable by the criterion and the 3 mm thickness was selected for further optimization as it provides the least part mass among other models.

Table 2. Results for different thickness values

Thickness (mm)	Max. Stress (MPa)	Max. Displacement (mm)	Fcritical (N)	Safety Factor	Stiffness (N/mm)	Specific Stiffness (N/g.mm)
3	4.81	1.09	5664	8.1	642	2.0
4	3.06	0.59	8898	12.7	1173	2.8
5	2.79	0.38	9765	13.9	1829	3.4
6	2.47	0.27	11020	15	2586	4.0

3.2 Material selection

In this case a solid fixator with 3mm of thickness is considered, and the effect of three different materials (PLA, ABS and PA) on the mechanical performance of the fixator investigated. Obtained numerical results are presented in Table 3. The results of maximum displacement and maximum stress are the two main criteria used for material selection. As observed all fixators met the criteria except the PA fixator, which met the maximum stress criteria, but the maximum displacement (3.05 mm) clearly exceeds the allowable displacement (2 mm). Therefore, PA is considered inappropriate. The maximum displacement of both ABS and PLA fixators are 1.09 and 0.75 mm, respectively showing displacements smaller than the maximum allowable displacement. Therefore, these two materials are suitable for the fabrication of the external fixation device. However, although both materials exhibit similar stresses, PLA fixators present smaller displacement than ABS. Therefore, PLA was the selected material.

Table 3. Results for different materials.

Material	Max. Stress (MPa)	Max. Displacement (mm)	Fcritical (N)	Safety Factor	Stiffness (N/mm)	Specific Stiffness (N/g.mm)
ABS	4.81	1.09	5664	8.1	642	2.0
PLA	4.86	0.75	7115	10.1	931	2.6
PA	5.08	3.05	3829	5.5	221	0.7

3.3 Geometry optimization

Three cut out structures were investigated (circular, hexagonal, and topological models). The effect of each option was assessed using structural FEA and the results are presented in Table 4.

Results show that the obtained fixators' designs after manual material removal (circles and hexagons) present larger displacement and stress values compared to both solid and topology optimized models. This is due to the random material removal from the model without considering the maximum stress regions which can affect the strength of the fixator. However, in the case of topology optimization, the maximum stress value was reduced compared to the solid model and the maximum displacement is relatively larger than the solid model. Therefore, results indicate that the topology optimized model is the stiffest and strongest model, whereas the hexagonal model is the least stiff and weakest model. The improvement obtained from the topology optimized model is obvious as the mass reduction of 15% was successfully achieved without sacrificing its integrity and the maximum stress reduced by 20%. Besides, the stiffness was also maintained as it is within the range of ± 10% of the original model.

Table 4. Results for different cut out configurations.

Design	Max. Stress (MPa)	Max. Displacement (mm)	Fcritical (N)	Safety Factor	Stiffness (N/mm)	Specific Stiffness (N/g.mm)
Circle	8.19	0.95	4230	6.0	736	2.4
Hexagon	12.72	0.99	2722	3.9	701	2.3
Topology	3.86	0.80	8975	12.8	865	2.9

4 Conclusion

A novel custom-made external fixator for lower limb injuries based on a patient's CT scan data, is presented. The fixator was optimized in terms of thickness, material and geometry focusing on its suitability for additive manufacturing and weight reduction. Manual and topological optimization strategies were considered for weight reduction and 15% of weight reduction was successfully achieved. This reduction also allows to reduce costs and fabrication time. The fixator was evaluated using finite element analysis under a static compressive load. Results show that the topology optimized model presents better mechanical performance.

References

1. A.D. Davis, U.S. Patent 5,288,286 (1994)
2. L.N. Solomin, *The Basic Principles of External Skeletal Fixation Using the Ilizarov and Other Devices*, 2nd edition, (Springer-Verlag, Italy, 2012), ISBN 978-88-470-2619-3 (2012)
3. J.R. Fox, W. Lovegreen, *Lower-limb orthoses, Chapter 22*, Fifth edition, 239-246, ISBN 9780323483230 (2019)
4. A.T. Fragomen, S.R. Rozbruch, HSS Journal **3 (1)**, 13-29 (2007)
5. Y.E. Lim, N.H. Kim, H.J. Choi and K. Park, J MECH SCI TECHNOL **31 (11)**, 5477-5483 (2017)
6. R.Y. Goldstein, C.J. Jordan, T.M. McLaurin, and A. Grant, *Bulletin of the NYU Hospital for Joint Diseases* **71 (1)** 96-103 (2013)
7. M.S. Alqahtani, A. Al-Tamimi, H. Almeida, G. Cooper and P. Bartolo. Progress in Additive Manufacturing **5** 85-94 (2020)
8. J.C. Arnone, A. Sherif El-Gizawy, B.D. Crist, G.J. Della Rocca and C.V. Ward, J Med Devices **7 (2)** (2013)
9. P. Gholawade, Simulation-based design of patient-specific femoral locking plate using topology optimization (Master dissertation, University of Missouri--Columbia) (2017)
10. S.H. Huang, P. Liu, A. Mokasdar and L. Hou, Int J Adv Manuf Technol **67** 1191-1203 (2013)

MATEC Web of Conferences **318**, 01026 (2020) https://doi.org/10.1051/matecconf/202031801026
ICMMEN 2020

Usability evaluation of mechatronic system by primary school children

Maria Chatzikyrkou[1,*]

[1] Aristoteles University of Thessaloniki, Department of Philosophy and Education Thessaloniki, Greece

Abstract. The paper examines the ease of use of a mechatronic system by children aged 7-12+ years living in the area of Thessaloniki. The research was conducted on students who participated in educational robotics seminars and as the results show, the trainees consider a mechatronic system very easy to use. The study concluded that older boys in particular were more familiar, than girls, in the construction and the programming of mechatronic systems. Children seem to be excited and this is evidenced by their strong desire to use it often. The area criterion does not make difference to the results. It seems clear that a mechatronic system is easy to use by primary school children. Through educational robotics, children can be taught in the different objects that combine this special tool. Learning takes the form of a game and escape from the narrow boundaries of conventional teaching that is now outdated and without interest.

1 Introduction

The aim of the paper is to investigate the usability of a mechatronic system by primary school children who use it during an extracurricular activity, educational robotics. But what does a mechatronic system mean? Any mechanical system that combines electro-mechanical parts with electronic devices and is controlled by microcontroller sytem, has been established to be called a mechatronic system. A mechatronic system includes the microcontroller, sensors, actuators and software. The sensors collect data from the process environment while the microcontroller together with the software take control of the variables and perform the control actions on the process. The software provides flexibility in changing the characteristics of the microcontroller, is able to easily and quickly perform calculations and shows better behavior in terms of sensitivity, noise. Finally, the actuators close the control loop of the process and the respective mission is implemented.

In order to be useful a mechatronic system, it must be easy to use. The usability of such a system is considered very important because, if the interface is not easy to the use, the user focuses on how to interact with it and not on its content [1]. In the literature, are different definitions of usability. In international standards, usability refers to efficiency and effectiveness in achieving specific goals and user satisfaction.

* Corresponding author: maria.chatzikyrkou@gmail.com

"Ease of use: the degree to which a product can be used by specific users to achieve specific goals with efficiency, effectiveness and satisfaction within a specific user framework" [2]

The main factor's of usability examined in the literature include ease of learning [3, 4], ease of use [3,4], usefulness [4], user satisfaction [5]. This study examines the "usability" of a mechatronic system in terms of usefulness, ease of use, ease of learning, and satisfaction.

2 State of the art

Usability refers not only to software and hardware products, but to any object or product that is designed and manufactured for use by people (users). Usability has been recognized as an important parameter in the design and evaluation of products, systems and services by researchers with different starting points in scientific fields such as Cognitive Psychology, Applied and Cognitive Applied and Cognitive Applied Economics, Computer Science and Industrial Product Design. Usability is described extensively in important relevant scientific books.

Child computer interaction (CCI) [10, 11] research has expanded greatly over the last decades, partly due to the realization that children's mental models are different from those of adults. Embodied interactions and body posture have also been analyzed to detect children's engagement and social interactions with robots [12, 13]. In another case [14] the interaction of children was analyzed with the help of gestures that showed a good percentage of acceptance of the new means of communication. The children were found to have great motivation for the activity and the ease of interaction was assessed as very positive [9].

The work of Nacher, Garcia-Sanjuan and Jaen showed that in young children and around the age of 6, it is difficult to interact with mechatronic systems. To be precise, most of them could not achieve the final goal, while in the parameter that had to do with the completion time, it showed that it is inextricably linked to the age group. The older children, are more successful on the goal of achieving time. In the same survey in response to (whether children can guide the robot on a predetermined path) and according to statistical analysis, the research question is answered in the affirmative to children aged three or older, as they were able to complete the work with success rates of about 100% [7]. In the work of Vasey et al. they compare the ease of use of an application in an android environment with a remote control, to control an Aldebaran Nao robot-type mechatronic system. The results were positive, pointing to the ease of use and organization of the interface [8].

Ranganathan et al. also studied the interaction with a robotic robot arm. Examining the ease with which children aged 9-12 were treated. Their results showed an impressive effect of age on motor learning. The 9-year-olds had great difficulty with the job, with only ~ 40% being able to complete the job, while the 12-year-olds had more difficulty controlling the robot arm as well as a low job completion rate, longer driving times and travel lengths [14].

3 Methodology

In this study, a sample of students consisting of 100 children aged 7-12 + years were asked to evaluate the usability of a remote control in relation to an autonomous robotic vehicle (Fig. 1). The evaluation was performed using a weighted questionnaire.

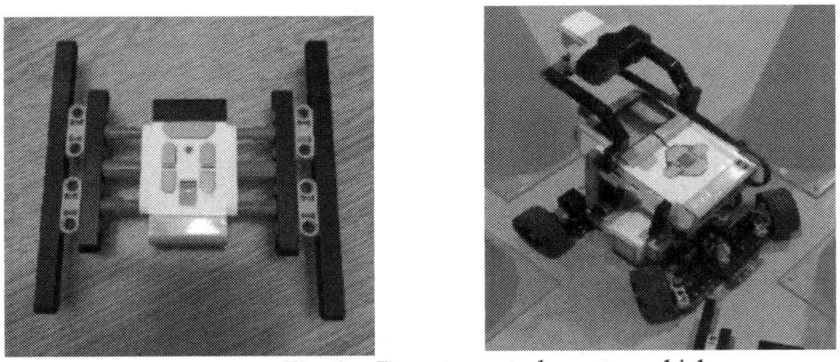

Fig. 1 - Remote control - motor vehicle

A Lego Mindstorms EV3 Infrared Beacon was used to build the remote control from the home edition version and some extra beams-bricks to achieve more stable handling. The Lego Mindstorms EV3 Infrared Beacon controller communicates with the Lego Mindstorms EV3 home edition's infrared sensor via infrared radiation. The commands send to a microcontroller on which are the wheels connected and the mechtronic system controlled. A Medium motor is also connected to an additional output, which is responsible for driving the vehicle by turning the front wheels either left or right. The system is controlled through appropriate programming that runs in the memory of the microcontroller and has been created in a Lego Mindstorms EV3 programmable language environment.

Having built the remote control of the robotic vehicle that interacted with it, it was deemed appropriate to define the research questions. In order to evaluate the overall usability, the following research questions were formulated.

- Did the trainees find the platform useful?
- Did the trainees find the platform easy to use?
- Did the trainees find the platform easy to learn?
- Did the trainees find the platform satisfactory?
- Did the trainees find the platform easy to use?

The research was conducted with the help of a structured questionnaire as a means of collecting information, with statistical analysis of the results and drawing conclusions as it was considered to be the most appropriate way to answer the above research questions. The 5-point Likert scale was used for the answers in the questionnaire.

Given that the object of the research is to evaluate the usability of interaction with an autonomous robotic vehicle (mechatronic system), the tool used after a thorough literature review is the structured USE Lund questionnaire (2001). The term USE is an acronym for the corresponding English words Usefulness (U), Satisfaction (S) and Ease of Use and Learning (E) and is a tool developed to evaluate software usability. The USE Questionnaire [6] measures the subjective utility (usability) of a product or service. This is a 30-item survey that looks at four dimensions of usability: usability, ease of use, ease of learning, and satisfaction.

The questionnaire, created electronically with the Google Forms tool, recorded the views of 100 children aged 7-12 + years living in the Thessaloniki area and using the mechatronic system as part of educational robotics seminars in which they were in groups of 4 -6 people.

Initially, a pilot survey was conducted in which the questionnaire was given to 5 students who were then asked about the ease and timing of completion as well as in case of misunderstanding or difficulty answering a question or a whole unit of questions. Their observations were useful as some minor changes were made, mainly in the vocabulary, and the average time to complete it was calculated. The USE questionnaire requires information such as age, gender and residential area. The structured USE questionnaire (Usefulness, Satisfaction, Ease of Use) has been proposed by Lund (2001) as a tool for categorizing user responses to the dimensions of usability, ease of use, satisfaction and ease of learning and is suitable for the effective measurement of the most important dimensions of the ease of use of an interface. It includes 30 LikertScale questions, in which users can choose one of the following agreement levels:

1 – I Completly Disagree, 2-Disagree, 3-Neutral, 4 – I Agree, 5- I Completly Agree

These levels correspond to the numbers 1 to 5, where "1" corresponds to "I completely disagree" and "5" to "I completely agree", in order to be able to quantify the variable response. These 30 questions are grouped into the following four categories:

- (A) Usefulness
- (B) Ease of Use
- (C) Ease of Learning
- (D) Satisfaction

The usability of the interface is based on the four categories, calculating the average number of user responses to the questions corresponding to each category.

4 Experimental results

The research was conducted to investigate the usability of the interaction of learners with a mechatronic system. The sample consisted of primary school children aged 7-12 + years living in Thessaloniki and its wider area and taking part in educational robotics seminars. The study involved 100 students, 61 of whom were boys (61%) and 39 were girls (39%). 9% are in the 7-8 age group, 58% are in the 9-10 category and 33% are 11-12 + years old.

The 30 statements were grouped into 4 categories and composed 4 new variables, "Usefulness", "Ease of use", "Ease of learning" and "Satisfaction". Cronbach's reliability index for each of the above variables was estimated. In order to be valid the answers given by the students, and therefore to be reliable the results and conclusions of the research and impartial, the estimated value of each factor a Cronbach should be higher than 0.6. The Cronbach's index for "Usefulness" was found to be 0.922, for "Ease of Use" 0.942, for "Ease of Learning" 0.902, and for "Satisfaction" 0.953, which shows high reliability, in the sense of internal consistency, of the 4 categories.

From the analysis of the statements that compose the "utility", its average was calculated. A total of 3,745 cases were found, suggesting that students' views on "usefulness" were mild. The standard deviation is 0.730 respectively. For ease of use, there were 3,313 cases, suggesting that students' views were neutral towards a soft agreement. The standard deviation is 0.860, respectively. From the analysis of the statements that make up the "ease of learning", its average was calculated and 3,698 facts were found, which suggests that the views of the students have a mild agreement. The standard deviation is 0.838, respectively. Finally, for the variable "Satisfaction" it appears that the average is 2,954, which suggests that the views of students show neutrality. The standard deviation is equal to 1,000.

In terms of overall usability, an average value of 3,427 indicates that learners tend to have a neutral to mild agreement on "Ease of Use" of the system as a whole. (Fig 2)

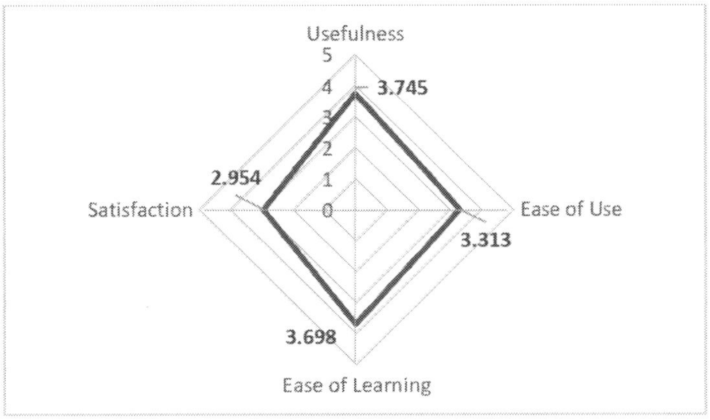

Fig 2.Usability Diagram

5 Conclusion

From the analysis of the results of the above research, important data emerged regarding the ease of use of the interaction of primary school students with a computer system.

Regarding the overall usability of the computer system, the majority of respondents are positive. They declare, on average, a mild agreement on overall usability, which is in line with other research. With the research of Pons (2019), Read (2013), Hourcade (2015), but also with the research of Sanghvi (2011), Henkemans (2017) and Ranganathan (2019) in the publications of which it seems that there is a positive interaction of the respondents with technology and electronics systems. It is also in agreement with the research of Vasey et al (2017), which is similar to the present research, which points out the ease of use and organization of the interface. Nacher's (2016) study found that children over 6 years of age reported ease of use and learning as opposed to younger children. In contrast to the results of these studies, Ranganathan's (2019) research is inconsistent because the findings of their research show difficulty in using and learning for children aged 9-12.
In summary, students 'views on "usefulness" and "ease of learning" are mildly consistent, while students' "ease of use" are neutral in terms of mild agreement. Finally, with regard to "Satisfaction", it appears that the views of students show neutrality.

References

1. G. H. Luo, E. Z. F. Liu., H. W. Kuo, S. M. Yuan, Design and implementation of a simulation-based learning system for international trade. The International Review of Research in Open and Distance Learning, 15(1) (2014)
2. ISO 9241. Ergonomics of Human-System Interaction. International Organization for Standardization.
3. J.Nielsen, Evaluating hypertext usability. In: D. H. Jonassen, & H. Mandl (Eds.), Designing hypermedia for learning(pp.147-168), Berlin: Springer-Verlag (1990a)
4. R. A. GuillemetteThe evaluation of usability in interactive information systems. In: J. M. Carey (Ed.).Human factors in information systems: Emerging theoretical bases.Norwood, NJ: Ablex, 207-221. (1995)
5. J. Nielsen, Usability Engineering, academic Press, London (1993)
6. A. Lund, "Measuring usability with the USE questionnaire", Usability and User Experience Newsletter, STC Usability SIG,vol.8, no.2,pp.1–4. (2001)
7. V. Nacher, F. Garcia-Sanjuan and J. Jaen, "*Evaluating the Usability of a Tangible-Mediated Robot for Kindergarten Children Instruction,*" *IEEE 16th International Conference on Advanced Learning Technologies (ICALT)*, Austin, TX, 2016, pp. 130-132. (2016)
8. E. Vasey, M. Fakhr Hosseini, Z. Zheng, C. H. Park, A. Howard, & M. Jeon. *Development and Usability Testing of a Remote Control App for An Interactive RoboT. Proceedings of the Human Factors and Ergonomics Society Annual Meeting*. 61. (2017) 808-812. 10.1177/1541931213601695.
9. P. Pons, J. Jaen, J Ambient Intell Human Comput (2019).
10. J.C. Read, P. Markopoulos, Int J Child-Comput Interact 1:2–6 (2013).
11. J.P. Hourcade, Child computer interaction. CreateSpace Independent Publishing Platform, North Charleston, (2015)
12. J. Sanghvi, G. Castellano, I. Leite, *Automatic analysis of affective postures and body motion to detect engagement with a game companion. In: Proceedings of the 6th international conference on human–robot interaction—HRI'11*. ACM Press, New York, pp 305–311,(2011)
13. O.A.B. Henkemans, B.P.B. Bierman, J. Janssen, Int J Hum Comput Stud **106**:63–76. (2017)
14. R. Ranganathan, M.Lee, M.R. Padmanabhan., Sci Rep **9**, 1960 (2019). https://doi.org/10.1038/s41598-018-38092-3

MATEC Web of Conferences **318**, 01027 (2020) https://doi.org/10.1051/matecconf/202031801027
ICMMEN 2020

An AM-oriented vehicle chassis' A-Pillar Design Approach

Agisilaos Kyriazis [1,*], *Dimitrios* Koulocheris[1,†], *Stamatios* Polydoras[2,‡], *Clio* Vossou[1,§]

[1]NTUA, School of ME, Vehicles Laboratory, 15780, NTUA Campus, Zografou-Athens, Greece
[2]NTUA, School of ME, Mechanical Design & Automatic Control Department, 15780, NTUA Campus, Zografou-Athens, Greece

Abstract. Design and production of highly demanding structural systems, such as the chassis, still rely on conventional forming and welding approaches, both because of their proven performance and the economies of scale achieved. Nevertheless, manufacturing of several chassis' segments is also expected to soon gradually switch towards AM, for increased design freedom and optimized performance. This paper proposes an alternative design approach for the A-pillar, a typical passenger car chassis segment; a design suitable in form for AM and equally capable in terms of its dynamic behavior, without undermining the chassis' safety. Prior A-pillar designs along with already published innovative AM-suited design approaches are reviewed. Moreover, these serve as a starting point for an inverse design towards the intended new AM-suited A-pillar alternative. Emphasis is given in the dynamic characteristics of the new structure, through proper modal analysis performed. Finally, the presented research concludes with a scaled-down assessment and verification prototype of the new design, planned to be built via FDM 3D Printing. The prototype is expected to demonstrate primary, as well as secondary/latent benefits from the use of AM in A-pillars, such as the increased diagonal visibility for drivers and passengers, arising from the redesigned, mesh-like form of the segment.

1 Introduction

Additive manufacturing (AM) technologies, commonly known as 3D printing, over the past decade have transformed the potential ways in which products are designed, developed, manufactured and distributed, [1]. For the automotive industry, these advances have opened paths for new designs, lighter, "cleaner", and safer products, shorter lead times and lower costs. Consequently, among other sub-systems and components in the automotive industry, manufacturing of several segments of the chassis is, at some point, also expected to gradually switch towards AM, mainly due to the increased design freedom and optimized performance AM has to offer. However, it is still difficult for AM to compete with

Corresponding authors:* agiskyri@hotmail.com
† dbkoulva@central.ntua.gr
‡ polsntua@mail.ntua.gr
§ cvossou@gmail.com

conventional techniques already long in use by automotive companies, that require the highest production rate and repeatability, with the lowest cost and rejection rates possible. Recent international standards (published and/or under development) by ISO & ASTM are clearly aiming in this direction, of appropriate reliability, repeatability, quality, safety and certification of modern AM processes, materials and systems, [2].

In terms of road safety, the occupant cell of a car, which consists of the chassis, is described as a 'survival box' suggesting that the survival of the occupants involved in the circumstance of a traffic accident depend highly on the mechanical response of the cell to the applied loads. The high strength of the occupant cell is a fundamental prerequisite for vehicle safety design, [3]. The chassis and other supporting parts of a vehicle should be designed so as to ensure safety and prevent loss of life in the event of a traffic accident. The chassis of an automobile differs according to its body-shape. Three typical body-shapes are the sedan that has four doors, the lift back that has a rear luggage compartment hatch type door and the station wagon whose roof extends straight back, [4].

As shown in Fig. 1, there are at least three main pillars supporting the frame work of a typical vehicle and these includes A-pillar, B-pillar and C-pillar. In the case of station wagons, that have large passenger compartments, an additional pillar, D-pillar, supports the rear side of the vehicle.

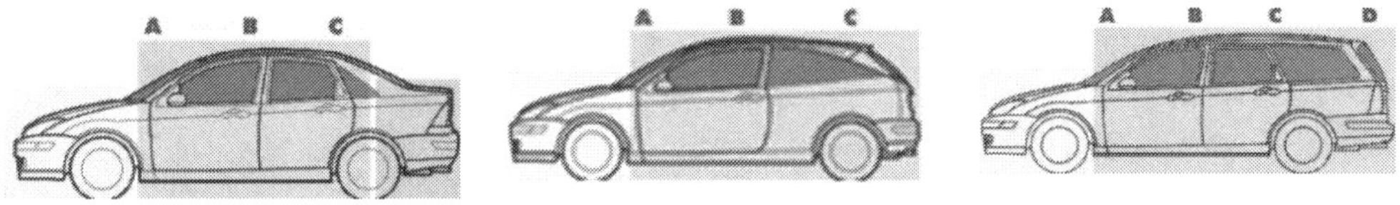

(a) (b) (c)

Fig. 1. Typical supporting pillars and their locations in (a) sedan (b) hatchback and (c) station wagon vehicles

A stage during the automobile design is the Body-in-White (BIW). In this stage, the car body is formed by assembled metal sheets, and the main components as chassis, power train, doors, etc. are not still mounted. The determination of the natural frequencies of the parts of the automobile body i.e. the A-Pillar, is crucial in reducing the probability of failure and expanding its life, [5]. An A-pillar is an important load carrying component of any automobile body, [6]. It is a thin-walled construction made of high strength steel alloys which is properly designed and dimensioned in order to successfully withstand the loads developed during a crash or a rollover and meet all the safety requirements. Industrial practice shows that A-pillars by design create large blind spots, blocking the driver's visibility. For this reason, some car-manufacturers attempt to design A-pillars slim and chamfered, to decrease the blind spots to a minimum and improve driver-visibility.

In this paper, three existing A-pillar designs are reviewed in terms of their dynamic behaviour, through modal analysis. Furthermore, an alternative AM-suitable design for the A-pillar is proposed, one that largely comes out of topology optimization and it is compared with a prior published innovative AM-targeted design. Both these designs adopt a perforated-lattice form that directly relates them with AM for manufacturing.

2 A-pillar Design and Modal Analysis

Three A-pillars, each corresponding to each type of an automobile body shape of Fig.1, are investigated and presented in Fig. 2. All 3D Computer Aided Models (CAD) models required were developed on Solidworks ®. It should be noted that they were not based on exact drawings of the vehicles' manufacturers, but on geometrical data retrieved from available online sources, [7, 8, 9]. In order to best approximate the actual geometry of each A-pillar in the simplest possible way in Solidworks, auxiliary lines on a proper reference plane were used, utilizing the overall dimensions of each car model and thus resulting in A-pillars forms of the real scale. The 3D CAD models developed are presented in Fig.2.

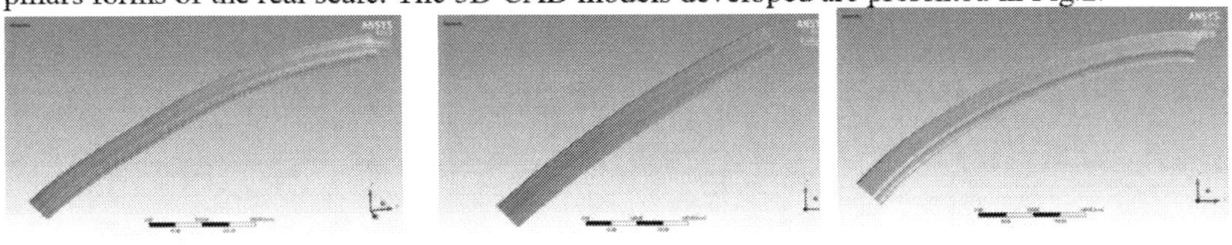

(a) (b) (c)

Fig.2. 3D CAD A-pillars of a (a) sedan (M1), (b) hatchback (M2) and a (c) station wagon (M3)

A surface finite element (FE) model has been created for each 3D CAD model in order to perform modal analysis of the three A-Pillar designs. Fig.3 shows the surface FE models.

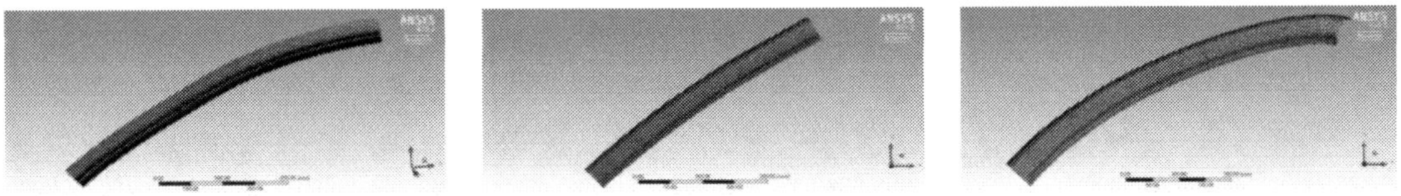

(a) (b) (c)

Fig.3. FE models of the A-Pillars of (a) sedan (M1), (b) hatchback (M2) and (c) station wagon (M3)

The mesh in all models was uniform, capturing curvature, consisting of FE of type SOLID181. SOLID181 elements are suitable for analyzing thin to moderately-thick shell structures. This element type is four-noded element with six degrees of freedom at each node: translations in the x, y, and z directions, and rotations about the x, y, and z axes, [10]. In all cases the thickness of the A-Pillars is considered equal 1.3 mm. In Table 1, the mass, the number of FE & nodes and the element quality of each FE model are presented.

Table 1. Models' Finite Elements (FE) and Nodes

Model	Mass (kg)	Area (m^2)	Number of nodes	Number of FE	Element Quality
M1	1.02	0.99	28590	28629	0.82
M2	0.57	0.56	4972	4890	0.78
M3	0.61	0.60	6977	6918	0.79

In Table 1 is obvious that M1 has almost double the surface of the rest, hence it is the heaviest and it consists of more FE and nodes. For all FE models, structural steel has been considered as material, with Young's modulus of 200 GPa, Poisson Ratio of 0.3 and Yield stress equal to 250 MPa.

Modal analysis has been selected in order to assess the dynamic behavior of the A-Pillars, since during the movement of an automobile, the engine cycle incentive and uneven road surface excitation will lead motor vehicles to the forced vibration. If the excitation signal frequency and the resonance frequency of the skin structure is close, the mechanical structure will produce local resonance and high deformation, [11,12]

For the modal analysis, both ends of each A-pillar have been considered fixed, in order to simulate the fact that with a vertical displacement of the chassis its top and bottom will remain untwisted and undeformed. The first six eigenfrequencies for each model are presented in Table 2.

Table 2. A-pillars Eigenfrequencies.

Model	Eigenfrequencies (Hz)					
	1st	2nd	3rd	4th	5th	6th
M1	107.67	147.41	190.92	275.61	312.88	389.10
M2	230.02	317.00	334.96	584.388	717.94	841.12
M3	139.16	248.89	338.63	449.63	564.26	650.92

In Fig.4 the contours of total displacement in the first three eigenfrequencies are presented for each FE model.

Model **Eigenfrequency**

1st 2nd 3rd

M1

M2

M3

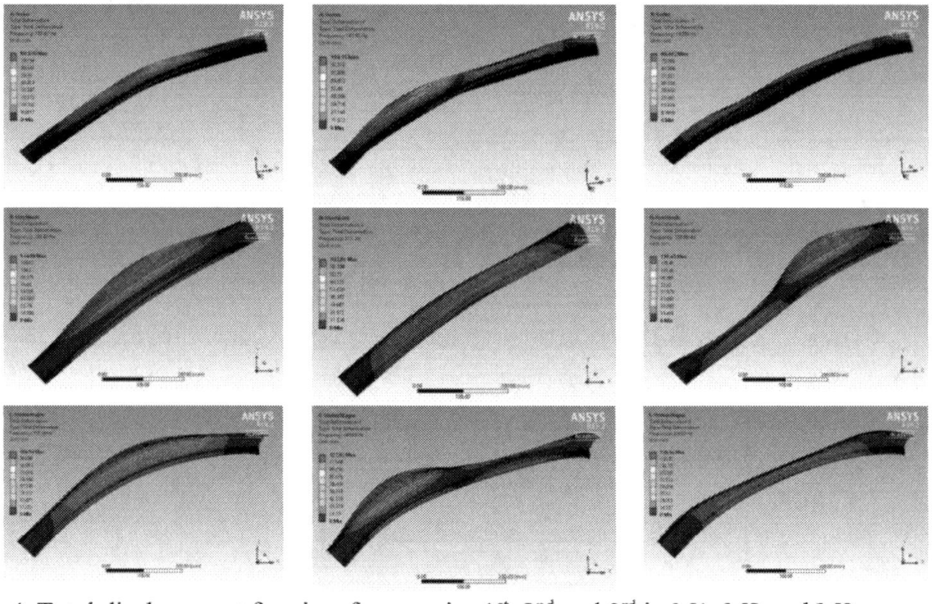

Fig.4. Total displacement for eigenfrequencies 1st, 2nd and 3rd in M1, M2 and M3

In Table 2 is obvious that M1 has the lowest eigenfrequencies, while M2 the highest. On the other hand, the displacement for the first eigenfrequency is highest for M2 (134 mm) while for the second and the third one is highest for M3 (128 mm and 131 mm, respectively).

3 M3 A-Pillar Topology Optimization

For the M3 model also an alternative, perforated, AM-suitable design exists. This design is compared to a design produced with topology optimization in a static structural loading simulating rollover. In more details, Ansys Workbench 2109R1 has been performed on a solid body corresponding to the M3 structure. The boundary conditions, along with the equivalent Von Mises Stress (VMS) contour are provided in Fig. 5. The force has been calculated in order to simulate rollover (vehicle's curb weight with one passenger 1500 kg).

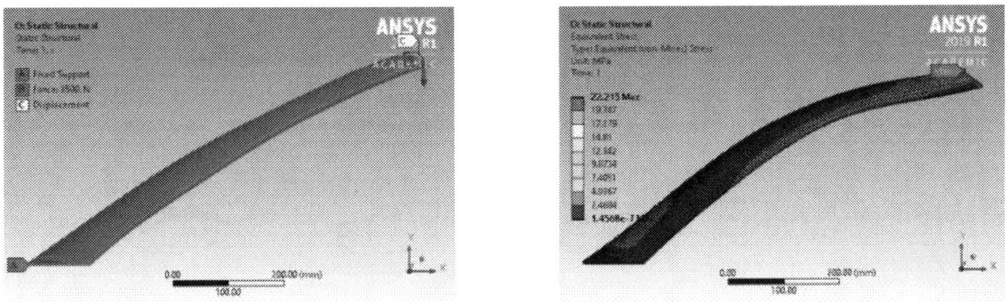

(a) (b)

Fig. 5. (a) Boundary conditions and (b) equivalent VMS contour of the static analysis of M3

In Fig. 5 is obvious that the highest value of equivalent VMS is 23 MPa, which is lower than the yield stress of the material.

For the Topology Optimization the Sequential Convex Programming solver has been used. Weight minimization has been set as objective function and maximum stress (yield stress of the structural steel) as the constraint. The optimization procedure converged after 75 iterations producing a structure with a weight of 4.80 kg. In Fig. 6 the result of the topology optimization is presented.

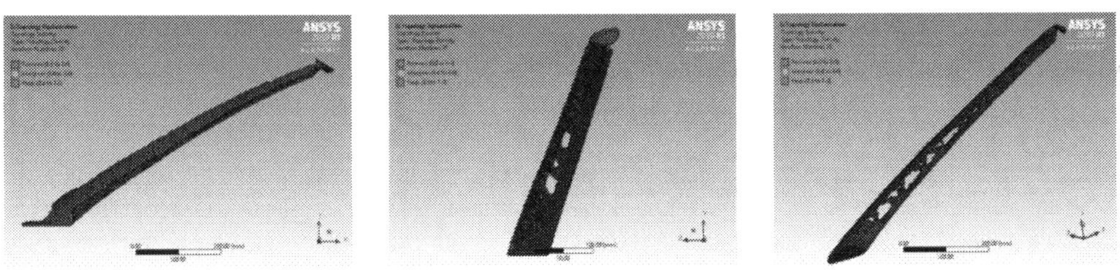

(a) (b) (c)

Fig. 6. Optimized structure for the M3 model in (a) front, (b) right and (c) arbitrary view

As mentioned above, the provided new structure is similar to a perforated A-pillar design (M3p) originating from M3 developed by the car manufacturer, in order to improve driver visibility. The perforated form characterizes both designs as candidate for AM. In Table 3 the result of the modal analysis of this A-Pillar (M3p) is presented.

Table 3. A-pillar Eigenfrequencies.

Model	**Eigenfrequencies (Hz)**					
	1st	**2nd**	**3rd**	**4th**	**5th**	**6th**
M3p	266.85	492.82	612.12	656.20	703.91	731.96

Comparing the values of the eigenfrequencies in Table 3 with these in Table 2 is obvious that M3p model has lower values than M3, but still, higher than M1 and M2.

In Fig. 7 the boundary conditions and the results of the static analysis simulating rollover for the existing perforated model are presented.

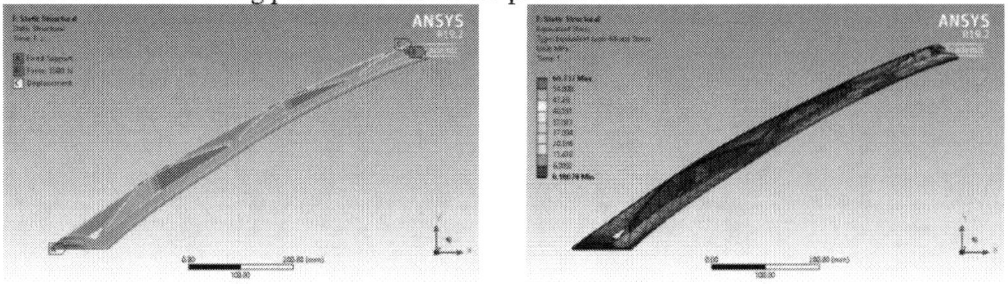

(a) (b)

Fig. 7. (a) Boundary conditions and (b) equivalent VMS contour of the static analysis of M3p

4 Conclusion

In this paper, three A-pillars (M1, M2, M3), each corresponding to a characteristic type of an automobile body shape, are designed in Solidworks and their dynamic behavior has been investigated through modal analysis in ANSYS. Furthermore, an alternative for the M3 model has been created using topology optimization. This design was compared to an already existing perforated design (M3p) in a static structural loading simulating rollover. The perforated form characterizing both designs makes them candidates for AM. In the future both these designs will used to create scaled down prototypes with AM in order to experimentally test their performance.

.References

1. M. Cotteleer, J. Holdowsky, M. Mahto, *The 3D opportunity primer: The basics of additive manufacturing*, Deloitte University Press, (2014).
2. S. Tranchard & V. Rojas, *Manufacturing our 3D Future*, iso.org/news , (2015).
3. A.E.Ikpe, E.K.Orhorhoro, A.Gobir, *Design and Reinforcement of a B-Pillar for Occupants Safety in Conventional Vehicle Applications,* Coventry University, (2016).
4. M. C .Venkataswamy, M. T .Naganna, & M. N. K .Mishra, Int. Journal of Advanced Engineering and Global Tech., (2015).
5. S. Joglekar, B. D. Patil, Int. Journal of Innov. Research in Science, Eng. and Technology, (2017).
6. K. S.Naik, & A. R. Patil, Int. Journal for Advance RnD, (2018).
7. G. H. Ruffo, *This is the technical presentation the Audi A5 and the S5 deserve,* (2016).
8. https://gr.pinterest.com/pin/419538521519806549/?lp=true
9. https://www.pinterest.ca/pin/471470654732646645/
10. P. Kohnke,. *ANSYS theory reference manual*. Ansys Inc., Canonsburg, PA, USA. (2013).
11. Y.L. Guoying,. F, Yongzhao, D., & Shouhuan, Z. Int. Journal of Engineering Research and Applications, (2014).
12. Y.Rashid, S.Aizzat, R.Ramli., S.M.Haris, & A.Alias. The Scientific World Journal, (2014).

MATEC Web of Conferences **318**, 01028 (2020) https://doi.org/10.1051/matecconf/202031801028
ICMMEN 2020

An ad hoc decision support method over additive vs. conventional manufacturing

Stamatios Polydoras[1,*], *Clio* Vossou[2], and *Dimitrios* Koulocheris[2]

[1]NTUA, School of ME, Mechanical Design & Automatic Control Department, 15780, NTUA Campus, Zografou-Athens, Greece
[2]NTUA, School of ME, Vehicles Laboratory, 15780, NTUA Campus, Zografou-Athens, Greece

Abstract. The mechanical design process considers numerous factors. Requirements related to performance and quality, limitations by legislation, standards, methods utilized or technological boundaries, urgency, cost, data preparation and preservation, design flexibility and organizational aspects. Successful design consists of proper decisions on form, geometry, materials, manufacturing methods, quality, reliability and more. Nowadays, a critical decision during design and realization of technological objects is whether they should be made conventionally or with Additive Manufacturing (AM)/3D Printing methods. Such a decision occurs under time-pressure or via a broader strategy for technological switch, is complex, multi-parametric and bears uncertainty and risk. A simple, effective and substantiated method to assist decisions for switching from conventional to AM could prove very useful. This paper refers to recent trends and activity in international AM-related standards, then presents and discusses preliminary work of the authors for an ad hoc decision method to be used upon specific "go/ no-go" decisions for AM. The method is largely based on the Pareto principle, to limit critical design factors contributing to this decision. All steps of the method towards a final decision are described. The method is demonstrated with a hypothetical, yet realistic example of a short run coolant vessel manufacture.

1 Introduction – Industry 4.0, Modern Design, Additive Manufacturing, Standards & Decisions

The technical world has recently seen a lot of change. Priorities have shifted, from maximum cost effectiveness and performance-maximized products, towards sustainability with optimized products. Design cycles nowadays tend to actually never meet a specific end point, as modern products, are "open" for change or improvement, throughout their full lifecycle, [1]. The same applies for priorities and strategic goals of production organizations, both large and small, where sustainability, longevity, effective operational structure, ecologic consciousness and social reputation, increasingly emerge above short- or mid-term financial results.

* Corresponding author: polsntua@mail.ntua.gr

Inevitably, this new technical environment affects and alters the nature and characteristics of modern products, in terms of design approach & process, production methods and industrialization. The role of electronics, computers, digitalization, CAD, CAE, PLM, the Internet, the "Cloud", Internet of Things (IoT) and other disruptive technologies, is crucial in this transformation, as they all formulate today's Industry 4.0, [2].

Equally important, Additive Manufacturing (AM), or simply 3D Printing is also correctly regarded as a key technology of this modern technical and industrial environment, already drastically acting as the critical link between the new "digital/virtual" universe of products and their "on demand" materialization. Most AM technologies have originally appeared in the late 1980's – early 1990's as Rapid Prototyping Techniques, [3]. Having almost immediately proven their value in product development, the past two decades they have gradually evolved into a vast future-capital for modern product design, development and manufacturing in the era of Industry 4.0, [4].

Design, both organizationally, as well as geometrically, is already severely affected by AM's special characteristics, adopting even more the concepts of individuality, customization & specialization, just-in-time & on demand production, minimization of inventories for products, parts and tooling as well, localization in production and product finalization, composite and multi-material components and of course, form-geometry-mass optimization, assembling minimization and parts consolidation, [5].

A reasonable question design & manufacturing-involved decision makers and industrial practitioners will soon have to often answer, is whether, when and why they should opt for an AM/3D Printing technology available, in place of a well proven conventional manufacturing method or process (e.g. casting, machining, forming etc.).

Modern mechanical products amass many of the following trends and characteristics:

- Global simplicity, individual/local complexity
- Interchangeability, physical and functional flexibility, upgradability
- "System-level" cross-platform compatibility, "system-level" modularity
- Increased "component-level" functional integration vs. modularity
- High-degree performance tunability/adaptability
- Easier assembly-disassembly; Reduced relevant time and effort
- Reduced materials' diversity, materials and components re-usability and recyclability
- Minimization of standardized machine elements and connecting elements
- Incorporation of "smart materials", "Built-in" Intelligence (AI)
- Easy-fast servicing, Easy-fast-cost effective repair

Obviously, incorporating the above characteristics into any modern product without a digital background would be from difficult to impossible. AM is directly involved with these when it comes to actual product realization. But, with AM still rapidly growing on capabilities, but equally on "inflated expectations", it is really difficult one to monitor current state of the art and at the same time objectively judge for, or against AM. On the other hand, conventional methods, when compatible with a current need or application, being fundamentally simpler, cheaper and usually already highly optimized, offer a high degree of security to industrial operators.

In such cases a simple, fast, effective decision support method could prove useful, especially under time-pressure. To develop, implement and support such an ad hoc decision method, members of the technical community need to acquire and retain objective knowledge of the actual capabilities and limitations of modern AM in every industrial branch and here is where standards step in.

In the field of AM, the American Society for Testing and Materials (ASTM) – Committee F42 and the International Standardization Organization (ISO) – Technical Committee 261 shortly after, have both recognized the need to develop and establish standards for the "frantic" AM technologies, covering aspects from terminology, technologies and systems, to proper practice, materials, design guides, tests – measurements – reports, or special requirements by specific industrial branches. They have already jointly published and are currently developing many new fundamental and targeted AM-related standards. Knowledge of these standards can lead to knowledge on the AM technologies themselves and their successful implementation.

The most important already published and effective standards that relate with and can be utilized as props to assist decisions on AM implementation, in the authors' opinion are: ISO/ASTM 52900:2015, ISO 27547-1:2010, ISO/ASTM 52901:2017, ISO/ASTM 52902:2019, ISO/ASTM 52910:2018, ISO/ASTM 52921:2013 and ISO/ASTM 52921:2013, [6]. Moreover, many more AM standards are imminent and currently under development, most important of which for decision makers should be: ISO/ASTM 52903-1, ISO/ASTM CD TR 52912, ISO/ASTM DIS 52921, ISO/ASTM DIS 52924, ISO/ASTM CD TR 52918, ISO/ASTM AWI 52908 and ISO/ASTM AWI 52909, [7].

A product is first born when expressed as an idea or need. Needs are translated into design requirements and later transformed into design specifications; the tangible goal every design has to meet, [8]. Requirements can be categorized into larger groups (e.g. performance, safety, economic, managerial, environmental, social, special etc.) containing specific design factors and each factor could also be further analysed into its constituent parameters, to make the design process both analytic and organized. The actual design outcome will comprise from appropriate techno-economic compromises that will best satisfy the mixture of attainable design factors, expressed within the design intent by preference and importance of each one.

It is a fact that the number of design factors and their parameters is very large, [9], making all decisions in the process more difficult to implement and harder to justify. A useful solution to this inefficiency comes from the Pareto principle, often applied in many technical problems and stating that only a 20% of crucial parameters in any problem is responsible for more than 80% of the outcome, [10]. This principle, in conjunction with "Miller's law", [11], regarding the maximum number of variables the human brain can successfully simultaneously handle, drastically helps all decision problems and is adopted by the authors in the proposed decision support method.

2 The proposed 7-step ad hoc decision method

In the proposed method, for the sake of speed and effectiveness, the "analysis" of design factors that determine AM fabrication over conventional manufacturing stops at the level of factors. Any design and manufacturing environment can formulate its own set of requirements and factors, overall, or per product. Not all requirements or factors would be applicable in every case and they could be selectively adapted for certain product or component types and categories. Apart from strictly product-specific design requirements, it is important for industrial operators to also include some strategic requirements, related to innovation and the dilemma of technological switch towards AM within an organization. The method is intended for decision makers closely related to the specific product's development and with a clear understanding on the capabilities of their own organization, as well as modern AM developments, where again current standards can serve as a prop.

Having properly combined and adapted well accepted and widely used design factors, parameters and criteria, [9, 12], with current industrial practices & state-of-art and the scope of the proposed method, for easier user-level implementation the authors present in Table 1 below some suggested indicative typical groups of requirements and their respective factors.

Table 1. Requirements and respective design factors applicable in AM decisions.

Requirement	Typical Factors
Performance	Strength, speed, hardness, flow, max/min temp, …
Functionality	Ease, ergonomy, readability, interface quality, …
Efficiency	Consumption, efficiency ratio, operation time, …
Durability	Fatigue, hardness, resistance to humidity, radiation, chemicals, …
Materials	Organic, technical, magnetic, conductive, precious, bio-compatible, …
Form-Aesthetics	Size, colour, texture, free forms potential, …
Manufacturability-Industrialization	Minimum part count, streamlined production, production rate, assembling time, assembly operations count, …
Reliability	Time to first failure, uses until failure, MTBF, …
Safety	Physical, electrical, toxicological, chemical, biological
Eco-friendliness	CO_2 impact, material recyclability, re-usability, raw materials efficiency, auxiliary equipment, …
Cost	Material cost, production cost, labour cost, tooling cost, running costs, ROI, …
Technological Availability	Domestic availability, local availability, in- house/outsource, …
Innovation	Modern technology, keep up with competitors, contribute to progress

A first step for the presented method is to select the most influential/applicable factors from Table 1, or any similar list of the design team. No "requirement" category is necessary. All non–applicable factors can be omitted.

In the second step the Pareto principle is applied on the selected factors, to narrow them down to a manageable number. This can be done e.g. by directly ranking their importance individually in a "1 (not important) to 5 (Very/extremely important)" qualitative scale. Depending on their initial number, important factors can be filtered by keeping only those scoring a "4" or "5", or just the latter; always trying to end up with a maximum of ten to fifteen "factors" for the assessment, to best satisfy "Millers's law".

A third - very important - step is to classify the remaining factors to those "in favor of" or "against" AM implementation, or else to those supporting – benefiting – justifying the use of AM and those that hinder – oppose – prevent it. If a factor is more or less equally served by both examined alternatives it can also be omitted as neutral. By the end of this step we should have two columns with the most dominant factors "for" and "against" AM.

Obviously, not all remaining important factors have the same impact or effect in the decision. For, or against AM, the preference, importance or utility of each factor on behalf of the decision maker must be registered in this fourth step as weights, in order to next reach a weighted aggregation. Pairwise comparison/ eigenvector method, [13], the "digital logic" approach, or even an arbitrary distribution of these weights must be implemented at this stage.

Some special handling is necessary for some critical factors that can act as "on-off" switches for the decision. We could name those factors "on-off" determinants or more aggressively "terminators". An example of such a factor for an assessed component would be under the "Materials" requirement group, the "Conductivity" factor. Given that an examined method, combined with a specific material would clearly lead to either a conductive, or a non-conductive part, such a factor – when required – will not be classified at all in the "for" or "against" columns for later scoring and weighted aggregation. In the method's fifth step, "terminators" are used more dominantly, as external multipliers of the weighted scores of each alternative, them always holding a "digital" value of "0" or "1" that allows them to totally terminate an option/alternative if not met by it.

During the method's sixth step, all already distributed important factors for AM or conventional manufacturing must now be ranked by the decision makers in a scale of only a few, positive-only, grades, e.g. "1: fair, 2: adequate, 3: good", corresponding to their satisfaction by their governing alternative, always with a specific method in mind and the available knowledge about their characteristics and performance (e.g. SLS of thermopolymers vs. Injection Moulding).

For the last seventh step, the product of corresponding weights with factors' grades is aggregated per column and multiplied with respective "terminators", ultimately leading either to the immediate elimination of an alternative (if zero-ed by a "terminator"), or to a final comparison between two positive total scores, one "for" AM and the other "against" it. The decision leans towards the greater number between the two.

An algebraic expression for the described method would include all the *n* important factors remaining after implementing Pareto filtering and "terminator" exclusion to be included in an **F** matrix, dimensioned 2xn, whose first column would contain values "1 to 3" for factors in favor of AM and "0" for factors against it. Respectively, the second column would include values "1 to 3" in all rows where column 1 had zeros and a zero in rows where in column 1 there was a value. **F** would be multiplied with a 1xn matrix **W** of weights, whose all elements sum up to 1. The result, driving the decision, would arise in a 1x2 matrix **R**, as expressed below,

$$\mathbf{R}=\mathbf{K}\times\mathbf{W}\times\mathbf{F}, \tag{1}$$

, where

$$\mathbf{K}=[\ \Pi k_i\ (i=1\ldots m)\quad 1] \tag{2}$$

, k_i = 0 or 1, m: the number of "terminators" applicable

If the 1st element of **R** is a number greater than the 2nd, AM should be opted for.

3 Discussion, conclusions & further work

Suppose an auto-industry subcontractor needs to manufacture a container for a relatively corrosive coolant, in a series of 50 to 500 pieces. The quantity of the coolant would be 0.5l and it could reach a temperature of app. 80°C. They need to examine if they should produce the part utilizing blow moulding of a proper plastic or use SLS in Nylon for the desired part. The Pareto-important factors for this specific case that falls in the category of a short-run production, as selected and filtered according to authors' prior technical experience, are listed in Table 2, and classified as "for" and "against" AM having followed the steps of the proposed method. The same applies for the importance weights (aggregating 1), that could of course also be derived by any valid relevant preference capturing technique, [13].

Table 2. Example important factors, distribution, "terminators" and weights.

Factors	pro AM	contra AM	"Terminator"	Weights	Grade
Flow	*			0.02	3
Water tightness			*	-	-
Transparency			*	-	-
Chemical resistance		*		0.1	3
Thermal resistance		*		0.1	2
Free Form potential	*			0.13	3
Impossible drafts	*			0.03	3
Minimum part	*			0.1	3

count					
Time to 1st Failure		*		0.02	1
Materials Efficiency	*			0.1	2
Auxiliary equipment	*			0.07	3
Tooling cost	*			0.2	3
Local availability		*		0.03	2
Competitors	*			0.05	2
Progress	*			0.05	3

Following the proposed method with the data of Table 2, we end up with a result vector of R = [2.1 0.58], namely AM is advisable for this specific case, provided that both "terminators" are satisfied (value=1) for the proposed AM solution. Nevertheless, examining more cautiously the characteristics of Nylon SLS, we can easily find out that there is no transparent Nylon powder in use for SLS. So, the correct result should eliminate this solution with the respective "terminator". Therefore, we should either relax the transparency requirement to an opaque vessel, or examine a different AM method, such as a UV resin-based method in order to proceed with the AM method.

The proposed decision support method is generic, fast, concise, simple, substantiated and relatively easy to implement, without the need of any special software package. It requires proper knowledge of the examined alternatives, thus probably calling for some collaboration of design teams and manufacturing experts during its implementation.

Several variations and improvements to the method could be applicable and are currently examined by the authors, as e.g. the simultaneous contribution of a single factor to both alternatives, the extension of the concept of "terminators" also to the conventional processes, the method's verification via sensitivity analysis etc.

References

1. M.E. Porter, J.E. Heppelmann, Harv. Bus. Rev., **93** (10) 96 (2015)
2. R. Y. Zhong, X. Xu, E. Klotz, & S. T. Newman, Engineering, **3**(5), 616 (2017)
3. C. K. Chua, K. F. Leong, & C. S. Lim, *Rapid prototyping: principles and applications,* World Scientific Publishing Company, (2010)
4. U. M. Dilberoglu, B. Gharehpapagh, U. Yaman & M. Dolen, Proc. Manuf., **11**, 545 (2017)
5. M. Kumke, H. Watschke & T. Vietor, V. & Ph. Prot., **11**:1, 3 (2016)
6. ISO, Standards by ISO/TC 261, Additive manufacturing, Published standards, Retrieved from www.iso.org on 28-04-2020
7. ISO, Standards by ISO/TC 261, Additive manufacturing, Standards und. Devel., Retrieved from www.iso.org on 28-04-2020
8. S. Eppinger, & K. Ulrich, *Product design and development*, McGraw-Hill Higher Education, (2015)
9. Y. S., Chen, P. P. Chong, & M. Y. Tong, Math. & Comp. Mod., **19**(9), 61 (1994).
10. S. Dowlatshahi, The Int. J. of Adv. Man. Tech., **9**(5), 324(1994)
11. G. A. Miller, Psyc. Rev., **63**(2), 81 (1956).
12. M.Sfantsikopoulos, *Mechanical Engineering Design Vol.I,* Athens, NTUA Publications (2005)
13. P. Sen & J. B. Yang, *Multiple criteria decision support in engineering design (Vol. 4).* London: Springer (1998)

Electrospun PCL-Surgihoney meshes for skin wound healing applications

Enes Aslan[1,*]

[1]Department of Mechanical, Aerospace and Civil Engineering, University of Manchester, Manchester, UK

Abstract. Skin is a complex and very important tissue, playing a significant protective and regulatory function. It is also prone to a large number of wounds and defects due to external factors such as temperature, chemical agents, and radiation. Accelerating the skin healing process and minimizing the risk of infection is a relevant research challenge. This paper investigates a novel wound dressing based on polycaprolactone (PCL), a synthetic biocompatible and biodegradable polymer, and honey-Surgihoney® (SH). Solution electrospinning was used to produce the wound dressing meshes. Different polymer solutions were prepared by mixing PCL and SH with acetic acid. Human dermal fibroblast were used to assess the biological characteristics of the electrospun meshes. Results show that the presence of SH1 has a positive impact on cell attachment and proliferation.

1 Introduction

Skin is the largest organ in the human body and has a crucial role in protecting the internal organs from the outside environment. When this tissue is disrupted by trauma and disease the natural self-regenerative capacity can be significantly compromised resulting in loss of function and serious health issues. According to the World Health Organisation (WHO), each year worldwide, nearly 300,000 people die as a result of burns and approximately 6 million people are injured [1–3]. Moreover, chronic skin ulcers affect over 6 million people [2,3], which poses a serious economic burden to healthcare systems worldwide and further societal costs due to loss of economic activity. Therefore, of the utmost priority is the development of cost-effective, efficient and effective wound treatment strategies.

Currently, the clinical gold standard to treat critical skin defects is based on the use of skin grafting which uses a healthy tissue taken from the human body (patient or other person) to treat the wounded area [4,5]. However, the use of autografts or allografts both present significant limitations [5]. To overcome the limitations novel wound dressings and artificial tissue engineered skin substitutes are being developed [6,7]. Hover, these dressings are able to protect the wound from the external environment, prevent wound contamination, and dehydration but present limitations in terms of both biological and antibacterial properties.

[*] Corresponding author: enes.aslan@manchester.ac.uk

Electrospinning is a versatile and simple technique to produce nanoscale fibers, highly relevant for the fabrication of wound dressings and meshes for tissue engineering applications [8,9]. In the case of tissue engineering, electrospun meshes presents high surface to volume ratios, which enables better attachment of cells. Moreover, the high porosity of such meshes also improves vascularization, oxygen and nutrients exchange [10]. Several studies reported the use of electrospun meshes for a wide range of tissue engineering applications [1,9]. Different synthetic and natural polymers have been investigated [1,11].

Honey has been used to treat wounds since ancient times due to its antiseptic, antibacterial and recuperative properties [12,13]. The bactericidal property of honey is due to the production of hydrogen peroxide and reactive oxygen species (ROS), high osmolality of sugar, and low pH [13–15]. Furthermore, ROS have been demonstrated to promote wound healing by encouraging cellular repair processes and tail regeneration in tadpoles [16,17].

This is paper investigates for the first time the use electrospinning to produce polycaprolactone wound dressings containing different levels of Surgihoney® (SH), a synthetic honey medically approved for skin applications and commercially available in the form of a cream.

2 Materials and Methods

2.1 Materials and Solution Preparation

Polycaprolactone (PCL) (CAPA 6500, Mw 50,000 g/mol) provided by Perstrop Caprolactones (Cheshire, UK), is a biocompatible polymer used to provide mechanical stability to the meshes. Surgihoney® (SH), provided by Matoke Holding (Abingdom, UK), was used as an antibacterial agent, and comprises different natural honey extracts developed by Matoke. Acetic acid (Fisher Scientific, UK) was selected as a solvent due to its low volatility, non-toxic behaviour and the capacity to dissolve both PCL and honey. Four different PCL-SH (0%, 10%, 20% and 30% w/w of SH) compositions were prepared. Firstly, PCL was dissolved in acetic acid by using a hot plate stirrer overnight at 40°. Then, SH was added into the mixture and stirred at room temperature until obtaining a homogenous solution.

2.2 Fabrication

Electrospinning system can be defined as a cheap and versatile fabrication technique, able to manufacture nanofibrous meshes and highly relevant for the fabrication of wound dressings. Usually, it involves ejecting a polymer solution from a syringe needle tip, by applying a high electric field. A drop of polymer material is formed at the end of the tip due to the effect of surface tension and viscoelastic stresses. The shape of the drop approaches a cone shape (Taylor cone) and a fluid jet is ejected towards the collector.

Electrospun meshes were produced using the Prefector system (Spraybase, Ireland). The electrospinning system consists of a high voltage power supply (from 0 kV to 30 kV), software to control a syringe pump system, stainless steel collector and a 1.8Gx25mm stainless steel emitter (needle). In this case, meshes were produced considering a voltage of 17 kV, flow rate of 0.1 mL/min and a distance between the needle and the collector of 18.7 cm.

2.3 Morphological Analysis

Morphological characterisation of electrospun meshes was conducted through scanning electron microscopy (SEM, Hitachi S-3000N, Japan) at an accelerating voltage of 15 kV. Samples were coated with gold prior to imaging. The images were analysed using Fiji software with the DiameterJ plugin to assess fibre diameter and length.

2.4 Degradation and Swelling Study

Degradation and swelling characteristics of produced meshes were investigated using two different media, Dulbecco's Modified Eagle Medium (DMEM) and simulation body fluid (SBF) which was prepared using Kokubo and Takadama's protocol [18]. Samples were cut as 1cm x 1cm square shape and placed into 24 well plates with DMEM and SBF and incubated at 37 °C during the experiment. Three replicates (n=3) were used to measure the weight as dry and wet. The experiment was conducted for 63 days and the DMEM and SBF were replaced once a week. Same samples were used to measure wet weight (Ww) when conducting swelling study for the first three days, after that different samples (n=3) were used for each time point (days 3, 7, 14, 21, 35, 49 and 63) for the rest of the study (degradation and swelling). Over the course of 9 weeks, samples were removed from the media and the wet electrospun meshes weights (Ww) measured. The swelling (Sw) was calculated according to the following equation:

$$Sw = W_w/W_i \quad (1)$$

where Wi is the initial dry weight and Ww is the weight of the wet electrospun mesh. After, the swollen meshes were dried in the incubator at 37 °C until completely dry and weighed again (Wd). The percentage of weight loss (Dl) was calculated as follows:

$$Dl = ((W_i-W_d)/W_i) * 100 \quad (2)$$

where W_d is the weight after drying.

2.5 Biological study

Electrospun meshes were assessed in terms of cell proliferation using human dermal fibroblast cells (HDF) (Cell Applications,Inc. Sigma Aldrich, UK). Cells were cultured with Dulbecco's Modified Eagle Medium (DMEM) containing 10% (v/v) fetal bovine serum, 1% (v/v) glutamine, and 1% (v/v) penicillin/streptomycin until 80% confluence and harvested using 0.05% trypsin-EDTA solution (Thermo Fisher Scientific, USA) at passage 2. The metabolic activity of cells, an indicator of cell proliferation, was evaluated using the AlamarBlue® assay kit (Sigma-Aldrich, UK) following the suppliers' protocol. The fluorescence intensity was measured on days 1, 3, 7, and 14 after cell seeding. Four replicates were considered. Cell viability was also assessed using a Live/Dead Assay Kit (ThermoFisher Scientific, UK) at days 3 and 14 according to the manufacturer's instructions.

3 Results and Discussion

3.1 Morphological Analysis

Figures 1a to 1d present the SEM images and inset fibre distribution graphs of PCL and PCLSH meshes. Results show that the mean fibre diameter decreases with the increase of Surgihoney® concentration (190 nm for PCL; 185 nm for PCLSH10; 168 nm for PCLSH20 and 165 nm for PCLSH30). PCL and PCLSH meshes exhibit a distribution of fiber diameters ranging from 50 nm to 600 nm. All samples present a uniform fibre distribution. SEM images demonstrate the ability to successfully electrospun both PCL and PCLSH meshes with nanoscale morphology mimicking the native tissue extracellular matrix (ECM).

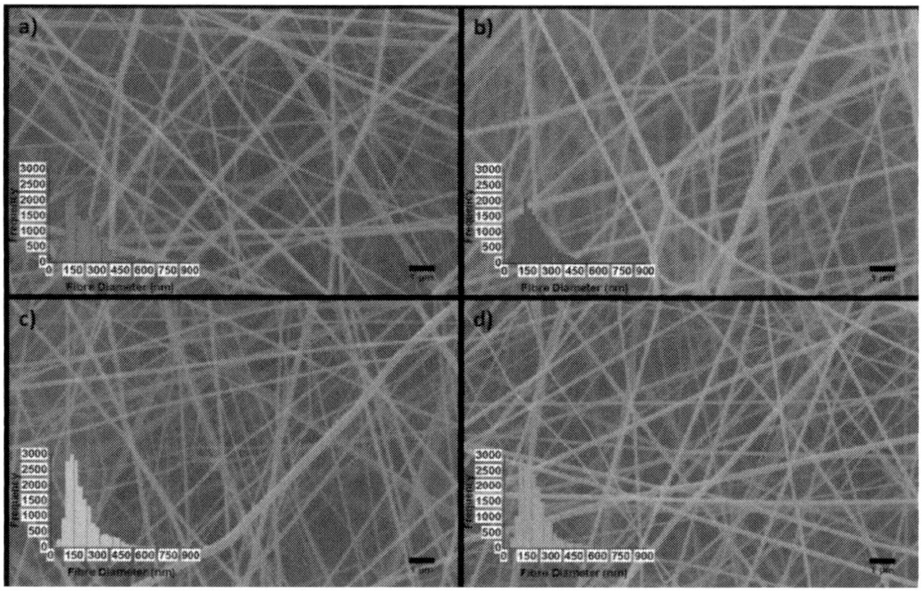

Fig.1. SEM images of electrospun meshes a) PCL, b) PCLSH10, c) PCLSH20, d) PCLSH30 (Scale: 1μm). Inset graphs of the distribution of fiber diameters

3.2 Degradation and Swelling

The swelling and weight loss values for PCL and PCLSH electrospun meshes are presented in Figure 2. Results show that PCL and PCLSH samples absorb DMEM and SBF until they reach an equilibrium state. It seems that all samples reached out the equilibrium state after day 7 in SBF solution and at day 14 in DMEM. It is also possible to observe that meshes in DMEM absorb more liquid than in SBF. This is due to the fact that meshes start to degrade in SBF solution earlier than in the DMEM. Results also show that meshes containing SH absorb more liquid than PCL meshes. It seems that water molecules containing hydroxyl groups form hydrogen bonds with other hydroxyl groups present in SH and consequently the liquid absorption is higher in PCLSH meshes than pure PCL meshes. It is also known that ideal wound dressings must be able to absorb the wound exudate. So, the meshes containing SH will be able to absorb more exudate during wound healing and keep the wound bed clean.

Figures 2c and 2d shows the weight loss of the electrospun meshes with time. Results show that PCL meshes present no degradation after day 63 in DMEM, while the meshes lose their half weight in SBF liquid at day 63. On the other hand, the meshes containing SH present a better degradation profile than pure PCL meshes in both liquids. It is also possible to observe that the increase of SH content on the meshes accelerates the degradation process due to the water solubility of SH. PCLSH30 meshes disappear completely at day 63 in SBF liquid while PCLSH20 meshes loses 86% of weight. Results also show a faster degradation in SBF than DMEM, and, as SBF is able to mimic the body fluids, indicate that PCLSH meshes could be completely degraded in the human body after two months.

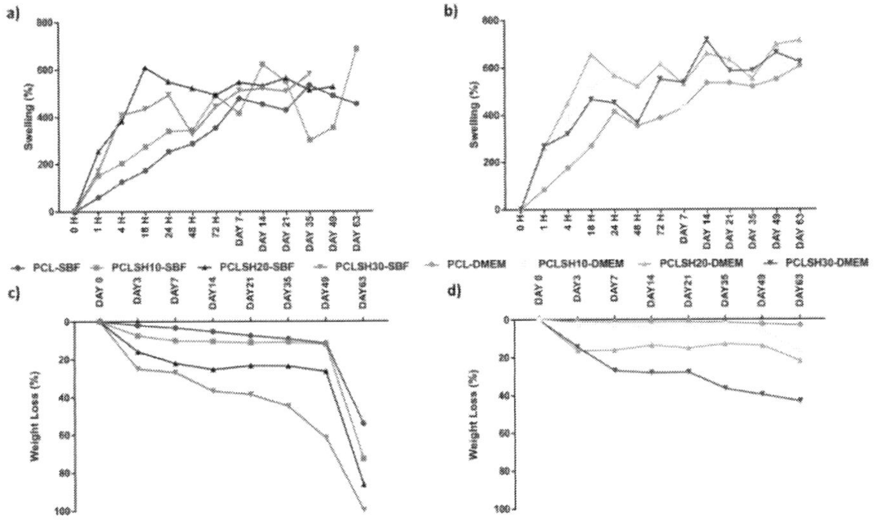

Fig.2. Swelling of the meshes in a) SBF b) DMEM and degradation of the meshes in c) SBF and d) DMEM as a function of time.

3.3 Biological Results

Fluorescence intensity values for different time points after cell seeding are presented in Figure 3a. As the fluorescence activity is an indicator of the metabolic activity of cells, results show that all meshes are able to sustain cell attachment and proliferation. Significantly better results are obtained for PCLSH30 meshes at day 14. Figure 3b shows the live cells on the attached meshes and their distribution on the meshes on day 3 and day 14. The results show that cells were able to spread on all samples, creating dense networks, particularly in the case of PCLSH30 meshes containing 30wt% of SH.

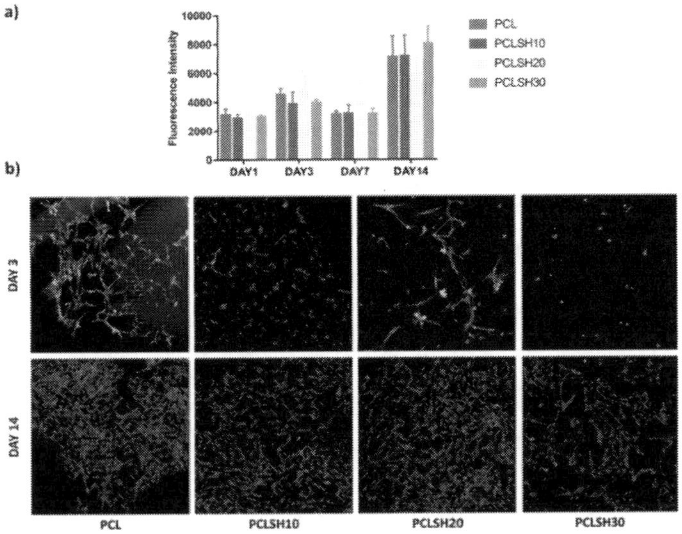

Fig.3. a) Cell proliferation with Alamar Blue Assay at day 1,3,7 and 14 b) Cell viability with Live/Dead staining at day 3 and day 14.

4 Conclusion

PCL and PCL/SH meshes were successfully produced using solution electrospinning. Meshes present fibers with diameters ranging from 50nm to 600 nm, mimicking the scale of the extracellular matrix (ECM. SH provides hydrophilicity to meshes which leads high water absorption. Meshes containing SH also degrades faster than pure PCL meshes. All meshes supported cell attachment and proliferation with best results obtained for PCLSH30 meshes.

The author wishes to acknowledge the funding provided by the Republic of Turkey Ministry of National Education, and also thanks to his supervisor Prof. Paulo Bartolo for his awesome and supportive guidance. This research was supported by Matoke Holdings.

References

1. D. R. Figueira, S. P. Miguel, K. D. de Sá, and I. J. Correia, Int. J. Biol. Macromol. 93, 1100 (2016)
2. R. F. Pereira, C. C. Barrias, P. L. Granja, and P. J. Bartolo, Nanomedicine 8, 603 (2013)
3. L. Yildirimer, N. T. K. Thanh, and A. M. Seifalian, Trends Biotechnol. 30, 638 (2012)
4. V. Jayarama Reddy, S. Radhakrishnan, R. Ravichandran, S. Mukherjee, R. Balamurugan, S. Sundarrajan, and S. Ramakrishna, Wound Repair Regen. 21, 1 (2013)
5. F. Groeber, M. Holeiter, M. Hampel, S. Hinderer, and K. Schenke-Layland, Adv. Drug Deliv. Rev. 63, 352 (2011)
6. K. Vig, A. Chaudhari, S. Tripathi, S. Dixit, R. Sahu, S. Pillai, V. A. Dennis, and S. R. Singh, Int. J. Mol. Sci. 18, (2017)
7. R. Augustine, Prog. Biomater. 7, 77 (2018)
8. J. R. Dias, P. L. Granja, and P. J. Bártolo, Prog. Mater. Sci. 84, 314 (2016)
9. P. Zahedi, I. Rezaeian, S. O. Ranaei-Siadat, S. H. Jafari, and P. Supaphol, Polym. Adv. Technol. 21, 77 (2010)
10. C. Vyas, R. Pereira, B. Huang, F. Liu, W. Wang, and P. Bartolo, Curr. Opin. Biomed. Eng. 2, 1 (2017)
11. S. Miguel, M. Ribeiro, P. Coutinho, I. Correia, S. P. Miguel, M. P. Ribeiro, P. Coutinho, and I. J. Correia, Polymers (Basel). 9, 183 (2017)
12. Z. H. Israili, Am. J. Ther. 21, 304 (2014)
13. M. D. Mandal and S. Mandal, Asian Pac. J. Trop. Biomed. 1, 154 (2011)
14. M. Dryden, G. Lockyer, K. Saeed, and J. Cooke, J. Glob. Antimicrob. Resist. 2, 168 (2014)
15. J. Cooke, M. Dryden, T. Patton, J. Brennan, and J. Barrett, BMC Res. Notes 8, 20 (2015)
16. N. R. Love, Y. Chen, S. Ishibashi, P. Kritsiligkou, R. Lea, Y. Koh, J. L. Gallop, K. Dorey, and E. Amaya, Nat. Cell Biol. 15, 222 (2013)
17. C. Dunnill, T. Patton, J. Brennan, J. Barrett, M. Dryden, J. Cooke, D. Leaper, and N. T. Georgopoulos, Int. Wound J. 14, 89 (2017)
18. T. Kokubo and H. Takadama, Biomaterials 27, 2907 (2006)

A CPS platform oriented for Quality Assessment in welding

Panagiotis Stavropoulos[1,*], *Alexios* Papacharalampopoulos[1] and *Kyriakos* Sampatakakis[1]

[1]Laboratory for Manufacturing Systems & Automation, Department of Mechanical Engineering and Aeronautics, University of Patras, 26504 Patras, Greece

Abstract. The major advantages of spot and seam welding are high speed and adaptability primarily for high-volume and/or high-rate manufacturing. However, this paradigm fails to meet the principles laid down by Industry 4.0 for real-time control towards Zero Defect Manufacturing for each individual product and intuitive technical assistance on the process parameters. In this paper, a Robust Software Platform oriented for a CPS-based Quality Assessment system for Welding is presented based on data derived from IR cameras. Imaging data are pre – processed in real-time and streamed into a module which utilizes Machine Learning algorithms to perform quality assessment. A database enables data archiving and post processing tasks along with an intuitive User Interface which provide visualization capabilities and Decision Support on the welding process parameters. The modules' IoT-based communication is performed with 5C architecture and is in line with Web Services.

1 Introduction

Welding is a well-established process involved in numerus manufacturing scenarios [1]. More than one welding techniques are often utilized in various industries at different manufacturing stages [2] for joining metal parts in high rate/volume production lines. Combined with robotized systems [2] these processes are offering high degree of automation and demanding on the other hand delicate and high level of control in order to address challenges related with the product quality. Eventually, states where weld's quality variability is maximized are configured due to factors such as complex part geometry, poor process parameters optimization, material variations and welding machinery and fixturing issues combined with the aforementioned aspects. The current Quality Assessment (QA) schema related with such welding applications is based mainly in periodical offline and partially on inline low-cognition-capability methods. That is, in the first case destructive tests [3] are incorporated along with costly NDTs methods [4] in a sample manner, creating a significant amount of scrap, introducing delays in production and configuring the product's quality stamp in a probabilistic manner. On the other hand, in-line methods [4] are limited by the onboard machinery sensor configurations, able to act in coarse-grained manner as per their defect cognitive and prediction capabilities. In this regard lately, several

* Corresponding author: pstavr@lms.mech.upatras.gr

attempts have been made on the development of software tools with dedicated interfaces for process modeling [5-6] and monitoring, knowledge extraction, cognitive quality control [7], machine state monitoring and user communication, in a non-unified way and with limited actual results in Laser Welding (LW) and Resistance Spot Welding (RSW). Therefore, the current study presents a web-based quality diagnosis platform for LW and RSW applications based on heuristic, Statistical and Machine Learning (ML) methods. Through the integration of two dedicated modules the platform is able to monitor the welding events, extract key features for classification and prediction and perform decision making for the overall part quality. Post processing functionalities are enabled with the utilization of historical data and they are available to the users along with the aforementioned ones though an intuitive Human – Machine Interface (HMI).

2 Architecture of Welding monitoring Cyber Physical System

In this study the concept of the Cyber-Physical System (CPS) [8] was adopted to describe the current approach within the context of LW and RSW applications based on the 5C architecture [9]. As shown in Figure 1 the physical counterpart is composed from the two robotized welding systems paired with the corresponding monitoring configurations. In cases of LW the LASER head is configured with a coaxial multispectral monitoring system including a MWIR high speed PbSe sensor-based camera (1 – 5μm) and a NIR CMOS camera (0.4 – 1μm) for monitor melt pool geometry, thermal distribution and surface dynamics of the workpiece. In addition, to suppress chromatic aberration a result of coaxial integration concept, narrow bandpass filters were aligned in front of the sensors. The RSW system was paired with a special mounting configuration to achieve magnetic protection and galvanic isolation of the SWIR camera which monitors the HAZ during and welding event. In order to bridge the gap from Physical into the Cyber domain an IoT module (Raspberry Pi) was utilized in order to acquire data from the sensors. This in turn, parses the generated metadata and data streams to the QA platform in real time. To this end the top-level architecture of the proposed QA platform (Figure 1) is structured into a back-end and a front-end component. The sensor data are captured from the IoT Data Acquisition (DAQ) module and passed to a server system in which they are preprocessed, clustered and distributed accordingly to three interconnected applications namely (HMI), LW QA module and RSW QA module paired with a Database element. The HMI application is offering visualization and QA functionalities that are partially supported from the two QA modules. The User Interface (UI) is located on the platforms front-end enabling the aforementioned features for the users.

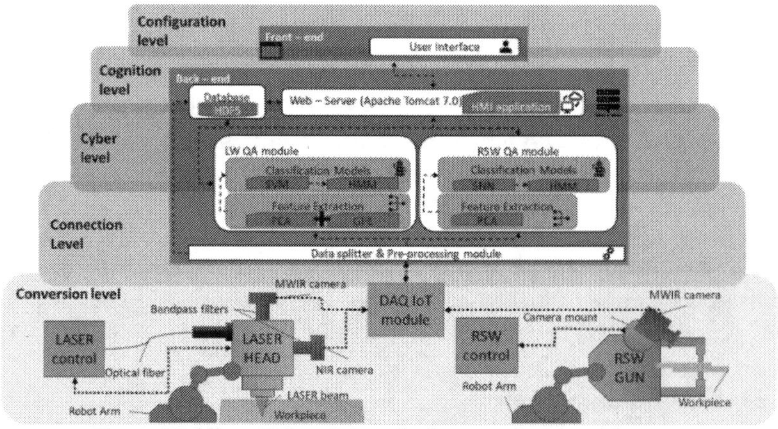

Fig. 1. Welding CPS and QA platform top level architecture

3 Quality Assessment modules

High process dynamics of LW [5] and RSW [10] and in general of the welding processes are making physical modeling not feasible for revealing the underlying high dimensional relationships of the input-output variables [7,11]. Two modules were developed in order to exploit such relationships by utilizing heuristic, statistic and ML methods. Their generic structure includes a data preprocessing, a feature extraction and a classification stage allowing the determination of the weld's quality (LW, RSW) and defects identification [12] (Figure 2). The development and initial implementation of modules was carried using MATLAB. Later on, were ported in Python scripts and packed into standalone modules.

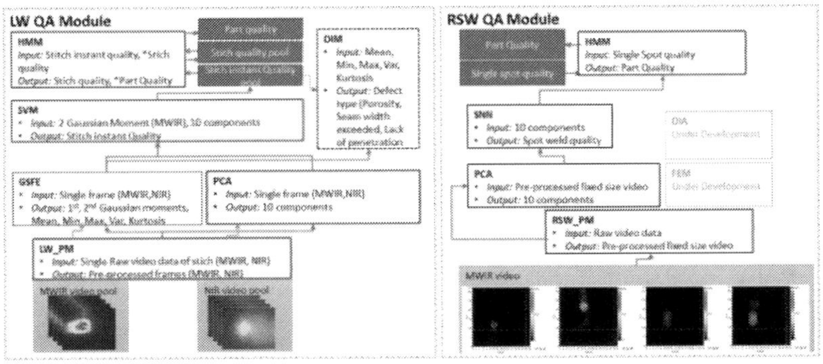

Fig. 2. LW & RSW QA module's top-level architecture

3.1 LW QA Module

The multispectral monitoring setup is configuring a high dimensional output feature vector not viable for the processing pipeline. In order to reduce its size, bunch of frames after subjected to filtering and normalization in the Pre-processing Module (LW_PM) are individually passed into the Geometrical & Statistical Feature Extraction (GSFE) module in which the 1st and 2nd order moments of the spatial temperature surficial profile are extracted along with statistical feature such as the mean, minimum, max, variance and kurtosis of the temperature field are calculated. As depicted in the Figure 2 along with the aid of Principal Components Analysis (PCA) [13] a set of 12 components in total are derived from the image pair and classified utilizing a Support Vector Machine (SVM) Model into GOOD or BAD stich instances. Finally, a Hidden Markov Model (HMM) deducing the overall quality of the stich (multiple SVM outputs) while decision in the same way can be made on the entire component's quality. In case of BAD stich instances, the Defect Identification Module (DIM) determines the defect type based on the output vector of the GSFE module.

3.2 RSW QA module

The nature of RSW process demands the analysis of spatiotemporal data. The incoming video data, although including a low spatial number of features, due to the high rate of process dynamics are demanding high acquisition rates configuring eventually a multidimensional output. In order to tackle this problem, as in the LW case PCA was utilized to extract a lower number of features from the video data that were previously subjected into the corresponding pre-processing module (RSW_PM) where thermal drift, background correction, median filtering and centering of the frames where carried out. With the implementation of a Shallow Neural Network (SNN) the reduced fixed-sized video vectors were subjected to binary classification (GOOD or BAD). The last step of this QA module utilizes an HMM in order to determine the overall quality of a part as its deduced from its single spot welds.

4 Platform requirements

As described in the Section 2 the development of the web-based platform was structured on a server-client logic with its back-end core hosting three real-time data generation modules (Figure 1). The generated data including data directly emerging from the sensor's out-streams and meta data as defined from the QA modules (quality labels, features). The Database element offering archiving/retrieving functionalities for the aforementioned meta/data enabling, along with the above, a series of potential functionalities for the HMI application and eventually for the UI, located at the platforms front-end. Therefore, in order to define the HMI's functionalities, the platform's requirements need to be specified. Platform's utilization from operators with different duties and responsibilities establishes a set of requirements regarding their access privileges into its different functions. Two type/categories of users namely "Workspace Operator" and "Shift Manager" can be identified with their access requirements to be defined base on their activities. Therefore, for the first category they are configured according to the specific nature of the machine and robot handling activities involved in specific tasks related with the production of specific products. On the other hand, the activities of the users which fall into the second category, are related with the optimization and of product and processes and their management (quality & product engineers). Based on the above and in combination with the information that must be available for the process, a set of requirements can be derived. One of its basic elements regards the real-time/historical data monitoring and visualization of a single process as well as of entire manufacturing sessions, which along with the analysis and processing of data/meta in real and post time can offer insights related with quality issues and correlating them with the status of the process. In the context of decision support towards process and product optimization another subset of requirements can be derived. More specifically, optimization and defect prevention of the process and part through parameter adjustment can be achieved based on communication, reporting and alerting actions among the production's line operators. In addition, collaboration between engineers and operators for overcoming quality issues can be implemented by offering direct recommendations on process parameters based on the analyzed data.

5 Results and Discussion

5.1 Platform functionalities

Based on the system and user requirements the platform's functionalities are implemented by utilizing the HMI application and the UI as presented below:

- User access rights: Two user type were defined (Workspace Operator, Shift Manager) enable through the platform's login screen, providing different access levels to the platform's functionalities for both LW and RSW case.
- Process stations: Two welding station were established to accommodate the different visualization widgets for two welding processes.
- Platform modes: Two modes (Real-time Monitoring, Post Process Elaboration) were implemented by utilizing two different screens in order to encapsulate separately the functionalities related with real-time and historical data for both LW and RSW cases.

- Data visualization: Real-time monitoring and historical data visualization of the data generated from the sensor and the QA modules, located on the corresponding screens. In the LW case these data including melt's pool evolution data, processed stitch, features evolution, and quality labels for each frame. In the RSW case the visualization of quality labels along with single videos corresponding to single spots are offered for both platform modes.
- QA model reconfiguration: Through the platform's API the reconfiguration of the QA modules is enabled for the classification and feature extraction modules with the utilization of historical data.

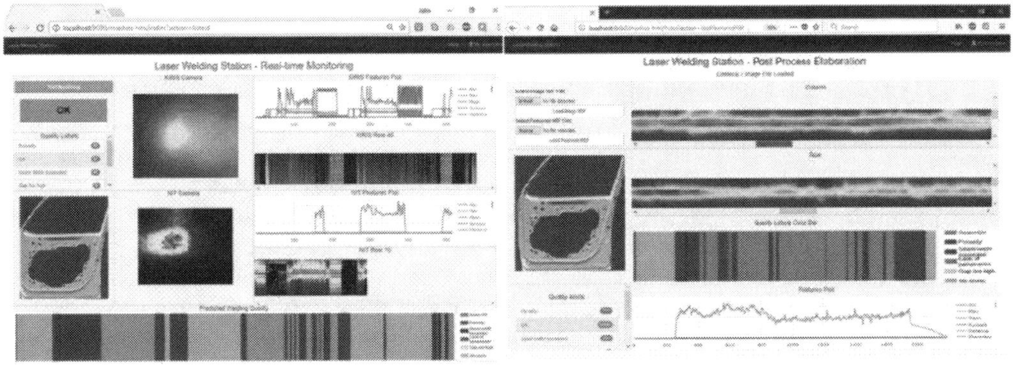

Fig. 3. LW UI station screens for Real-time Monitoring and Post process Elaboration

5.2 System's software specifications

Based on the system requirements that obtained from the architecture of the platform as it is analyzed previously the stand-alone QA modules were implemented into Python scripts (CGI) and paired with the appropriate API to enable their utilization from the rest of the back-end's components. The Database system using the HDF5 file format was developed along with a dedicated script for synchronizing, archiving and reading operations of large amounts of data, in order to reduce the volume and exchange time of critical tasks. The last element of the platform's back end is the HMI application which is hosted on an Apache Tomcat 7.0 web server, offering functionality to tasks related with data visualization, processing and data management. For the web framework the Bootstrap v3.3.7 library was utilized, enabling the creation of the UI (Figure 3) for different type of devices.

5.3 QA modules performance

In the LW case the collected data processed and classified by studying separately each video frame. This fact along with the absence of extreme welding phenomena (metal expulsion, sparks) had as a result the pixel value's range to be relatively limited allowing this way the PCA algorithm to extract a small number of components (10) with a significant explanation of the total variance (98%). Consequently, the SVM classifier, by utilizing a Linear Kernel, was able to reach a prediction accuracy of 92%. On the other hand, during the RSW Monitoring extreme welding phenomena were captured. Combined with the high dimensional nature of the video data, resulted the thermal footprint per spatial dimension to be described, across the different videos, from quite different ensembles. Due to these factors the PCA was not able to extract a low number of features (10) that can describe data's variability (87%) adequately, forcing the creation of a classification model with high number of tunable parameters. In addition, the small dataset used for training and testing along with information constrains resulted in a "small" classification model with mediocre prediction performance (85%).

6 Conclusions and future work

This study presented the development and implementation aspects of a quality diagnosis web-base diagnosis platform in the context of the LW and RSW applications. Based on the CPS architecture the elements and functionalities of the platform are enabled insights regarding the quality of the welds and products, allowing the users to perform immediate evaluation of them as well as of the involved process while enhancing actions related with the optimization of their KPIs. The platforms modular architecture can be easily adapted and scaled for different welding processes and complex monitoring setups by adding the appropriate QA modules and creating the corresponding stations in the HMI application. Challenges for the future including the improvement of the RSW QA module by studying a number of feature extraction methods, utilizing genetic algorithms for selecting the best suited NN architecture and experimenting with different classification models to improve and extend its classification performance and capabilities. Except of these the integration of additional sensors predicts that will lead, in general, to an improvement of the platform's defect cognitive capabilities, whilst the development and integration of Digital Twins can be integrated in a later phase for adding process control automation features.

This work is under the framework of Project REALM. This project has received funding from Region of Western Greece under the framework of Partnership Agreement ESPA.

References

1. G. Chryssolouris, N. Papakostas, D. Mavrikios. CIRP-JMST, **1(1)**, 45-52 (2008)
2. T. A. Barnes, I. R. Pashby, J. Mater. Process. Technol. , **99(1-3)**, 62-71. (2000)
3. M. Pouranvari, S. P. H. Marashi, Sci. Technol. Weld. Join., **18(5)**, 361-403, (2013)
4. J. Stavridis, A. Papacharalampopoulos, P. Stavropoulos, nt. J. Adv. Manuf. Technol., **94(5-8)**, 1825-1847, (2018)
5. G. Pastras, A. Fysikopoulos, C. Giannoulis, G. Chryssolouris, nt. J. Adv. Manuf. Technol., **78(5-8)**, 723-736, (2015)
6. A. Fortunato, G. Guerrini, S. N. Melkote, A. A. G. Bruzzone, CIRP. Ann. Manuf. Technol, **64(1)**, 189-192, (2015)
7. J. Günther, P. M. Pilarski, G. Helfrich, H. Shen, K. Diepold, Procedia Technology, **15**, 474-483, (2014)
8. D. Mourtzis, E. Vlachou, J. Manuf. Syst., **47**, 179-198, (2018)
9. J. Lee, B. Bagheri, H. A. Kao, Manufacturing letters **3**,18-23, (2015)
10. Y. B. Li, Z. Q. Lin, S. J. Hu, G. L. Chen, J. Appl. Phys., **101(5)**, 053506, (2007)
11. M. Schmidt, A. Otto, C. Kägeler, CIRP annals, **57(1)**,213-216, (2008)
12. N. M. Nasrabadi, J. Electron. Imaging, **16(4)** ,049-901, (2007)
13. F. Song, Z. Guo, D. Mei, IEEE: System science, engineering design and manufacturing informatization, **1**, 27-30, (2010)

Experimental Analysis and Soft Computing Modeling of Abrasive Waterjet Milling of Steel Workpieces

Panagiotis Karmiris-Obratański [1,2], *Nikolaos* E. Karkalos[1] , *Anastasios* Tzotzis[3], *Panagiotis* Kyratsis[4] , and *Angelos* P. Markopoulos[1,*]

[1]National Technical University of Athens, School of Mechanical Engineering, Heroon Polytechniou 9, 15780 Zografou, Athens, Greece
[2]AGH University of Science and Technology, Faculty of Mechanical Engineering and Robotics, al. Mickiewicza 30, 30-059 Cracow, Poland
[3]University of Zaragoza, Department of Design and Manufacturing, C/Mar a de Luna 3 -50018 Zaragoza, Spain
[4]University of Western Macedonia, Dept. of Product and Systems Design Engineering, Kila Kozani, GR50100, Greece

Abstract. Conventional machining processes such as turning, milling and drilling have long been prominent in the metalworking industry but alternative processes which do not require the use of a cutting tool in order to conduct material removal have also been proven to be sufficiently capable of achieving high efficiency in various cases. In particular, Abrasive Waterjet (AWJ) machining can be regarded as a rather appropriate choice for cutting operations, taking into consideration that it involves no heat affected zones, is able to process all material types and create a variety of complex features with success. In the present work, a comprehensive study on the effect of four process parameters, namely jet traverse speed, stand-off distance, abrasive mass flow rate and jet pressure on the width and depth of machined slots on a steel workpiece is conducted. The results are first analyzed with statistical methods in order to determine the effect and the relative importance of each parameter on the produced width and depth of the slots. Finally, these results are used to develop soft computing predictive models based on Artificial Neural Networks (ANN), which can efficiently relate the process parameters with its outcome.

1 Introduction

Abrasive Water Jet (AWJ) machining is an unconventional process used for material removal, as an alternative to traditional cutting methods. High-pressure water-abrasive cutting technology is commonly termed as cold, due to its association with the absence of a heat-affected zone on the workpiece. Although this method is used for cutting through the materials, researchers are also trying to establish the right combination of the jet pressure

* Corresponding author: amark@mail.ntua.gr

(P), stand-off distance (h), traverse speed (v_t) and the abrasive mass flow rate (m_a) in order to achieve the desired depth, when cutting is not performed through the entire workpiece. If the depth of the cut is uniform and controlled, then this process is known as AWJ Milling.

Goutham et al. [1] tried to establish a monitoring system for AWJ milling using acoustic emissions and cutting forces during a controlled depth cutting of pockets. They found that the depth of cut and the Material Removal Rate (MRR) increase as the jet pressure increases and decrease as v_t and h increases. Pal and Tandon [2] conducted a similar experiment by AWJ machining of blind pockets on different materials, including stainless steel 304 and tool steel M2 Rc 20. They investigated the milling depth, materials characteristics, surface roughness and the milling time during controlled depth milling. The experiments showed that there is a non-linear correlation between milling time and milling depth. Also, the machinability of each material influences significantly the milling time of pockets of the same depth. Akkurt et al. [3] machined AISI 1030 steel and AISI 304 stainless steel, among other materials, in order to evaluate the influence of v_t and the workpiece thickness on surface roughness. The experimental results explained that the "cutting wear" mechanism results in better surface quality than the "deformation wear" mechanism. Moreover, surface roughness was higher for the 5mm specimens than in the 20mm specimens for the steel-based materials; that could be explained as high strength materials require higher cutting forces and thus deformation is higher for the workpieces with thinner cross section. For the same stainless steel, Yuvaraj, Pavithra, and Shamli [4] carried out experiments in order to achieve surface patterns through the controlled milling. In this experimental study, a multicriteria technique was used to evaluate the effect of input parameters, namely jet pressure, traverse speed, stand-off distance and abrasive flow rate in the process. The results showed that based on selected parameters the surface morphology on stainless steel implants would be able to be predicted. On the other hand, in their investigation, Hlaváč et al. [5] tried to investigate the characteristic phenomena of the taper inside the cut during AWJ milling. This experimental research was performed also to examine the influence of brittle/ ductile behavior during liquid nitrogen cooling on plates of various steel types with thickness close to 30mm. The results were compared with the theoretical relation between the traverse speed and taper angle and difference between the inlet and outlet width of kerf up to 35.4% and 44.9% was found at lower and higher traverse speed values, respectively.

Yuvaraj and Pradeep Kumar [6, 7] have studied the AWJ cutting of AISI D2 steel by different abrasive mesh sizes and different jet impingement angles. Grey Relationship Analysis and Analysis of Variance (ANOVA) test was performed in order to determine the influencing parameters such as pressure, jet angle and mesh size on the different response parameters, namely, the MRR, taper angle, taper ratio, surface quality and the penetration depth. The results of the aforementioned tests revealed that there was an improvement during the AWJ cutting by setting the jet angle at 70°, jet pressure at 225MPa and by using the #100 abrasive mesh size. Finally, the purpose of the investigation of Ankush and Lalwani [8] was to explore the machinability of AISI H13 die steel by using AWJ. An ANOVA test for the surface roughness revealed that, the traverse speed is the most significant factor with contribution almost 77%, followed by the jet pressure and the stand-off distance in order to achieve smoother surfaces.

In the present work, a comprehensive experimental study on AWJ milling of steel is conducted in respect to four major process parameters, namely jet pressure, traverse speed, abrasive mass flow rate and stand-off distance in order to determine their effect on depth of penetration and kerf width. For these parameters, a relatively wide range of values is employed and after statistical analysis is conducted on the experimental outcome, Artificial Neural Network (ANN) models are developed in order to effectively establish the correlation between input parameters and responses of the AWJ milling process.

2 Methodology

AWJ milling experiments were performed on a low carbon S355 steel workpiece under three different levels of traverse speed, abrasive mass flow rate, stand-off distance and jet pressure. In each case, a straight, non-through slot was created with its depth (d) and width (w) considerably varying according to the process conditions. The experiments were designed according to Taguchi L9 orthogonal array for traverse speed, abrasive mass flow rate and stand-off distance, with each sequence of experiments repeated for three different jet pressure values; the levels for each process parameter are presented in Table 1.

Table 1. Process parameters levels.

Parameter	Level 1	Level 2	Level 3
Traverse speed (mm/min)	100	300	500
Abrasive mass flow rate (g/s)	2	5	8
Stand-off distance (mm)	1	3	5
Jet pressure (MPa)	233	317	400

The AWJ milling experiments were carried out on a H.G. RIDDER – Automatisierungs GmbH model HWE-1520 machine with 80 mesh size garnet as abrasive. In any case, the jet impingement angle was 90^0, perpendicular to the workpiece surface, focusing tube diameter was 1 mm and waterjet nozzle diameter was 0.3 mm. After the experiments were carried out, depth of penetration and kerf width were measured based on high resolution images captured on an optical microscope which were subsequently processed by ImageJ software.

The neural networks used in the present study are multi-layer perceptron (MLP) models with a single hidden layer. Two different neural network models will be created by using data from the 27 experiments, one regarding depth of penetration and one for the kerf width with different input parameters and their evaluation will be performed based on Mean Squared Error (MSE) values.

3 Results and Discussion

In order to determine the influence of process parameters to the depth of penetration and kerf width, statistical analysis of the experimental results is necessary. In Fig. 1, the main effects plot for the depth of penetration is depicted. Fig.1 reveals that a considerable drop in depth occurs when traverse speed increases, whereas an increase of abrasive mass flow rate and jet pressure leads clearly to higher depth. These trends are anticipated as the abrasive flow rate and jet pressure affect the kinetic energy of the jet [9], thus creating deeper slots at high values, whereas higher traverse speed leads to shorter time for the jet to remove material from the workpiece at a certain region. Moreover, from these graphs it can be observed that traverse speed, abrasive mass flow rate and jet pressure lead to the highest variation of depth of penetration values, whereas stand-off distance seems to have only a small effect on depth. The observations regarding the influence of each process parameter on depth are confirmed by ANOVA results, as it was found that the relative importance of v_t, m_a, h and P were 53.1%, 23.9%, 3.4% and 13.6%, respectively. All process parameters except for stand-off distance were found statistically important, indicating that they can be used to effectively regulate the depth of penetration during AWJ milling.

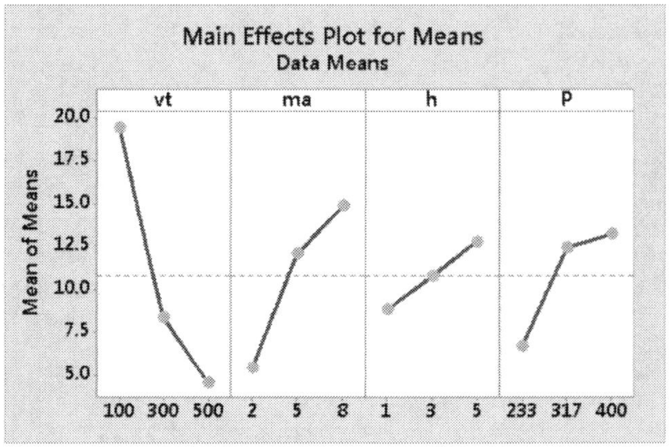

Fig. 1. Main effects plot for depth of penetration.

In Fig.2, the main effects plot for kerf width is presented, in which stand-off distance is shown to have the most noticeable impact on kerf width, clearly increasing it with increasing values. Kerf width is also positively affected by abrasive flow rate, negatively affected by traverse speed and a small positive effect is caused by the jet pressure. Higher stand-off distances lead to a larger divergence of the jet above the workpiece [9] and larger dispersion of the abrasive particles, which remove material at larger distances. Moreover, higher abrasive mass flow rate leads to more intense impact of the particles with the surface and higher traverse speed does not allow for widening of the slot due to shorter time of impact of the jet with the workpiece surface [10]. ANOVA results again confirm the observations on Fig.2, as stand-off distance is proven to be most important factor with 56.8% relative importance, whereas traverse speed and abrasive mass flow rate follow with 20.1% and 14.6%, respectively. For the selected range of jet pressure values, this parameter is considered statistically not important for kerf width with a relative importance of 1.13%.

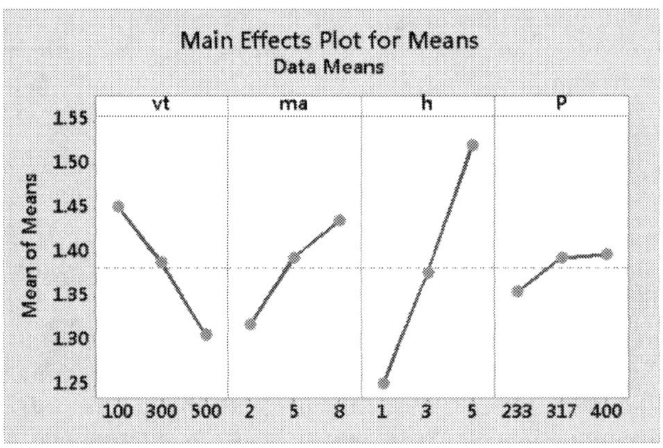

Fig. 2. Main effects plot for kerf width.

After the experimental results were analyzed with statistical methods, soft computing models using Artificial Neural Networks are developed in order to predict depth of penetration and kerf width in respect to process parameters. ANN were selected, as they have been proven to be capable to predict the outcome of even highly non-linear processes with high accuracy compared to other traditional methods, such as regression models. As the depth of penetration is not fixed during AWJ milling, it is important to be able to predict it for any desired combination of process parameters. Furthermore, kerf width considerably higher than the desired value is a defect of the AWJ milling process and it is crucial to select process conditions that lead to acceptable values of kerf width.

Different MLP models will be developed for each response and in each model only the statistically significant parameters are employed, as depicted in Fig.3a and b. Given the amount of available data pairs and number of input and output neurons, the number of the neurons in the hidden layer will be varied between 2 and 8. Each network will be tested 50 times and the optimum network architecture will be determined according to MSE values during training and test procedures.

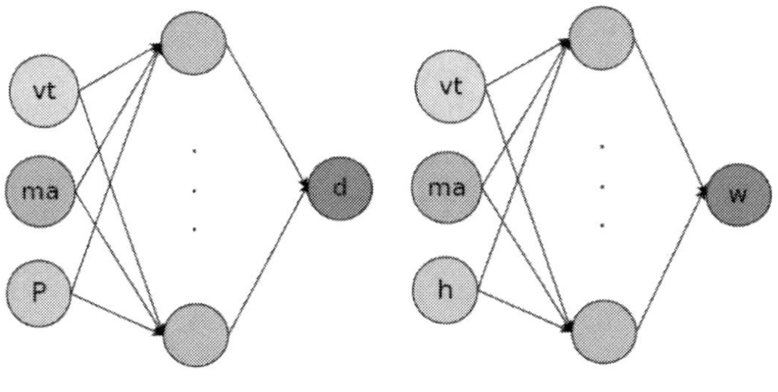

(a) (b)

Fig. 3. Schematic of ANN models for (a) depth of penetration, (b) kerf width.

In the case of depth of penetration model, the values of MSE for each model during test procedure are depicted in Fig. 4a. Usually during the development of an ANN model, training can lead to overfitting, especially when an unnecessary large number of hidden neurons is employed, which causes a lack of generalization abilities, something that can be evaluated during test procedure. Thus, from the results of Fig. 4a, it becomes obvious that the network with 5 hidden neurons is the best performing one as it has the lowest MSE test value of 1.26×10^{-4} and at the same time, it has a sufficiently low MSE train value of 7.35×10^{-5}. In the case of kerf width model, the MSE test values for each employed network are displayed in Fig. 4b. The results of Fig. 4b indicate that the model with 6 hidden neurons is the most suitable in this case as it exhibits the lowest MSE test value of 1.52×10^{-4} and its training performance is also acceptable with MSE value of 1.63×10^{-4}. Thus, it was proven that in both cases, involving highly non-linear correlations, the ANN models which were selected exhibited sufficient accuracy and generalization capabilities.

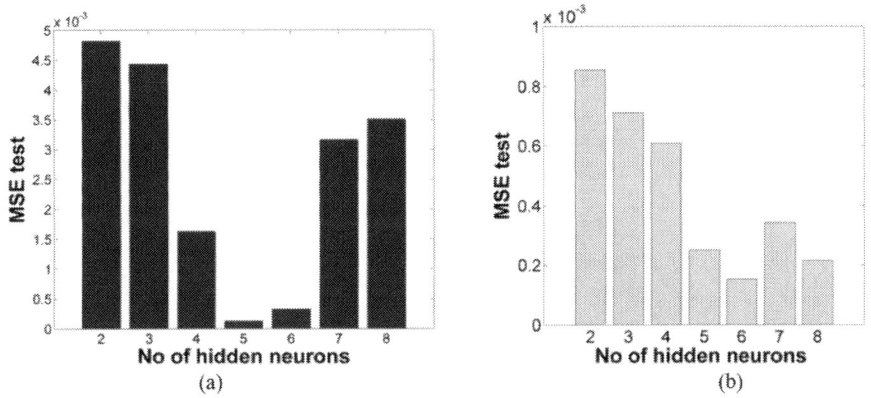

(a) (b)

Fig. 4. MSE test values for different ANN models regarding: (a) depth of penetration, (b) kerf width.

4 Conclusions

In the present work, the effect of process parameters during AWJ milling of steel was investigated through experiments and statistical analysis, whereas the correlation between process parameters and its outcome, namely depth of penetration and kerf width was established by the use of artificial neural network models. From this work, various conclusions were drawn.

At first, the contribution of different parameters on depth of penetration was determined. Traverse speed was identified as the dominant factor, with a relative importance of over 50%, followed by abrasive mass flow rate and jet pressure. Depth of penetration can be increased with lower traverse speed values, as it leads to a longer interaction time between jet and workpiece; also, increased abrasive mass flow rate and jet pressure lead to higher depth as the jet impacts the workpiece with higher kinetic energy.

Then, the effect of process parameters on kerf width was investigated. In this case, the stand-off distance was shown to play a major role, whereas traverse speed and abrasive mass flow rate followed. The reduction of kerf width was shown to be achieved at lower stand-off distance, higher traverse speed and lower abrasive mass flow rate values.

Finally, ANN models were developed in order to establish the correlation between process parameters and the two responses of the experiment. After an investigation regarding the number of hidden neurons in the ANN models was carried out, the best performing models, exhibiting both high accuracy and generalization capability, for depth of penetration and kerf width prediction were determined, having 6 and 5 neurons, respectively, in a single hidden layer.

References

1. U. Goutham, M. Kanthababu, S. Gowri, K. R. Sunilkumar, M. Mathanraj, J. J. R. Jegaraj, R. Balasubramanian, *Advances in Forming, Machining and Automation, Lecture Notes on Multidisciplinary Industrial Engineering*, 153-164 (2019)
2. V.K. Pal, P. Tandon, Int. J. Adv. Manuf. Technol. **66 (5-8)**, 877-881 (2012)
3. A. Akkurt, M.K. Kulekçi, U. Şeker, F. Erkan, J. Mater. Process. Technol. **147(3)**, 389-396 (2004)
4. N. Yuvaraj, E. Pavithra, C. Shamli, J. Test. Eval. **48 (4)** (2020)
5. L. M. Hlaváč, I. M. Hlaváčová, V. Geryk, Š. Plančár, Int. J. Adv. Manuf. Technol. **77**, 1811-1818 (2015)
6. N. Yuvaraj, M. Pradeep Kumar, Mater. Manuf. Process **32(2)**, 151-161 (2017)
7. N. Yuvaraj, M. Pradeep Kumar, Mater. Manuf. Process **32(2)**, 162-170 (2017)
8. S. Ankush, D.I. Lalwani, *Proceedings of Conference on Intelligent Robotics, Automation and Manufacturing*, 333-341 (2013)
9. S. Saravanan, V. Vijayan, S.T. Jaya Suthahar, A.V. Balan, S. Sankar, M. Ravichandran, Mater. Today Proc. **21(1)**, 116-122 (2020)
10. P. Gudimetla, J. Wang, W. Wong, J. Mater. Process. Technol. **128(1-2)**, 123-129 (2002)

MATEC Web of Conferences **318**, 01024 (2020) https://doi.org/10.1051/matecconf/202031801024
ICMMEN 2020

A Pedagogical Methodology for Introducing CAD Modeling Tools and 3D Printing Technologies to Adult Trainees

Maria Chatzikyrkou[1,*], *Athanasios* Manavis[2], *Prodromos* Minaoglou[2], *Nikolaos* Efkolidis[2]

[1] Department of Philosophy and Education, Aristoteles University of Thessaloniki, Thessaloniki, Greece
[2] Department of Product and Systems Design Engineering, University of Western Macedonia, Kozani, Greece

Abstract. This paper presents a four-step training model based on pedagogical methodologies for introducing CAD (Computer-Aided Design) modelling tools and 3D printing technologies to adult trainees. Nowadays, 3D modeling techniques and 3D printing technologies are emerging core competencies due to the increasing popularity of STEAM education frameworks. The proposed training model has these four stages for trainees' team: a) theoretical background of program based on traditional educational techniques, b) e-learning tools and parallel interaction with the trainer, c) micro-teaching frameworks and case studies and d) brief questionnaires with motivation and satisfaction criteria completed by the trainees. The present paper attempts to identify the pleasure and satisfaction of participating in New Technologies Education through CAD (Computer-Aided Design) modelling tools and 3D printing technologies program.

1 Introduction

Nowadays, CAD (Computer-Aided Design) modelling tools and 3D printing technologies are used on educational procedures, progressively. This novel direction of modern educational frameworks created the need for *training-the-trainers* in the field of CAD modelling techniques and 3D printing issues according to basic pedagogical and educational principles. Furthermore, trainer's training focuses on basic pedagogy principles including the three notable rules: knowledge, skills, and attitudes. Moreover, it is considered necessary the training of the new advisers in eLearning platforms which offer a great number of advantages in distance learning models (i.e. electronic classroom and environments, asynchronous educational programs and evaluation of participation and performance).

After completing the training program, trainees of novel digital tools are able to explain the main principles of CAD modelling tools and 3D printing technologies, to show the Additive Manufacturing capabilities and technology in general, to support educational

* Corresponding author: **maria.chatzikyrkou@gmail.com**

activities based on CAD exercises, and finally to design basic digital applications for 3D printing.

2 Literature Review

Literature review provide a foundation for future empirical research. Furthermore, it can help reveal the status of knowledge related to a main scientific topic, describe the quality of current research, situate research findings, and provide rationale for future research directions [1]. From pedagogical point of view, Kapsalis and Papastamatis (2000) define five stages of planning a training program. [2] Specifically, the first stage concerns the evaluation of the existing situation, the investigation and recording of the training needs of the target group, the analysis of important data and problems. In the second stage, are defined the general educational goal and the individual objectives of the program. The selection of the contents and the layout of the program follows. The fourth stage is the design of teaching strategies and the selection of training methods and practices. At the final stage, the project is being built and implemented. [3]

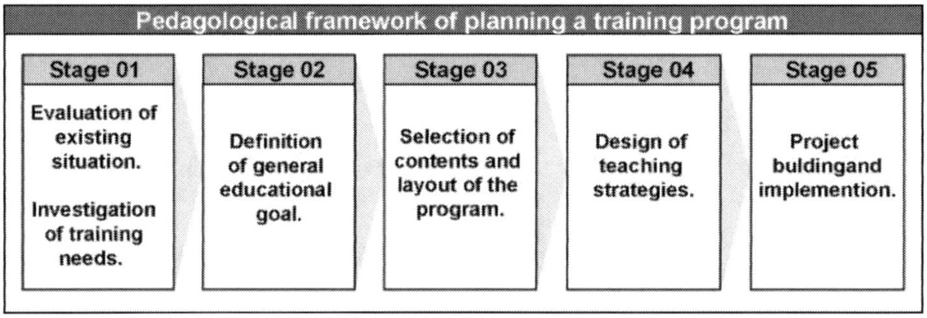

Fig 01. Pedagogical framework of planning a training program.

There is a large number of educational methodologies, depending on the type of training frameworks, they all share common steps, referring to purpose, content, methods, and evaluation. Shaikh (2017) proposed in his research is a systematic approach to the management of educational programs and is therefore also applied to knowledge subjects of a particular technological nature. It consists of five phases: Analysis, Design, Development, Implementation and Evaluation. [4]

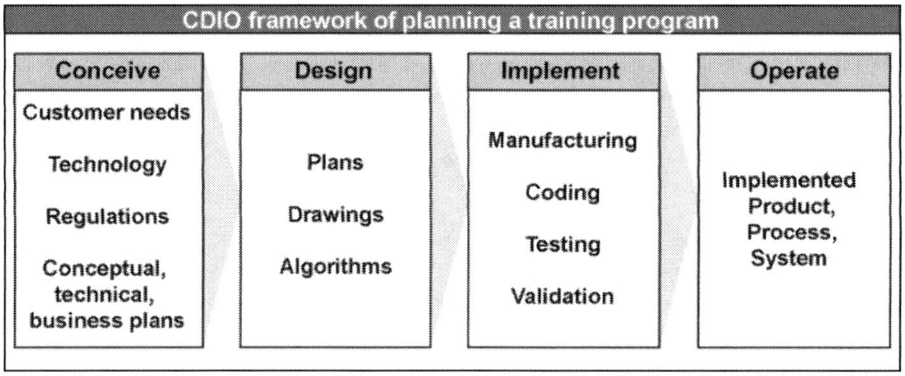

Fig 02. CDIO framework of planning a training program.

Bankel et al. (2005) define four stages of planning a training program. CDIO (Conceive – Design – Implement – Operate) is an innovative, educational framework. (Fig.02) As trainees develop professional skills, they also cultivate practical thinking skills. [5] The CDIO educational process is used to describe a lifecycle of any product, process, project, system, software, material development. Specifically, *Conceive Stage* contains notions like customer needs, technology, enterprise strategy, regulations, and conceptual, technical, and business plans. Second phase is *Design Stage* that focuses on creating the plans, drawings and algorithms that describe the product, process or system that will be implemented. Next, *Implement Stage* is about manufacturing issues, coding, testing and validation. The final phase is *Operate Stage* that uses the delivered, implemented product, process, or system to satisfy the intended value. The CDIO framework is widely used in engineering education and it has been found to enhance learners' technical knowledge and improve leadership and problem-solving abilities. [6]

2.1 CAD Modelling Tools and Education

Digital modeling is a process in which a CAD (Computer Aided-Design) system is used to assist in defining the geometry and visual appearance of design [7]. According to Pavel & Omid (2015) CAD is taught in different scenarios. It can be taught by software distributors who organize courses as a course in design study curriculum. It could also simply be left to students to implement it in problem-based learning design projects. [8] The main intention of using CAD systems in training programs is the increase capacity of a learner to form connections between sources of information, and thereby create useful information patterns. [9]

2.2 3D Printing Technologies and Education

Over the years, novel technological systems (i.e. robotics and Computer Aided-Design) approaches have emerged to play an important role in enhancing the education procedure for 3D printing techniques. Furthermore, the concepts of Science, Technology, Engineering, and Mathematics (STEM) have driven the integration of 3D printing technology in education. [10]. Thus, this paper explores the impact of integrating 3D printing into the 3D modelling program on learning outcomes among adult learners with different background.

2.3 e-Learning and Education

The educational potential offered by e-learning in adult education is supported by an increasing number of arguments related to the initiatives taken at the level of educational systems-especially in higher education, within universities that have developed an infrastructure which favours the use of virtual learning systems. According to Ingvarson & Gaffney (2008), digital learning systems "can provide a more responsive, personalized, effective, equitable and efficient learning experience for each student. [11]

3 Training Methodology

The methodology described in this study is based on the fundamental rules of adult education. Specifically, a training program has been designed that combines novel techniques with proven educational methodologies from pedagogical point of view. This integrated type of training is based on the advantages of traditional education combining them with the benefits of the use of e-learning techniques. The proposed training methodology for CAD (Computer Aided-Design) and 3D printing technologies is consisting of four specific stages: lessons 1, 2, 3 and 4. Figure 03 illustrates the workflow for the training programme. The first lesson was organised by educators in an open software e-learning platform. The content of the course was about the theoretical background of CAD technology. Specifically, the two-hour course included special units from the CAD theory. The issues that it was presented by educators were: a) history of CAD/CAM development, b) CAD/CAM contents and tools, c) definition of CAD/CAM tools, d) introduction to CAD/CAM softwares and e) all types of geometric modelling representations. Through the platform the trainees attended the necessary lectures about CAD theory and interacted in real time (synchronous e-learning) with the instructors. The second lesson (four-hour course) took place in a computer room. Trainees participated in the course in physical presence. Instructors used traditional teaching methods to train the adult trainees in CAD software. Specifically, instructors presented the 123D Design© software. Autodesk© 123D Design© is a freeware CAD designer software app filed under 3d design and made available by Autodesk© for Windows.

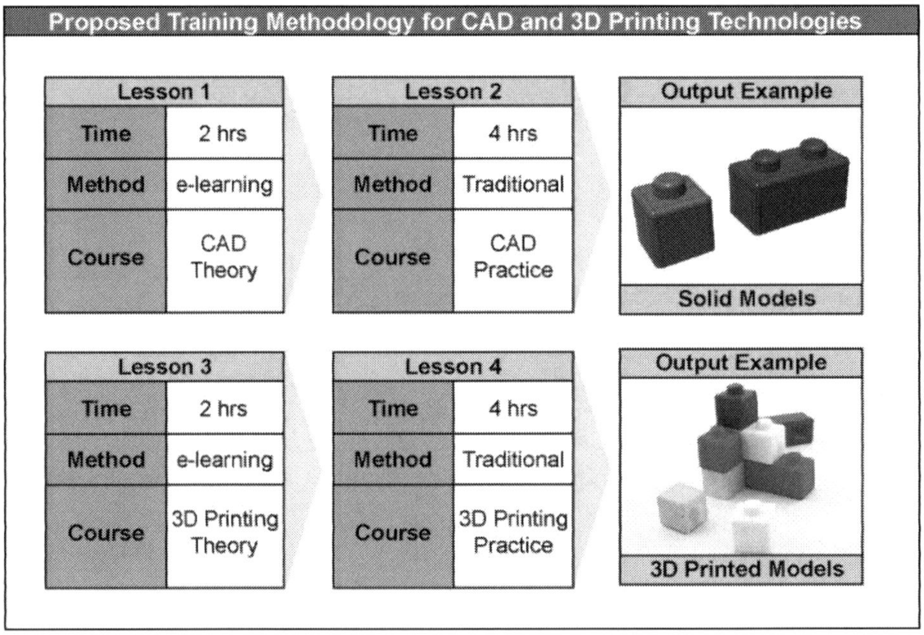

Fig 03. The proposed training methodology for CAD and 3D printing technologies.

Furthermore, trainers presented a great number of technical CAD issues from 3D printing point of view. Some of these issues were: a) support for opening, inserting, and saving meshes in STL and OBJ formats, b) perform Combine, Subtract and Intersect operations between meshes and solids, c) option for combining objects on STL export in order to support printers that read first body only and d) importation of SVG files and the usage of them as sketches or as simple extrusions. Finally, the trainers presented a series of exercises, which included the following: a stencil ruler, a Lego-type cube, and a baby blocks shape sorter toy. The third lesson (two-hour course) was organised by instructors in an open software e-learning platform, again. The content of the third meeting was about the Additive Manufacturing theory. Specifically, the two-hour online lesson included issues from the 3D printing technologies. The issues that it was presented by trainers were, a) additive manufacturing processes, b) additive manufacturing technologies, c) additive manufacturing materials, d) Additive manufacturing applications and e) educational and industry case studies. Subsequently, the fourth lesson was classroom based and trainees participated with a physical presence. The group of trainees was trained in the use and operation of FDM 3D printer and finally, they all printed out the objects they had designed in previous lessons (Fig.04).

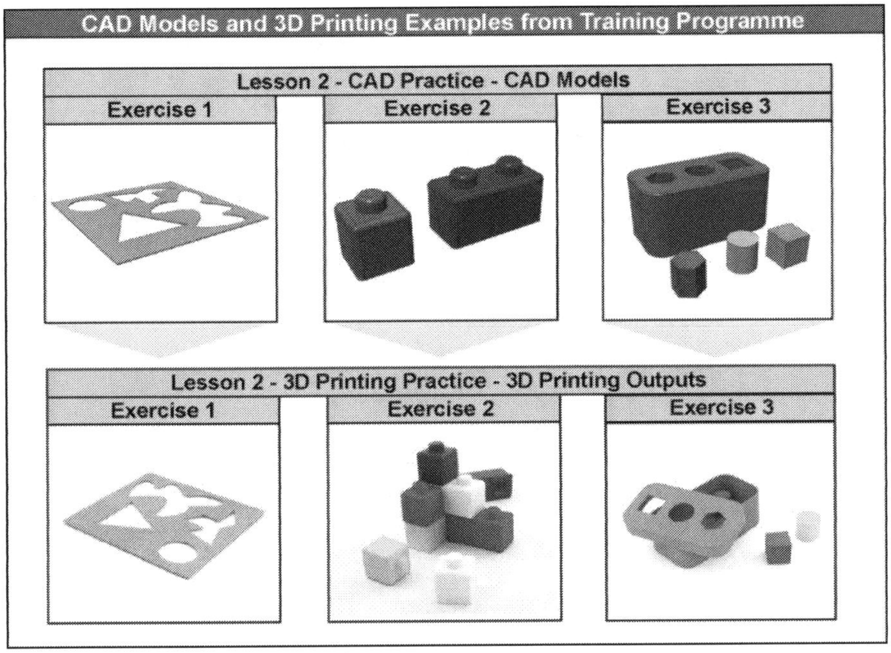

Fig 04. CAD models and 3D printing examples from training programme.

4 Experimental Results

At the end of the fourth meeting is also given an evaluation questionnaire for the programme where the results of the educational process and the impressions of the trainees are recorded. The program has been applied so far to 17 trainees who completed a short evaluation questionnaire that was given to them. At the beginning of the questionnaire, a brief reference is made to the nature of the survey and to the responsible organization that supports it, as well as information confidentiality issues, as well as completion instructions. The questions of the second section are listed below and form part of a balanced survey of satisfaction with participation in educational programs based on bibliography (Fig.05). As a scoring scale, a four-scale, with four possible answers, two positive (Very much and Much) and two Negative (Little and Very little) were selected. The processing of the questionnaires shows that the overwhelming majority of participants in the program, is pleased. Furthermore, the concepts of CAD (Computer Aided-Design) and 3D printing had been explained well. Overall, a percentage exceeding 90% responds that it would highly recommend the program to a colleague or friend.

5 Conclusions

This paper presented a four-step training model based on pedagogical methodologies for introducing CAD (Computer-Aided Design) modelling tools and 3D printing technologies to adult trainees. The methodology that was described follows the fundamental rules of adult education (traditional and e-learning formats). Specifically, the proposed training program combines novel techniques from CAD and 3D printing scientific fields. He results of the methodology are spectacular, since, as can be seen from the sample, the overwhelming majority is absolutely satisfied with the seminar and would of course recommend it to a colleague or friend.

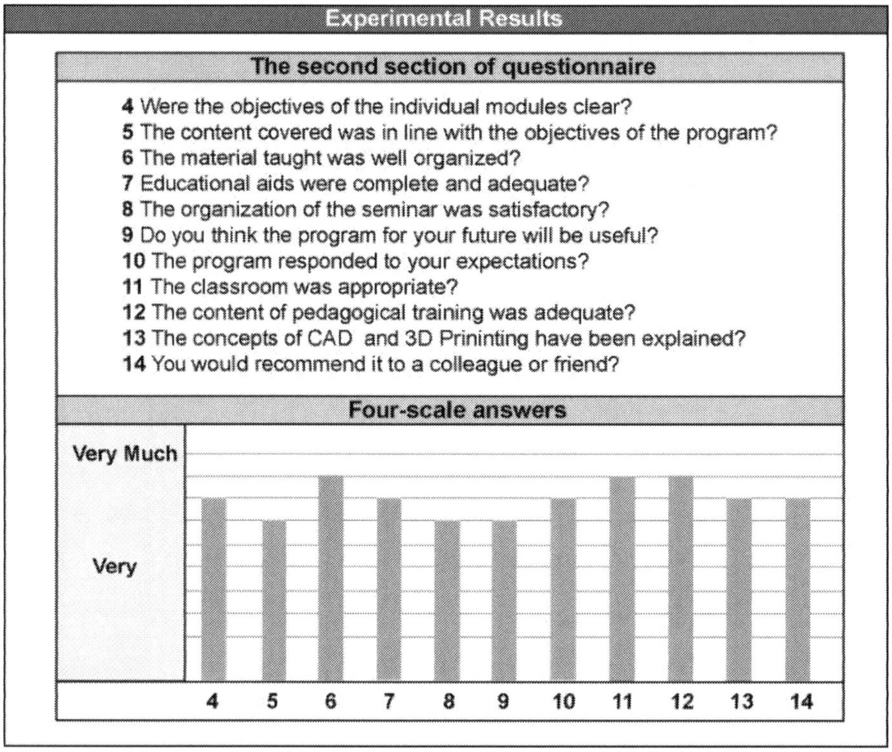

Fig 05. Experimental results.

References

1. A. Fink, *Conducting Research Literature Reviews: From the Internet to Paper* (Sage Publications, Thousand Oaks, CA, 2005)
2. A. Kapsalis, A.Papastamatis, Education of Adults B: Instructive Adults, (Publications of University Macedonia, Thessaloniki, Greece, 2000)
3. A. Tsagaris, M. Chatzikyrkou, I. Simeli, *JCETR* **2**, 38 (2018)
4. H. Shaikh, Discuss the Role of Instructional Design and how beneficial is Instructional Design in teaching & learning (http://eclipse.mu.ac.in) (2017)
5. J. Bankel, K.F., Berggren, M. Engström, I. Wiklund, E.F. Crawley, D.H., Soderholm, S. Östlund, Int. J. Eng. Educ. **21**, 121 (2005)
6. A. Babuscia, J.L. Craig, J.A. Connor, Acta Astronautica **77**, 138 (2012)
7. R. Coyne, H. Park, D. Wiszniewski, Design Studies **23**, *263* (2002)
8. N. Pavel, M. Omlid, Inter. Con. Eng Prod D Edu 144 (2015)
9. G. Siemens, A. Connectivism, ITDL **2,** 3 (2005)
10. V. Kostakis, M. Papachristou, Telemat. Inform. **31,** 434 (2014)
11. D. Ingvarson, M. Gaffney, Developing and Sustaining the Digital Education Ecosystem: The Value and Possibilities of Online Environments for Student Learning (M. Lee & M. Gaffney Eds. Acer Press, 2008)

MATEC Web of Conferences **318**, 01033 (2020) https://doi.org/10.1051/matecconf/202031801033
ICMMEN 2020

Non-destructive testing of welded fatigue specimens

Sofia Papanikolaou[1,*], *Dimitrios* Fasnakis[1], *Andreas* Maropoulos[2], *Dimitrios* Giagopoulos[1], *Stergios* Maropoulos[1] and *Theodoros* Theodoulidis[1]

[1]Department of Mechanical Engineering, University of Western Macedonia, Kozani, Greece
[2]Department of Mechanical Engineering, Aristotle University of Thessaloniki, Greece

Abstract. Non-destructive tests were conducted on welded fatigue specimens prepared using the same steel material and welding method as the one used in the manufacture and repair procedures of a KRUPP SchRs 600 bucket wheel excavator to reveal any defects present. The chemical composition, the mechanical properties, tendency to cracks and the microstructure of the bucket wheel material were determined using appropriate tests. The initiation of cracks and their subsequent growth during fatigue testing of the welded specimens was studied using ultrasound testing (UT) and a metallographic examination in order to investigate the causes of failure during service and predict fatigue life of the bucket wheel welded parts. It was found that the welding method used produces welds with numerous discontinuities that can only be detected using ultrasound techniques.

1 Introduction

Various non-destructive testing methods have been used to inspect welded joints of bucket wheel excavators BWE's [1-2]. Radiographic testing (RT) of a butt welds is used to identify the gas pore flaws and slag inclusions which are not within the limits of acceptability for a welded joint [1]. Ultrasonic testing (UT) is conducted to reveal inhomogenities of a local character and inhomogenities with depicted frequency [1]. The chemical composition and mechanical properties, the impact toughness, hardness, tendency to cracks and the microstructure of bucket wheel materials were determined using appropriate tests [3]. High values of residual stresses, as well as the cold cracking observed on the welded joint of the knife and the bucket body of a BWE, suggest that the defects also play a significant role in the failure [3,4]. The present case study is concerned with investigating the reasons for failure of welded joints in the rim and supporting diaphragms of a KRUPP SchRs 600 bucket wheel excavator. Non-destructive testing was conducted to investigate for defects present in welded tensile and fatigue specimens prepared using the same steel material and welding method as that used in the manufacture and repair procedures of the excavator. In addition, the initiation of cracks and their subsequent growth during fatigue testing of the welded specimens was studied using ultrasound tests. A

* Corresponding author: spapanikolaou@uowm.gr

metallographic examination was carried out in order to study the microstructural changes during welding that may be the cause of failure during service of the bucket wheel welded parts.

2 Experimental

The steel material, S355J2G3 EN 10025-2, was provided by the Kardia Lignite Mine, Western Macedonia Lignite Centre and is the same as that used in the manufacture and repair procedures of the Bucket Wheel (BW) under investigation. Six steel plates, 300mm x 300mm x 8mm, each were produced by welding together two steel plates as shown in Figure 1. The joint design and the welding sequences are also shown in Figure 1. The electrodes used were Bohler-FOX EV50 EN ISO 2560-A: E42 5B 42 H5 2.50x350mm for the root and 3.2x350mm for runs 2 to 4. The welding details are shown in Table 1.

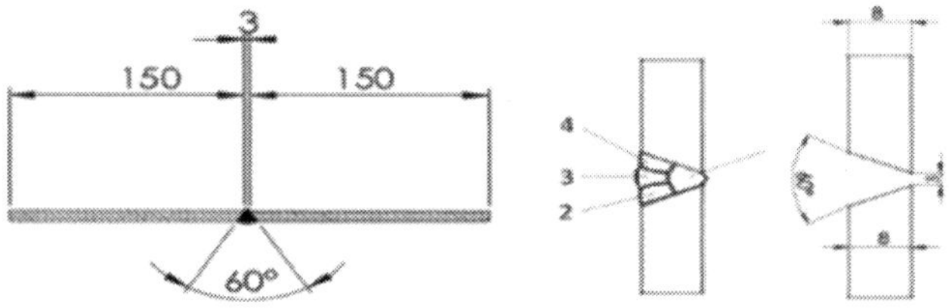

Fig. 1. Welding of steel plates, joint design and the welding sequences.

Table 1. Welding details.

Run	**Welding Process**	**Size of filler material**	**Current (A)**	**Voltage (V)**	**Type of current/ Polarity**	**Run out length/Travel speed(mm/min)**	**Heat input (kJ/mm)**
1	111MMA	2.5	85/90	24	DC+	80-100	1.2
2-4	111MMA	3.2	110/120	24	DC+	80-100	1.5

Seven, 300mm x 40mm x 8mm, specimens were machined from each of the six welded steel plates as shown in Figure 2. Thus, forty two in total specimens were produced as shown in Table 2. Steel plates E and Z were welded using the Backstep technique.

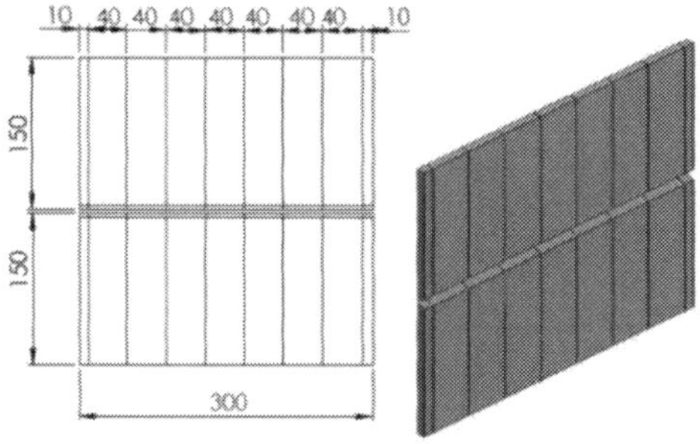

Fig. 2. Tensile and fatigue specimens prepared from the welded steel plates.

High cycle fatigue tests were conducted under axial tension loading according to ASTM E 466 using a 100kN capacity Instron machine until cracking or reaching the boundary number of cycles of $2 \cdot 10^6$ [5]. Prior to testing any excess weld material was removed by milling using a 45mm diameter 3 cutting edge T25 tool so as to obtain a smooth surface

finish and thus prevent the initiation of fatigue cracks from the specimen surface during testing [5,6].
In order to reproduce the effect of the very high tensile residual stresses that are expected to be present in real welded structures [6] in a small-scale specimen, the testing was carried out under a high stress ratio (R = 0,8) at a high tensile mean. This stringent condition is expected to yield conservative results of the weld fatigue strength [7,8].

Table 2. Specimen codes.

Steel Plate	Specimen set
A	A1-A7
B	B1-B7
Γ	Γ1-Γ7
Δ	Δ1-Δ7
E	E1-E7
Z	Z1-Z7

The maximum (F_{max}=80kN) and minimum (F_{min}= 64kN) loads applied were calculated with a maximum stress of σ_{max}=250 MPa (0.8*Yield Strength) and a minimum stresses of σ_{min}=200 MPa. Since the cyclic loading waveform shape and a frequency, of up to 100Hz, have no significant effect on fatigue performance in passive environments [5, 6] sinusoidal loading at 50 Hz was selected to reduce the testing time and to avoid undesirable out-of-plane bending stresses due to resonance and excessive heating of the specimen.
Hardness and tensile tests were carried out so as to determine the as received, un-welded material properties. The chemical composition of the material was determined using optical emission vacuum spectroscopy and the steel microstructures were studied using optical and scanning electron microscopy (SEM) to investigate the presence of defects or inclusions that could lead to premature failure during testing. The fracture surfaces of the mechanical tests were further studied with scanning electron microscopy to obtain a better picture of the failure process. The specimens were also tested with the magnetic particle method using a Karl Deutsch yoke electromagnet and appropriate sprays according to the ISO 17638 [13] and ISO 23278 [14] standards. The Karl Deutsch Digital Echograph and the Olympus Epoch1000i ultrasound devices were used to investigate for subsurface defects. The audit was conducted in accordance with ISO 17640 [15] and ISO 23279 [16]. The initiation and development of cracks during fatigue testing was studied using ultrasound non-destructive testing.

3. Results

The chemical analysis of the material showed that the material was a plain carbon steel (0.17 %C, 0.3%Si, 0.45% Mn, 0.04% P, 0.04% S), common in the manufacture of BW structures. The tensile and hardness properties of the as received un-welded material are shown in Table 3. The tensile and hardness properties of the welded material are also shown in Table 3.

Table 3. Tensile and hardness properties of the un-welded and of the welded material.

Specimen type	Tensile Strength (MPa)	Yield strength (MPa)	% Elongation	Hardness VPN
Un-welded	430	290	42	360
Specimen set A-Δ	435	298	36	360-460-550
Specimen set E-Z	446	302	30	362-445-520

As expected, the un-welded material exhibited a ferrite - pearlite microstructure and no martensite or other phases were present that could lead to premature failure. The microstructures present along the length of a welded specimen showed an initial increase in grain size followed by an area of small grain size leading to a dendritic microstructure in the weld. The microstructures observed are typical of such steels, and do not indicate reasons for premature failure during service. The fracture surfaces observed using scanning microscopy revealed the presence of manganese sulphide inclusions but as this very common in steel [18-20] it is not expected to cause premature fracture.

Table 4. Discontinuities present in specimen set B.

Specimen	VT	PT	MT	UT
B1	overlap (lid)	overlap (lid)	overlap (lid)	pore (root)
B2	overlap (lid)	overlap (lid)	overlap (lid)	pore (root)
B3	misalignment	misalignment	misalignment	misalignment
B4	-	pore (lid)	pore (lid)	pore lid and lack of fusion (root)
B5	-	overlap (lid)	overlap (lid)	pores (pool) and incomplete fusion (root)
B6	-	-	-	crack (pool)
B7	-	pores (root)	pores (root)	pore (pool) and incomplete fusion (root)

All non-destructive testing methods employed methods indicated the presence of various types of discontinuity for both specimen sets A-Δ and E-Z. The findings of the non-destructive tests for specimen set B are shown in Table 4. Although most specimens contained some type of discontinuity only a small number of them failed to reach the boundary of $2{\cdot}10^6$ cycles during the fatigue tests. The fatigue test results for specimen sets B and E are shown in Table 5.

Table 5. Fatigue Tests of specimen set B.

Specimen	Failure (yes/no)	Specimen	Failure (yes/no)
B1	no	E1	no
B2	no	E2	no
B3	no	E3	yes 1,556,340 cycles
B4	yes 1,918,221 cycles	E4	no
B5	no	E5	no
B6	yes 1,256,355 cycles	E6	no
B7	no	E7	no

The metallographic and fractographic investigations carried out showed that the microstructures present and the inclusions contained cannot account for the frequent service failures of the welded joints. However, it seems that poor quality welding may be the cause of the problem. As can be seen, for both specimen sets A-Δ and E-Z, almost all non-destructive testing methods indicate the presence of some type of discontinuity. For example, for specimens B1-B3 and B5, VT, PT and MT showed the presence of considerable overlapping and misalignment and would have prevented such welds from going into service. However, for specimens B4, B6 and B7 UT showed the presence of pores, incomplete fusion and a crack which were not detected by the VT, PT and MP.

In practice, when rim replacement takes place, visual inspection is performed forty eight hours after welding is completed and liquid penetrant inspection is used only where defects

are suspected [21]. This means that, for example, welds containing discontinuities such as those in specimens B4, B6 and B7 would not be detected and the repaired BW would go into operation. Although all samples, even those containing defects performed well under tension specimens B4, B6 and E3 failed to achieve the required number of cycles. The initiation and development of cracks, which led to fracture, during fatigue testing in specimens B4, and E3 was studied using ultrasound non-destructive testing. The development of the discontinuities in specimen B4 is described below. The ultrasound test showed a lack of fusion (LOF) defect starting at 5mm-10mm width at a 6mm depth. After 500,000 fatigue cycles the discontinuity did not progress. However after 1,000,000 cycles the discontinuity progressed from 5mm-10mm to 5mm-14mm width. Finally, at 1,918,221 cycles there was sudden fracture of the specimen and the test ended.
Failure occurred suddenly, which is characteristic of fatigue [22], usually initiating from a preexisting defect and after progressing to a critical length. The question that arises from this observation is why some defective welds fail and some do not and how one can decide which "usual" quality welds are to be allowed to go into service. The issue has been addressed by a number of researchers [23-25] who suggest various statistical evaluation methods for censoring of failed and non-failed welds for fatigue strength assessment of welded structures. The use of high frequency mechanical or ultrasound impact, TIG re-melting of welds and weld grinding has been recommended for fatigue strength improvement of welded joints [25-27] and will be studied by the authors in future work.

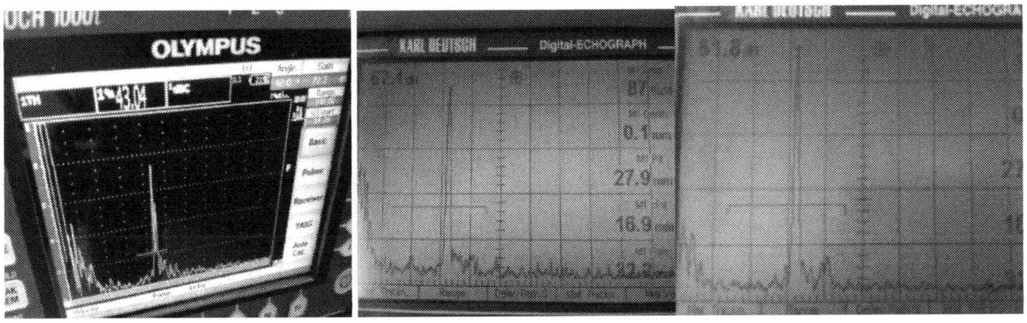

Fig. 3. The development of the discontinuities in specimen B4.

4. Conclusions

The manual metal arc welding process used for the welding repairs in the bucket wheel excavator needs improvement as most welds contain numerous discontinuities. Even though they may not always lead to fracture they will in some cases cause premature failure under fatigue loading. Welding with electrode type EN ISO 2560-A: E42 5B 42 H5 produces better quality welds. Visual testing (VT) and liquid penetrant (PT) testing is not sufficient to detect serious weld defects and ultrasound testing (UT) should be applied in all cases. Steel material of higher yield strength is suggested to be used as the first step towards improving fatigue strength. Additionally, weld grinding, TIG remelting and mechanical impact of welds would result in longer fatigue life.

References

1. S. M. Bosnjak, M. A. Arsic, N. D. Zrnic, M. P. Rakin, M. P. Pantelic, EngFail Anal, Vol.**18**, pp. 212-222 (2011)

2. S. Bosnjak, M. Arsis, S. Savicevic, G. Milojević, D. Arsić, Maint and Rel, Vol.18, No. 2, pp. 155-163 (2016)
3. S. M. Bosnjaka, M. A. Arsicb, N. B. Gnjatovica, I. L. J. Milenovica, D. M. Arsic, Eng Fail Anal, Vol. **84**, pp. 247-261 (2018)
4. M. Arsic, V. Aleksic, Z. Radakovic, Struct Int and Life, Vol. **8**, No 1, pp. 13-22 (2008)
5. ASTM E 466-2002. 'Standard Practice Conducting Force Controlled Constant Amplitude Fatigue Tests of Metallic Materials'
6. ISO/TR 14345-2012, 'Fatigue testing of welded components'
7. N. H., S. Nisshijima, A. Ohta, Y. Maeda, K. Uchino, T. Kohno, K. Toyomasu , I. Soya, *J. Test. Eval.*, **16**, pp. 280–285" (1988)
8. S.J.Maddox, "Key developments in the fatigue design of welded constructions", Proc. IIW Int. Conf. *Welded Construction for Urban Infrastructure*, ISIM Timisoara (2003)
9. ISO 17637:2016, Non-destructive testing of welds-Visual testing of fusion-welded joints
10. ISO 5817:2014, Welding-Fusion-welded joints in steel, nickel, titanium and their alloys (beam welding excluded), Quality levels for imperfections
11. ISO 3452-1:2013,"Non-destructive testing-Penetrant testing Part 1: General principles"
12. ISO 23277:2015, "Non-destructive testing of welds – Penetrant testing of welds"
13. ISO 17638:2016, "Non-destructive testing of welds – Magnetic particle testing"
14. ISO 23278:2015, "Non-destructive testing of welds Magnetic particle testing of welds"
15. ISO 17640:2010 "Non-destructive testing of welds-Ultrasonic testing-Techniques, testing levels and assessment"
16. ISO 23279:2017 "Non-destructive testing of welds-Ultrasonic testing-Characterization of discontinuities"
17. T.J. Baker, "Sulphide Inclusions in Steel", *American Society for Metals*, Metals Park, OH, USA, pp. 135 -158 (1958)
18. T. Gladman, "Sulphide Inclusions in Steel", *American Society for Metals*, Metals Park, OH, USA, pp. 273-276 (1975)
19. S. Maropoulos, N. Ridley, Mat Science and Eng, vol. **384**, pp. 64-69 (2004)
20. A. Zarkas, A. Maropoulos, S. Papanikolaou, S. Maropoulos, "Rim replacement of a bucket wheel excavator", *1st International Conference on Welding and Non Destructive Testing* (1st ICWNDT), HSNT and WGI, Athens, Greece (2018)
21. M. A. Maleque and M. S. Salit, Mat Sel and Des, Springer Briefs in Materials, pp. 17-38 (2013)
22. S. J. Maddox, 'Fatigue assessment of welded structures', Proc. Int. Conf. *Extending the life of welded structures*, Pergamon, Press, Oxford, pp. 33-42 (1993)
23. K. Wallin, "The probability of success using deterministic reliability", *Fatigue Design and Reliability: ESIS Publication 23*, G. Marquis and J. Solin, Eds., Elsevier Science Ltd., Amsterdam, pp. 39–50, (1999)
24. G. Marquis, T. Mikkola, Weld in the World, Volume **46**, Issue 1–2, pp. 15–22, (2002)
25. P. J. Haagensen, S. J. Maddox. *IIW recommendations on post weld fatigue life improvement of steel and aluminium structures*. Paris: International Institute of Welding, (2011).
26. H. C. Yildirim, G. Marquis, Int Journ of Fat, Volume **57**, pp. 803–822, (2013)

MATEC Web of Conferences **318**, 01034 (2020) https://doi.org/10.1051/matecconf/202031801034
ICMMEN 2020

Safety Management of Hazardous Materials - Orphan Radioactive Sources: Contribution of STRASS Project

Stelios Xanthos[1,*], *Alexandros* Clouvas[2], *Fokion* Leontaris[2], and *Nikolaos* Liampas[1]

[1]Department of Industrial Engineering and Management, International Hellenic University, Greece
[2]Nuclear Technology Laboratory, Department of Electrical & Computer Engineering, Aristotle University of Thessaloniki, Greece

Abstract. STRASS project is an INTERREG project, collaboration between Greece and North Macedonia that includes the following aims a) Discovery and identification of radioactive materials (especially orphan radioactive sources and materials that are transferred mainly accidentally) during check in cross border area, b) Location of dangerous points of the roadway Thessaloniki-Skopje. c) Investigation of radiological risk after a traffic accident d) Establishment of common emergency response protocols for both countries. The project deals with circular economy and sustainability and its main challenge that is presented here is to minimize any risk of accident (traffic accident during transportation, dispersion, loss etc) and pollution when handling and transferring willfully or accidentally hazardous radioactive materials.

1 Introduction

Global challenges like health, safety and environmental issues require transnational and interdisciplinary approaches. An effort towards this direction is being done by STRASS project. STRASS project is an INTERREG project, collaboration between Greece and North Macedonia that focus on two axes. One that has to do with radiation protection and a second one that deals with safe transportation. The combination of above-mentioned fields is the main objective of the project which is about safe transportation of hazardous materials and especially radioactive ones.

For the first field Radiation Portal Monitors (RPM) are commonly used to detect and intercept unauthorized movement of nuclear and other radioactive materials both at borders and within States [1]. However, under certain circumstances, detection portals can fail to detect radiation (eg due to shielding). After this step of detection in the border area of a hazardous load that may fail or not the next challenge is to minimize traffic accidents with trucks carrying these hazardous materials.

For the second field it is well known that traffic accidents are the leading cause of death for people aged under 50 [2] and the second-leading cause of death worldwide [3] for children aged 5–14 years and young people aged 15–29 years. Greece and especially Central Macedonia are negative pioneers in traffic accidents counting thousands every year. Among

* Corresponding author: xanthos@autom.teithe.gr

38 European countries, Greece is ranked in the seventh highest position in road crash fatalities, (24 men and 5 women per 100,000 citizens) [4]. To eliminate them and reduce their consequences requires a systematic analysis.

The purpose of the present study is to examine a) safety and detection of radioactive materials provided by the installation of Portal Monitors in the Evzoni border station and b) the traffic accidents on roadway from Thessaloniki to Evzoni border station and therefore the safety of transported products.

2 Methodology

As mentioned before, this research is oriented towards two branches: a) radiation detection and determination of Minimum Detectable Activities of RPM in Evzoni and b) traffic accidents on roadway from Thessaloniki to Evzoni. They are discussed in detail bellow.

2.1 Radiation Detection

In the Greek-North Macedonia border, Radiation Portal Monitors have been installed since 2004 for detecting and intercepting unauthorized movement of nuclear and other radioactive materials. The detection system is a TSA PM 700 AGN polystyrene portal monitor with two scintillating detectors per pillar.
The calculation of the threshold over which the alarm will be triggered is directly dependent on the background of natural radioactivity and is given by the following formula

$$TL = background + c \times \sqrt{background} \qquad (1)$$

TL : The alarm threshold in counts per second
background : background of natural radioactivity in counts per second
c: Constant that determines the confidence level

Obviously, the $\sqrt{background}$ indicates a Gaussian (normal) background distribution. In fact, the distribution of the background in counts per second is not a normal distribution, but it is very similar to it [5]. Therefore, constant *c* determines the confidence level of this distribution (eg for $c = 3$, we have a confidence level of 99.73%).

In order to estimate Minimum Detectable Activities for different radionuclides, Monte Carlo simulations and measurements were used in combination.

Simulation of the scintillating detectors of the Radiation Portal Monitors in the Greek-North Macedonia border was carried out with a set of Monte Carlo simulations. The tool used in order to simulate spatial response to gamma radiation of these detectors was the MCNP a general-purpose Monte Carlo N-Particle code of the Los Alamos National Laboratory.

In order to test the scintillating detectors of the RPMs, Monte Carlo simulations were performed in combination with in situ gamma radiation measurements. Measurements were performed using Eu-152 and Cs-137 radioactive point sources located 70 cm above ground and in half distance between pillars of the portal (3m) for Eu-152 and 30 cm from the detector for Cs-137. This is due to the fact that Cs-137 had very low activity (0.35 μCi).

2.2 Traffic accidents

The following procedure was followed for this research on the part that deals with traffic accidents:

- Examination of traffic accidents data on roadway Thessaloniki to Evzoni was carried out.

- It was decided to exclude from the investigation the section of the highway E1 from Thessaloniki to the intersection of Chalastra, because most of the vehicles in this section of the highway travel for south, western and northern Greece.
- The data, which were used for this research, was the recorded traffic accidents by the traffic departments of Thessaloniki and Kilkis, which are responsible for the section of the PATHE E75 highway from the intersection of Chalastra to the Evzoni border station.
- Because of its completely different morphology the part of motorway that is under consideration is divided in two sections. The first one is a single carriageway 47 km long, begins from intersection of Chalastra (485 km) and stops nearby the exit of Nea Kavala (532 km) and the second part is a continuation of the first part extends up to the Evzoni border station, (549 km) is 17 km long, is dual carriageway having two lanes per direction.
- The road was examined for locating dangerous points of truck accidents. Because the number of truck accidents were limited, only 11 occurred in the last decade, we could not draw safe conclusions. For this reason, we used all the recorded traffic accidents (77).
- The percentage and impact of truck accidents were compared with all the other accidents.
- A correlation study was performed between weather conditions and traffic accidents involving a truck.
- Furthermore, it was examined if the months of the year, the days of the week, the daylight and the hours of the day were correlated to increased number of traffic accidents.

3 Results and discussion

3.1 Simulation Results of RPMS

The following numerical “experiments” were performed: By Monte Carlo simulations were calculated the number of photons per second (cps) detected by the scintillator per μCi. Cs-137 and Co-60 point sources were placed in different distances from the scintillator (0 up to 3 meters). The radioactive sources were located 70 cm above soil. Three meters is the half distance between the two pillars of the RPM. The installation is shown in figure 1.

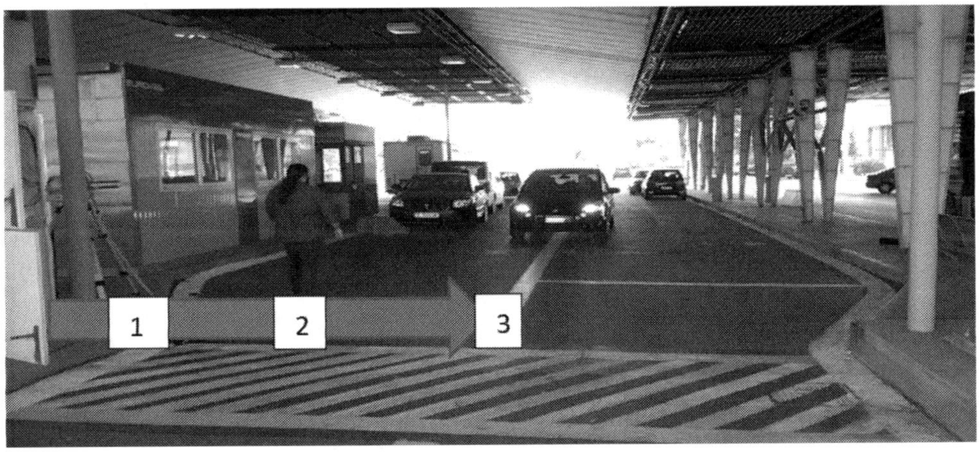

Figure 1. Distance in meters from the scintillator.

As mentioned before calculation of the threshold over which the alarm will be triggered is directly dependent on the background of natural radioactivity and is given by formula (1)

For the calculation of the minimum alarm activity we assume that the cps measured by the detector due to the source is 4 sigma of the cps of the Background. In the Evzoni area the counts per second due to background is 303 cps for the total window (22-1595 keV) of the scintillator. Following the above remarks:

- the threshold over which the alarm will be triggered in the RPM is:
  $TL = 303 + 4 \times \sqrt{303} = 373\ cps$
- the minimum activity of the source for triggering the alarm corresponds to $4 \times \sqrt{303} = 70$ photons per second detected in the scintillator due only to the source (background subtracted).

In Table 1 are presented the calculated (by MC simulations) cps per μCi for Cs-137 and Co-60 point sources placed 70 cm above soil and 300 cm away from the scintillator. As mentioned above three meters is the half distance between the two pillars of the RPM. Dividing the value of 70 cps by the tabulated values (second column of Table 1) we deduce the minimum activities for triggering the alarm which are presented in column 3.

Table 1. Calculated cps per μCi for Cs-137, Co-60 point sources placed 70 cm above soil and 300 cm away from the scintillator and the corresponding minimum activities for triggering the alarm of the RPM.

POINT SOURCE	**cps/μCi**	**Minimum activities for triggering the alarm of RPM**
Cs-137	11.9	5.9 μCi
Co-60	38.0	1.8 μCi

3.2 Measurement Results of RPMs

As mentioned before, in order to validate results from Monte Carlo simulations of the scintillation detectors of the RPMs, in situ gamma radiation measurements were performed in combination with simulations. Experimental measurement results are presented below.
Estimation of the minimum activity of Eu-152 source for Alarm triggering of the RPM.

- The Eu-152 source (unshielded) is located 70 cm above soil and in the half distance between the two pillars of the RPM (3 meters).
- The activity of the source is 42.5 μCi.
- The measurement in the RPMs due to the source is 604 cps more than 5 times higher the background (BKG =116 cps). The energy window of the RPMs is the default one (22-144 keV). For the calculation of the minimum alarm activity we assume that the cps measured by the detector due to the source is 4 sigma of the cps of the Background.
- A minimum alarm activity of about 3 μCi (111 kBq) is estimated for Eu-152.

Estimation of the minimum activity of Cs-137 source for Alarm triggering of the RPM.

- The source Cs-137 (unshielded) is located 70 cm above soil and in the half distance between the two pillars of the portal (3 meters). Unfortunately, we do not have high activity Cs-137 source in order to perform such measurement at 3 meters.
- We have performed, measurements with a Cs-137 point source of activity 0.35 μCi up to 0.3 meters from the detector.
- The measurement in the RPMs due to the source at 30 cm is 161 cps. In order to deduce the cps at a distance d = 3 meters we use the $d^{-1.7}$ distance dependence factor found experimentally for the Eu-152 source. With such dependence the estimated cps for a Cs-137of 0.35μCi source at 3 meters is estimated at 3.2 cps.
- For the calculation of the minimum alarm activity we assume that the cps measured by the detector due to the source is 4 sigma of the cps of the background which was 144 cps.
- A minimum alarm activity of about 5.2 μCi (192 kBq) is estimated for Cs-137.

Similar results concerning minimum alarm activities for Cs-137 sources were taken from 19 radiation portal monitors in Spain [6].

3.3 Investigation for dangerous points and traffic accidents in the roadway Chalastra - Evzoni

The research showed that in the last decade (2007-2018), seventy-seven (77) traffic accidents were recorded, the impact was 28 deaths, 12 serious injuries and 94 slightly injured.

The highest proportion of accidents had involved cars, which is quite normal due to the majority of traveling the road axis. Trucks and motorcycles were following.

In the following figure 2, comparison of the number of traffic accidents and their severity in the motorway Single carriageway 47 km long (485 – 532 km) and Dual carriageway 17 km long (532 – 549 km) is depicted.

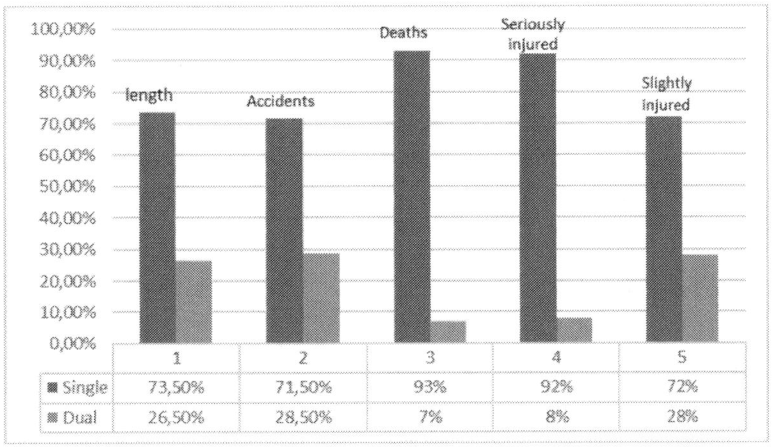

Figure 2. Comparison of traffic accident rates and their severity

After examining the data, it was found that the percentage of the number of recorded traffic accidents in the above two parts is approximately proportional to their lengths. Although the same percentage of accidents is reported, their severity was four times less in the second part than in a first one.

Traffic accidents involving trucks were recorded on the first part of the roadway (single carriageway). They were uniformly distributed throughout this part of the motorway and it includes long straight sections.

In order to characterise a km of the roadway as a dangerous point three repeatable traffic accidents should have been occurred in the last decade. Afterwards the topography and the type of accident that occurred in each point were investigated.

The examination had shown that there are no road points that can be described as dangerous. Most recurring accidents had taken place in perfectly straight sections, without dangerous causes compared with other sections of the whole roadway.

3.4 Causes of truck accidents and impacts.

The main reason of serious traffic accidents is due to the movement in the opposite traffic lane, that causes often a head-on collision. This is also confirmed in our study, where all the deadly truck accidents caused by movement in the opposite traffic lane. In the following table 2, causes of truck accidents and impacts are presented.

Furthermore, it was investigated if weather conditions and night hours were associated with the occurrence of truck traffic accidents. The conclusion is that there is no correlation between bad weather or limited light and truck accidents. Finally, a relationship between morning hours (07:00 - 12:00) and truck accidents due to the increased traffic in those hours was observed.

Table 2. Causes of truck accidents and impacts

Causes of accidents	Movement in the opposite traffic lane	Lack of attention	Uknown	Priority	Are being investigated
Number of accidents	5	2	2	1	1
Accident impacts	4 deaths & 1 slightly injured	3 slightly injured	3 slightly injured	1 severe injured	2 slightly injured

4 Conclusions

Based on the above we come to the following conclusions after this study:

For Radiation Detection part of this study:

Minimum activities of Cs-137, Co-60 sources for alarm triggering of the RPMs were calculated by Monte Carlo simulations and found in good agreement with those (Cs-137) deduced experimentally. They are also in good agreement with the literature. Minimum activities for triggering the alarm of the RPM were estimated for Cs-137 and Co-60 radioactive sources for 70cm above ground and 300cm away from pillar (half distance between both pillars) were estimate and found 5.9 μCi and 1.8 μCi respectively.

For Traffic Accidents part of this study:

- The percentage of traffic accidents involving a truck is 12%, with the same percentage of impacts compared with all the other traffic accidents.
- It was observed that the severity of all the vehicle accidents in the dual carriageway part from 532 km up to 549 km with two lanes per direction is four times less.
- There is a relationship between all serious and fatal truck accidents with the movement in the opposite traffic lane and head-on collisions.
- It was observed that all truck traffic accidents were recorded in the single carriageway part from 485 km up to 532 km which is 47 km long.
- There are no road points that can be described as dangerous. Most recurring accidents had taken place in perfectly straight sections, without differences if compared with other points of the roadway.
- There is no correlation between bad weather or limited light and truck accidents. There is a relationship between the morning hours (07:00 - 12:00), with increased number of truck accidents and their severity with increased traffic load. There is a relationship between the weekend and increased number of truck accidents which cannot be reasoned.

Project SRASS co-funded by the European Union and national funds of the participating countries.

References

1. S. Apikyan, D. Diamond, R. Way, *Prevention, Detection and Response to Nuclear and Radiological Threats* (Springer, 2008)
2. P. Saukko, B. Knight. *Forensic Pathology*. 281–300 (Arnold, 2004)
3. M. Peden, R. Scurfield, D. Sleet, *World report on road traffic injury prevention*, (World Health Organization., 2004)
4. WHO Regional Office for Europe, *Atlas of health in Europe*, 56 (WHO, 2008).
5. J. Ely, R. Kouzes, J. Schweppe, E. Siciliano, D. Strachan, D.Weier, , Nucl. Instr. and Meth. Phys. A 373-387 (2006)
6. C. Correa, T. Ortiz, O. González, J.I. Serrano, J.A. González., IRPA12, (2008)

Code development for the implementation of the Part Conditional Scale Feature in 3D CAD software platforms

Konstantinos Vogiatzis[1] and *Georgios* Skordaris[1]

[1]Laboratory for Machine Tools and Manufacturing Engineering, Mechanical Engineering Department, Aristotle University of Thessaloniki, Greece

Abstract. Most contemporary 3D CAD software has implemented the Scale feature as a plain multiplication or division of an object's standard dimensions. A variety of options exists; sometimes the Scale process can be applied to an origin or a user-defined coordinate system, or it can be applied to one or two axes. No software so far has offered the option of setting the rules as to how the Scale feature can be applied to one or more components of an object. This article is about the development of a code that adds this functionality to a popular 3D CAD software. As a first step, the code is only applied to holes that lie on a planar face.

1 Introduction

In the process of Product Development, engineers often rely on the design of a product that has already been successfully manufactured, which they modify, with the aim of improving it [1]. Many times, a successful product needs to be modified so that it can be used in different applications, industries, consumer categories etc. from which it was originally developed. In order for a product to acquire additional or different properties, one of the most common modifications used by engineers is to change its dimensions. But the shape in which a successful product has been designed is almost always an important factor of its success. Thus, when attempting to improve it, the requirement is to change it as little as possible [2]. Given the above limitations, regarding the complexity of the product's geometry, the redesign process can be quite laborious and time consuming [3]. In order to solve this problem, all modern CAD software programs have implemented the Scale feature. The Scale feature can uniformly change the shape of an object, by a certain scale factor [4], which at a first glance seems to be the solution to the problem. However, in the real world the product is redesigned so that it has "almost" the same shape as the original. It is never "exactly" the original product multiplied by a number. A typical example of geometry that cannot be changed this way is the holes. Usually the holes are used to assemble the product components together or to fasten the product. The assembling as well as the fastening is done with standard connection elements (e.g screws, pins, etc.). However, all these connection elements also have standard dimensions. Therefore, by applying the Scale feature to the entire product, the holes should be excluded and modified individually. This article is about the development of an algorithm that applies the Scale command to the whole object, but then applies exclusion rules and individual scale rules to the holes that lie on planar surfaces, defined by the user. The algorithm was developed as an add-on to the DS SOLIDWORKS commercial software.

2 Methods

At first, a C++ dynamic link library (dll) project was created inside the Visual Studio Community 2017. Then, a code has been developed in order to embed the Add-on to the SOLIDWORKS software. More specifically, using the functions of the SOLIDWORKS Automated Program Interface (API), all necessary connections (Events) between the SOLIDWORKS software and the Add-on have been developed, toolbars, menu items and buttons have also been created inside the SOLIDWORKS Graphical User Interface (GUI). Then, the main dialog (Fig. 1) of the Add-on has been designed.

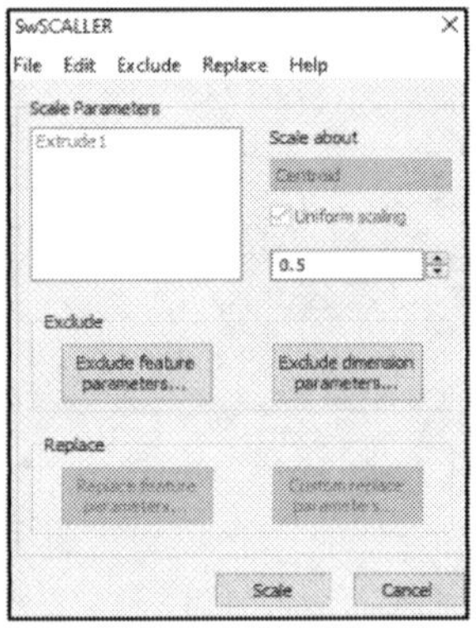

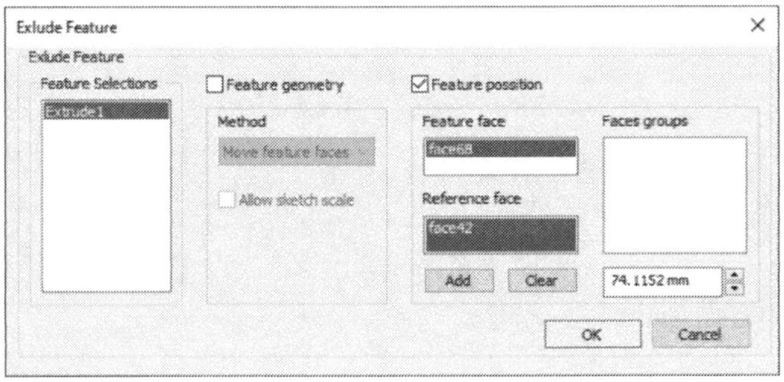

Fig. 1. Main dialog **Fig. 2.** Exclude feature dialog

Scale feature is firstly applied on bodies; the main dialog bodies are selected and scale rules are defined (scale factor and scale type). The next step is to exclude holes from being scaled. For this purpose, an Exclude Feature Dialog (Fig. 2.) has been developed.

All hole features that must be excluded from the Scale operation are selected. For each hole feature, a reference face must be selected (e.g the inner face of the hole) and the face from which the hole distance must remain constant after the Scale process. There is also an option to input the distance value. For each hole, two position rules must be set, in order to determine the hole's final position on the planar surface (Fig. 3). All those rules are saved in memory.

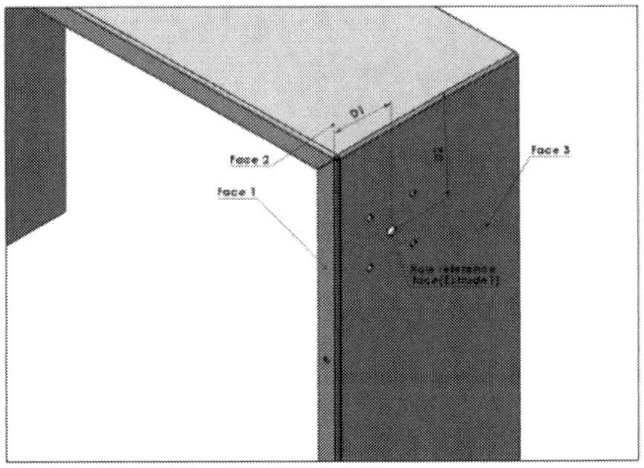

Fig. 3. Illustrating the defined distances D_1 and D_2, of Hole feature from Face 1 and Face 2 respectively.

Now it is possible to calculate the coordinates of point X (x_x, x_y, x_z), that is the center of the hole that we want to exclude from Scale process. Point X must preserve its distance (D_1) from Face1, Face2 (D_2) and Face3 D_3, after the Scale process will have taken place. Firstly, the coefficients of the Face1, Face2 and Face3 planes equations must be calculated. SOLIDWORKS API has a function that returns a face's normal vector and the vector's root point. For reference face, Face1, the normal vector and the root point are

$$\vec{\mathbf{A}}\,(a_x, a_y, a_z) \tag{1}$$

$$\mathrm{P}\,(p_x, p_y, p_z) \tag{2}$$

for reference face Face2 are

$$\vec{\mathbf{B}}\,(b_x, b_y, b_z) \tag{3}$$

$$\mathrm{V}\,(v_x, v_y, v_z) \tag{4}$$

and for the face on which the hole lies are

$$\vec{\mathbf{C}}\,(c_x, c_y, c_z) \tag{5}$$

$$\mathrm{N}\,(n_x, n_y, n_z) \tag{6}$$

The equation of a plane is [5]

$$a_1 x + b_1 y + c_1 z + d_1 = 0 \tag{7}$$

because the normal vector of this plane is

$$\vec{\mathbf{n}}\,(a_1, b_1, c_1) \tag{8}$$

The plane equation can be written as follows:

$$a_x x + a_y y + a_z z + d_1 = 0 \tag{9}$$

And because the root point P, lies on the plane, its coordinates must satisfy the plane's equation:

$$a_x p_x + a_y p_y + a_z p_z + d_1 = 0 \tag{10}$$

From the above equation d_1 can be calculated as follows:

$$d_1 = -a_x p_x - a_y p_y - a_z p_z \tag{11}$$

And by using the same methodology, d_2 *and* d_3 are calculated as follows:

$$d_2 = -b_x v_x - b_y v_y - b_z v_z \tag{12}$$

$$d_3 = -c_x n_x - c_y n_y - c_z n_z \tag{13}$$

The distance of the point

$$\mathrm{X}\left(x_x, x_y, x_z\right) \tag{14}$$

from the plane of Face1

$$a_1 x + b_1 y + c_1 z + d_1 = 0 \tag{15}$$

is given by the following formula [6]

$$D_1 = \frac{\left|a_1 x_x + b_1 x_y + c_1 c_z + d_1\right|}{\sqrt{{a_1}^2 + {b_1}^2 + {c_1}^2}} \tag{16}$$

The distance of the point X from the plane of Face1 is given by:

$$D_2 = \frac{\left|a_2 x_x + b_2 x_y + c_2 c_z + d_2\right|}{\sqrt{{a_2}^2 + {b_2}^2 + {c_2}^2}} \tag{17}$$

The distance of the point X from the plane of Face3 is given by:

$$D_3 = \frac{\left|a_3 x_x + b_3 x_y + c_3 c_z + d_3\right|}{\sqrt{{a_3}^2 + {b_3}^2 + {c_3}^2}} \tag{18}$$

where $D_3 = 0$, because point X is a point on Face3.

By solving the system of equations (16), (17) and (18), point X coordinates x_x, x_y *and* x_z can be calculated. The solution to this system of equations, are four points, X_1, X_2, X_3 *and* X_4(Fig. 4).

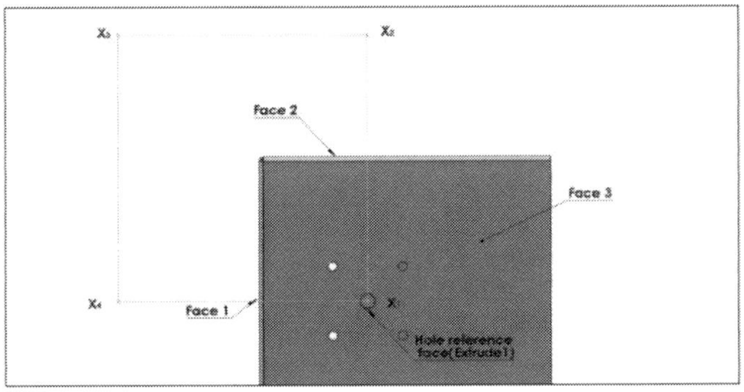

Fig. 4. Illustrating the positions of the four points, that abstain distance D_1 from Face1 and distance D_2 from Face2.

Next, by checking which of the four points lies on Face3, the coordinates of point X can be determined.

Now, by using the Hole Reference Face, the center of the hole can be calculated. If,

$$C_H(x_H, y_H, z_H) \tag{19}$$

is the center of the hole coordinates after the Scale process, then Hole Face must be moved as follows:

$$dx = x_x - x_H \tag{20}$$

$$dy = y_x - y_H \tag{21}$$

$$dz = z_x - z_H \tag{22}$$

Finally, by using the SOLIDWORKS API Move Face command, the face hole translation that has been previously calculated can be applied.

3 Typical example

The process explained in this article can be applied as illustrated below, in the example of a stainless-steel sheet metal part which is the outer cover of an electrical panel.

The electrical panel has a main power switch, which is fastened on the sheet metal cover by four bolts. The installation of the switch also needs a hole at the center for the switch button (Fig. 5). In this example, the electrical panel must be scaled at half the size, but the same main switch must be installed. The code described in this paper will be used in order to exclude the main switch holes from the Scale process.

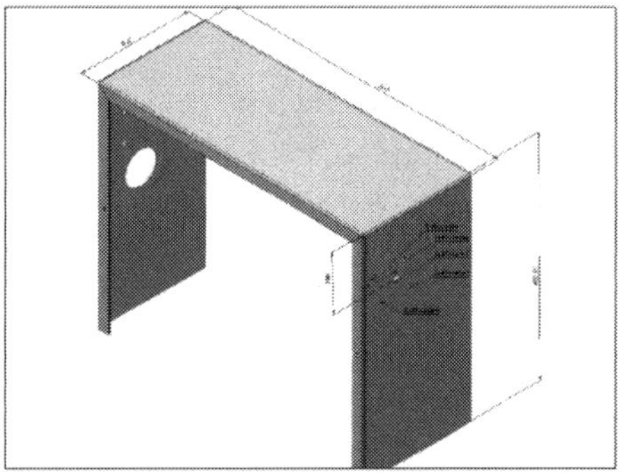

Fig. 5. Illustrating, all hole features, that must be excluded from the Scale operation.

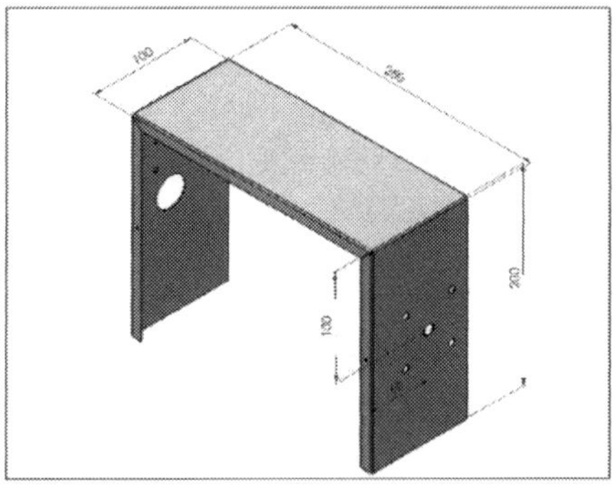

Fig. 6. Illustrating the upper electrical panel cover, after the Part Conditional Scale code has been applied. The cover has been scaled half size, but the main switch holes distances from Face1 and Face 2 are unchanged.

The code is executed inside the SOLIDWORKS user interface. In the main dialog of the application the scale factor is set.

The next step is to exclude all the hole features of the main switch (CutExtrude, CutExtrude9, CutExcrude8, CutExtrude7 and CutExtrude6).In this step, all Hole features are selected, one by one. For each hole face, two reference faces are selected. The distance of the hole face and the reference faces will be preserved after the Scale process. The resulting sheet metal part, after running the code, is shown In Fig. 6.The upper cover of the electrical panel has been scaled, using a scale factor of 0.5, but the holes where the main switch will be installed remains unchanged and at the same distance from the reference faces.

4 Conclusions

This paper introduced how a code has been developed, for the implementation of the Conditional Scale feature. Conditional Scale feature is extending the functionality of the Scale feature that is included at all contemporary CAD software, by adding the option of setting the rules as to how the Scale feature can be applied to one or more components of an object.

References

1. J. C. YU, S. Krizan, K. Ishii, Journal of Intelligent Manufacturing, **4,** 199–208, (1993)
2. H. Rui, Z. Shusheng, B. Xiaoliang, X. Changhong, H. Bo, Computers in industry, **67,** 38-53, (2015)
3. C. L. Tan, M. A. Vonderembse, Journal of Operations Management, **24,** 494-510, (2006)
4. SOLIDWORKS API Help
5. https://brilliant.org/wiki/3d-coordinate-geometry-equation-of-a-plane/
6. https://mathinsight.org/distance_point_plane

3D Printing Assisted Product Design Addressing Refugees Needs

Antigoni Panagiotidou[1], and *Dimitris* Karalekas[2*]

[1]International Hellenic University, School of Science and Technology, 57001 Nea Moudania, Greece
[2]University of Piraeus, Department of Industrial Management and Technology, 18534 Piraeus, Greece

Abstract. With one of the largest refugee crises of modern times currently occurring, the need for technology-based solutions to address related pressing issues is eminent. In the past few years, 3D printing has attracted considerable attention as a technology that could help to address specific refugees needs in a customized way. The aim of the presented work was to investigate and demonstrate the contribution of 3D printing to the design of specific products that could easily and rapidly manufactured to assist the refugees integration into the host country. Specifically, the undertaken study focused on facilitating the integration of refugees' children into the Greek society through the design and 3D printed toys for educational purposes. It is demonstrated in this preliminary study that such 3D printed toys can be proven a powerful tool for improving the integration process of displaced people by making the learning of a new language a pleasurable experience.

1 Introduction

In a rapidly and radically changing world, there are many complex societal problems. According to a U.N. refugee agency report [1], Europe is facing the worst refugee crisis with the number people who are displaced exceeding the one of the post-World War II. Many European organizations have tried to solve living and societal challenges of refugees by using new innovative methods and design tools. Considering that the design of ordinary products is user oriented, a more human-centered approach should be used for the development of products that could satisfy the refugees needs. The solutions offered can be products for daily use, services, garments and toolkits to education and training.

Children have been the worst affected by the refugee crisis, facing the lack of proper hygiene, nutrition and education. In the camps of the host countries usually there is no access to learning environments, literacy or basic education. One of the most effective learning techniques is through play because enhances creativity, invention and communication skills. Learning the language of the host country would be a skill for education and integration. Through education the refugee children can gain a sense of self value and reawaken their talents.

* Corresponding author: dkara@unipi.gr

Additive Manufacturing, or widely known as 3D printing, could help to address such issues related to important refugees needs. It has been reported that non-governmental organizations (NGOs) have been exploring the possibilities with 3D printing [2-6]. They have explored different ways of localizing the manufacturing of items needed in crisis such as items for water access, shelter, protection and prosthetics. Recently, attention has been drawn in developing robust and easily deployable 3D printers for humanitarian response works where the printers' operational environment is not well controlled [7].

This study investigates the design and fabrication, using a commercial 3D printer, of a simple educational toy product to facilitate the integration of refugee and migrant children into the Greek society through a playful experience.

2 Methods and materials

2.1 Product design requirements

An educational toy was selected to be designed and 3D printed models to be fabricated. In general, educational toys are specially designed to provide stimulation and learning for young children. By using such toys children can develop skills and abilities through the playing process. For instance, they can help them in problem solving, developing language skills, encouraging the communication with each other and expanding both their mental and physical abilities.

A simple structured toy in the form of a cubical building block was designed. Such shaped wooden toys of various sizes are commercially available and intended for constructive playing that natures spatial and logic-mathematical intelligence, develops logical thinking and creative problem solving. The selected cubical block shape satisfied the following design requirements:

- To encourage the refugee children to learn and develop their Greek language skills by learning the alphabet's letters and forming words.
- The overall size to be as minimal as possible to be easily handled by the young children.
- To not have sharp edges and be safe during playing.
- To be made of Acrylonitrile Butadiene Styrene (ABS) material which is harmless and suitable to prototyping.
- To be lightweight to be easily handled and carried.
- To be easily manufactured and painted with different colours.

2.2 Product design sketches and CAD models

Initially, a series of sketches were created in order to evaluate various ideas about the new toy block designed for the refugee children. These toy blocks had various shape forms such as cuboid, cylindrical cube, triangular prism and others as shown in Figure 1. Since the main function of the toy was to assist the children to the learning of the Greek language a cubical shaped block was selected having dimensions of 40 x 40 x 40 mm. The cube as a three-dimensional object has six square faces where different Greek letters, either capital or in small lettering, could be positioned. The toy set would be consisted of ten cube-shaped blocks in total while the simple form of the blocks provides numerous word formations since they allow easy revolution and horizontal or vertical placement during the playing process. Representative CAD models of the alphabet blocks are presented in Figure 2.

Fig. 1. Sketches of considered shape ideas.

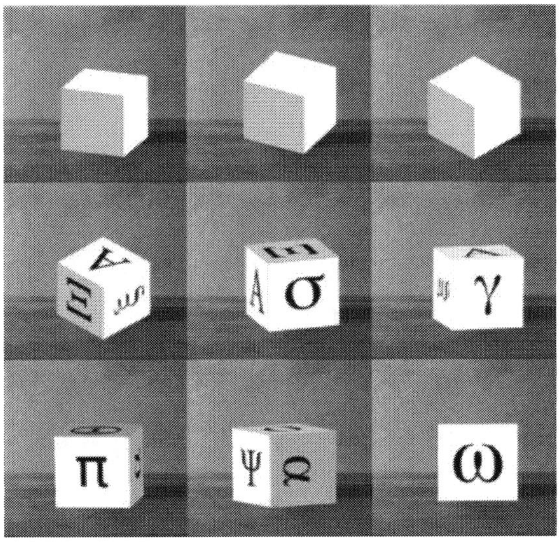

Fig. 2. CAD models of alphabet blocks.

2.3 Physical models fabrication

After having designed the blocks on the computer using a commercial CAD software, a desktop BCN3D Sigma R17 printer was used to fabricate the three-dimensional physical models. A layer thickness of 0.254 mm, infill density 25% and printing speed of 60 mm/sec were selected while the building time for each cube was 2.5 hours. For purposes of minimizing material consumption, the toy blocks were internally hollow. The 3D printer constructed simultaneously four cubes by adding ABS material layer-by-layer, fusing each layer together to make the model. Figure 3 shows four fabricated models at the end of the 3D printing process while in Figure 4 the final 3D printed blocks are presented with the Greek lettering positioned on their sides.

Fig. 3. 3D printed models on the build platform.

Fig. 4. Final alphabet physical models.

3 Discussion – Printed blocks evaluation

By testing the final 3D printed cubical models, valuable feedback and insight were obtained. Compared to adults, small children are unlikely to give any well documented feedback to asked questions. They do not have the same experiences as adults do, so they cannot express themselves in clear and precise way. Moreover, it must be considered that some of the involved children hardly spoke the Greek language. As a result, the kids' genuine impressions and reactions were observed, based on their body language, as they played with the printed cube. The children had no previous experience with such a product while no additional explanations were given.

The process was both unpredictable and demanding as the children needed time to feel comfortable and to start playing with the printed blocks. Since the children are distracted easily and cannot keep their focus to a certain object for a long period of time, the process was of short duration to hold their interest. The evaluation test took place in a playground, a familiar environment for them, and they were already in a playful and relaxed mood. Some children were more excited to participate while others were unwilling to engage.

The group that was selected was consisted of children between six to nine years old (see Figure 5). At the beginning the children started experimenting with the printed lettered blocks by touching them. The next step was to try to put the letters in a formation next to each other. If they did not know any letter they could ask for help. Later on they were asked to find a specific letter and to participate in a quick game of words. They were asked to find the letters of a specific word, or to put some letters in a row to form a word or other word game variations. After that, they were left unguided and free to play with the blocks the way they wanted to.

Fig. 5. Children are playing with the alphabet blocks.

The process with the word games seemed to hold their attention and to stimulate their interest. Judging from their reaction, at the end of the playing process was concluded that they found playing with the printed blocks pleasant and fun. They enjoyed stacking the blocks up, knocking them down and clacking them together. The blocks proved to be right sized as they were big enough to read the alphabet letters on them and not huge to have difficulties in manipulating them. In order to form words, they collaborated often with each other discussing about the right letters. Although not all children responded the same way, it was concluded that the quality of the printed model satisfied the initially set design and functional/educational requirements.

4 Conclusions

This proof-of-concept work explored the prospect of a 3D printed toy to assist to the preparatory education of young refugee children. For this purpose, cubical blocks with alphabet letters on their sides were designed and built using a commercial low-cost 3D printer. Based on this study observations, although the related comments from the observed children were broad to some extent, it was found that the simple in geometry solid blocks had a positive impact on their leaning experience. The children were engaged in a process of forming Greek words, helping them initiate conversations and exchange ideas during the words' formation exercise. Overall, this preliminary study provides encouraging information to researchers and designers to investigate further the development of additively manufactured products that can satisfy important refugees needs related to their easy integration into the educational system and labor market of the host country.

The authors would like to acknowledge the valuable assistance of Mr. Manolis Tzimtzimis during the fabrication process of the physical models at the Digital Manufacturing and Materials Characterization Laboratory of the International Hellenic University. The presentation of this work at the 7th International Conference on Manufacturing and Materials Engineering - ICMMEN 2020 Conference has been supported by the University of Piraeus Research Center.

References

1. The UN Refugee Agency, Available online: http://www.unhcr.org/globaltrends2017/
2. Field Ready, available online: http://www.fieldready.org/
3. Refugee Open Ware, available online: http://www.row3d.org/
4. P. Tatham, J. Loy, U. Peretti, Journal of Humanitarian Logistics and Supply Chain Management, **5**, 188-208 (2015)
5. E. James, D. Gilman, *Hyperlocal Manufacturing and 3D printing in Humanitarian Response* (OCHA Policy and Studies Series, 2015)
6. S. Saripalle, H. Maker, A. Bush, N. Lundman, *Proceedings of IEEE Global Humanitarian Technology Conference* (Seattle, USA, 205-208, 2016)
7. B.L. Savonen, T.J. Mahan, M.W. Curtis, J.W. Schreier, J.K. Gershenson, J.M. Pearce, Technologies, **6** (2018)

Prospective analysis of the impact of a pandemic in Industry 4.0

Georgios Andreadis[1] and *Ana Isabel* Quirós Gámez[2,*]

[1]Laboratory for Machine Tools and Manufacturing Engineering, Mechanical Engineering Department, Aristoteles University of Thessaloniki, Greece
[2]Escuela Técnica Superior de Ingeniería. University of Seville, Spain

Abstract. The world population in 2020 was estimated at 6.070 million and is projected to grow to around 9 billion by 2050. The evolution and transformation of society in the technological field is a challenge due to its rapid evolution and the social consequences it triggers. The pursued aim is the prospective analysis of diverse scenarios directed by Industry 4.0 macro-drivers, breeding cornerstones for the purpose to presage future pandemic backdrop. In accordance with the evolution of the force for change due to the analyzed factors, a series of recommendations and future work is elaborated to create a weapon to confront this possible context.

1 Introduction

The analysis of miscellaneous probable scenarios predictions impeached to the evolution of the industry or the setback of the latter is the focus of this research. Taken hand in hand with the social evolution, these progressions will entail in the adaptation of Industry 4.0 or even the evolution towards a fifth industrial revolution linked to building a bridge to the Japanese concept of society 5.0. It ought to be emphasized that the proposed scenarios are predictions based on a prospective analysis, not measurements.

The structure of the document consists of an initial section explaining the methodology for the classification, where engines of changes related to Industry 4.0 will be examined. Likewise, influential actors will be identified in this epigraph. Consecutively, the substantial point of the analysis is found in the codification of different situations governed by the development of the aforementioned mega drivers. To conclude, recommendations in a pandemic prospect would divert the circumstance with the subsequent improvement are assessed

2 Methodology

One of the uncertainties sought for noteworthy prediction is what is to come. Topicality some authors and theories attest that the future can be built and not merely be predicted. Under this prism, there are two types of futures within the discipline: presumable future and potential future [1].

* Corresponding author: anaquirosgamez@hotmail.com

Everything is happening at the present has an explanation in the past. Along these lines, predictive analysis can be accomplished in the event of a pandemic based on COVID-19 and consider its repercussion in possible scenarios. Thus, leaving evidence of the importance of good decision making.

2.1 Driver identification

The purpose of this section is to identify the macro drivers and micro drivers considered as change motors that are skilled to prompt the different scenarios. These megatrends variables are systematized and schematized in Table 1, which will be explained in the subsequent subsections.

Table 1. Analytical resources

Macro drivers	Micro drivers
Risks	Technological risks
	Natural disaster
	Human risks
IoT	Low consumption
	Device and connexion compatibility
	Network traffic problems
Artificial Intelligence	Biosensors
	Encryption and Data Integration
	Machine learning
Big Data	Quality of the data
	Secure storage
International Political Agenda	R&D
	Centralization and Inter-community agreements for innovation
Cloud Computing	Remote medical assistance and telework development
	Social Networks
	Online commerce
Society 5.0	Adaptation of new technologies and technological education
	Access to privacy

2.1.1 Risks

Risks are substantial for impact analysis. The two great threats confronting the course of humanity are incorporated, likewise including a third pivotal factor:
-Nature, where it is enclosed both natural, climatic, and biological catastrophes. It is a decisive engine of change for any conceivable future.
-The human itself. It is considered the risk of social crisis as financial. Extraordinary improvement of the industry could infer the vanishing of manual work and the non-adjustment of society. Nuclear disasters or warfare are also enveloped, which would be a strong blow and would undoubtedly be determining variables respect any prediction.
-The technological risk, both the hardware and software ones, it is contemplated cybersecurity, cloud computing, or infrastructure servers' risks.

2.1.2 Internet of Things

The improvement of the Internet of Things is also a megatrend. The possibility of gathering data through sensors on mobiles and the interconnection of devices is a fundamental piece to fit in the development of probable futures. In the case of pandemics, it could provide valuable data such as information on confinement and population movement. Tracing the origin of the disease, the spread and the contagion of the virus through patient zero are factors that could be controlled because of the development of this engine of progress.

2.1.3 Artificial Intelligence

Progress in this field is an underlying element. The development of smart algorithms for useful samples collection to use in future pandemics or the advance in machine learning are imprescriptible components of metamorphosis. The capacity to encrypt and protect this information is one of the security challenges against cyberattacks. Notwithstanding within artificial intelligence, it could be listed an enormous volume of drivers for change, but centering in a pandemic scenario, it will be distinguished the development of biosensors, detectors, and forecasters of the disease to contain the spread of the virus. The propulsion of 3D printing in addition to scanning technologies with the noteworthy capacity to cover demand will also compound a weapon to combat dearth. The virtual reality will allow the flow of teleworking and the development of cloud computing macrotrends.

2.1.4 Big Data

The enormous volume of information that is dealt with today assumes a crucial role in society. Useful data as referenced above is critical to forestall and has the power to predict future occasions. The real-time integration of the number of infected people from the official sources, both medical and security, is essential for good dynamic decision-making against the battle with the virus.

2.1.5 International Political Agenda

Government entities have an essential role in managing the ailment. GDP is a hazard factor that can cause a global collapse of the bourse and financial exchange; therefore, a monetary emergency would emerge. The results of a crisis can trigger the stoppage of the influencer drivers considered. Institutions need to oversee food and personal protective equipment deficiencies, as well as the possible appearance of mafias. Investment in R&D for the development of Industry 4.0 will be unequivocal in the evolution towards a modern revolution and the intra-community agreement towards progress will make the difference between scenarios.

2.1.6 Cloud Computing

Amidst isolation, life goes on, albeit in a digitally and not personal way. Owing to this macro driver, the new normality can be lived ceaselessly in time without venturing out from home. The connectivity between people through social networks, the development of the telecommuting, or computerized business will make it conceivable to maintain a strategic distance from monetary insolvency and market stoppage. Moreover, because of the advancement of remote clinical care, a large mass of undiagnosed patients is avoided in hospitals and can be followed up on those already diagnosed in a non-dangerous phase of the disease. This is a fundamental service.

2.1.7 Society 5.0

Finally, it is mentioned one of the most determining factors in those different situations: the evolution of the population. This Japanese concept of Society 5.0 is referenced owing to the development of the pandemic has accelerated various patterns and could trend a bridge to this new intelligent lifeway. The objective pursued is to situate individuals at the focal point of all ongoing technological transformation, taking the German concept of Industry 4.0 to a higher level and offering a path to a coordinated and interconnected society. As per this vision, the sophisticated integration of cyberspace (data) and physical space (genuine world) will be a reality and will shape human-centered socio-economic and technological models. One of the keys to ensuring success in its execution is the change in mindset and education in innovations. Thusly, we would move from Information Society 4.0 to coordinated, integrated, and sustainable sophistication [2].

2.2 Actors identification

In this section, the actors that are part of the evolution and financing within the Industry 4.0 paradigm are distinguished. In this concise portrayal of them, it is considered the European Union as the focal point of the investigation.

2.2.1 Internal Actors

It is differentiated between economic operators, such as influential investment and development companies, and politician agents made up of every nation's governments and the European Union as their coordinator. The administration of research and development (R&D) is an overriding factor to emphasize, whose burden is on the internal actor's management. Inside this gathering, it is additionally incorporated specialists' confederations, researchers, and engineering designing advancement entities.

2.2.2 External Actors

The actors who influence the transformation of the object of study but have no direct impact on the development of Industry 4.0 in Europe are englobed in this paragraph. It is incorporated supranational entities, such as the UN, specifically UNIDO (United Nations Industrial Development Organization). Emerging markets, where it is included China, Japan, and Russia (which is also a border with Europe). It is also considered the US, whose Industry 4.0 market between 2016-2023 will develop at a CAGR of 12.9%. PR Newswire said, "The United States, the world's second-largest manufacturer, with a 2017 industrial output reaching a record level of $ 2.2 trillion, will apply Industry 4.0 technologies to replace China as the world's largest manufacturer." [3].

3 Analysis of potential future scenarios

Three possible future situations depending on the combination of furtherance of the recognized drivers are introduced. The + and - symbols imply the level of positivity and negativity of the drivers.

Table 2. Codification of drivers.

	Risks	IoT	AI	Big Data	IPA	Cloud Computing	Society 5.0

Severe Scenario	+	-	-	-	-	+	-
Status Quo	-	+	-	-	+	+	-
Positive Scenario	-	+	+	+	+	+	+

3.1 Severe Scenario

The uncontrolled pandemic situation, where the attack rate is 80% [4]. The lasting decline in the labor flow is around 0.8% and along these lines, the demand for labor decreases considerably. A case like the Spanish flu of 1918-1919 is confronted. Containing the disease becomes impossible and sanitary breakdown occurs (health personnel contagion due to lack of PPE and the uncontrolled number of patients in an advanced stage of the disease). Deficiencies and inadequacy of manufacturing are reached due to the majority of non-automatization of the business and the insufficiency of adaptation. Inconceivability advancement of online work because of the requirement for an administrator to operate a machine occurs. The economy and the stock market plummet due to confinement and governments do not have the capacity to finance technological development. The GDP of nations falls significantly. In a vulnerable circumstance, the risk from cyberattacks enhances.

Solidarity movement is generated with 3D sanitary material printing however it is insufficient performance due to the poor development of this technology. The data (the contagions detected) are stored in the cloud and they tend to be considered for preventing a future pandemic, despite they are not genuine in order of magnitude. Communication among individuals and nations would likewise be secured covered because of correspondence communication networks and mobile telephony networks. Industry 4.0 standard covers a super communicated but not integrated and intelligent society. The technology exists yet is not executed or financed.

3.2 Status Quo

McKibbin & Fernando consider attack rates (share of the population who become sick) in the range 1-30% and case-fatality rates (share of those infected who die) in the range 2-3% [5]. Industry 4.0 technologies have the capability of providing better-computerized solutions for our daily lives during this crisis. Online commerce and the voluptuous industry fulfill demand. There is no lack due to the transformation of operability in smart factories (environments where are used to make perfumes are transformed to manufacture hydroalcoholic gel during COVID-19 pandemic period). The possibility of teleworking is limited, and remote medical assistance or control of population movements are a reality thanks to the IoT. Exists a pure state of the Industry 4.0 concept neutral (neither in advance nor in reverse bias). It could be argued in this circumstance an acceleration of the process of adoption of Industry 4.0 to a situation of abnormality, passing to a more mature state of the German concept.

3.3 Positive Scenario

Verity et al. evaluated for this situation a disease casualty proportion of 0.145% (0.08-0.32) for individuals younger than 60, and 3.28% (1.62-6.18) for individuals matured at least 60 [4]. Pandemic regulation is practiced rapidly because of valuable data store in the cloud from past pandemics and legitimate preventive and control of this infection. The breakdown is avoided owing to the investment funds for Industry 4.0 and the development of new technologies, which is the first step towards communication integration and distribution.

Society would thus be able to get closer to Industry 5.0 and therefore it would be immersed in awareness mentality getting closer to the super-intelligent population, almost involved in the Japanese concept of society 5.0.

4 Conclusion

Going forward, the medical industry would grow and the requirement of digitization of certain sectors is indispensable. An intelligent, centralized, and digitized healthcare system mitigate the risks of collapse and guarantees coverage of remote populace areas. Emphasis is placed on the development of smart manufacturing in order to assemble, conveyance, archive, and analyze data framework. Advances in Industry 4.0 will provide infrastructures to control the proper isolation of the patients, accelerating the medication fabricating and treatment process control. Along with, "formulation of national strategies and integration of the government's promotion system". Counting the architecture of a 'practical IoT system' and a think-tank function [2]. Regulated and centralized administrative digitization and the breaking of the digital cleavage are essential challenges to face the situation.

The consequences will rely upon the implication of the different elements and circumstances. The prerogative society is submerged in is a pandemic scenario, so consequently, recommendations so as to fortify the entrepreneurial framework in case of a resurgence of similar situations, are designed. Industry 4.0 is positioned at the epicenter of the study to improve production capacity and trace a channel to future developments. Given this circumstance, it is anticipated a noticeable requirement to fortify the execution of the existent technology (it has been evidenced that the impact of the crisis on the pandemic has been substantially less aggressive in automated companies due to their ability to production adaptation). The need to reduce the risk factors mentioned above as far as possible so that to provide stability for a favorable sustainable environment is pivotal. Hence, persuading the cooperation of the public and private sectors is indispensable in a progressively and successful environment. To end up, and parallel that, reinforce intra-community network financing projects is a vigorous motor of changes in order to further the potential of the individual-technology relationship in fostering the enhancement of the quality of life of all people through a super-smart society [6].

References

1. Espejo, V. *Megatendencias: un análisis del estado global (Análisis de tendencias del mundo)*. C.N. P. E (2016).
2. Keidanren. Society 5.0: The big societal transformation plan of Jap.
3. HSRC. The U.S. and China Invest Heavily in Industry 4.0 Technologies to be the World's Largest Manufacturer.
4. Verity, R., Okell, L. C., Dorigatti, I., Winskill, P., Whittaker, C., Imai, N., Cuomo-Dannenburg, G., Thompson, H., Walker, P. G. T., Fu, H., Dighe, A., Griffin, J. T., Baguelin, M., Bhatia, S., Boonyasiri, A., Cori, A., Cucunubá, Z., FitzJohn, R., Gaythorpe, K., … Ferguson, N. M. Estimates of the severity of coronavirus disease 2019: A model-based analysis. The Lancet Infectious Diseases. (2020)
5. Del Rio-Chanona, R. M., Mealy, P., Pichler, A., Lafond, F., & Farmer, D. Supply and demand shocks in the COVID-19 pandemic: An industry and occupation perspective. (2020).
6. Serpa, S., & Ferreira, C. Society 5.0 and Social Development [Preprint]. (2018)

3D finite element analysis of Al7075-T6 drilling with coated solid tooling

Anastasios Tzotzis[1], *Angelos* Markopoulos[2], *Nikolaos* Karkalos[2] and *Panagiotis* Kyratsis[3,*]

[1]University of Zaragoza, Department of Design and Manufacturing, C/Maria de Luna 3, 50018 Zaragoza, Spain
[2]National Technical University of Athens, School of Mechanical Engineering, Heroon Polytechniou 9, 15780 Zografou, Athens, Greece
[3]University of Western Macedonia, Dept. of Product and Systems Design Engineering, Kila Kozani, GR50100, Greece

Abstract. Due to the fact that simulation of drilling was added in commercial finite element analysis (FEA) software only recently, 3D finite element modelling is an invaluable asset during related researches. The present study employs 3D FEA to model the drilling process of Al7075-T6 alloy with solid carbide tooling, investigates important phenomena that occur during drilling and finally compares the simulated results with experimental data. A number of simulations were performed with DEFORM3D™ software at different cutting conditions; cutting speed of 50m/min, 100m/min, 150m/min and feed of 0.15mm/rev, 0.20mm/rev, 0.25mm/rev. The proposed model takes into consideration certain aspects like damage initiation and evolution of the material, contact interface between the drill bit and the workpiece and standard boundary conditions. Eventually, the acquired numerical data for thrust force were compared to the experimental results for the same cutting conditions and parameters. To obtain the experimental data, a series of nine drilling tests were performed. Upon validation of the numerical data, the temperature distribution on the tool tip – workpiece interface, as well as the chip morphology (shape and curling radius) were determined. Results showed a good agreement between the numerical and the experimental data. Specifically, thrust force and chip morphology exhibited an agreement of about 95% and 90% respectively, which confirms the potential of 3D FEA implementation on machining investigations.

1 Introduction

Since drilling is one of the most used machining process, its study draws the attention of many researchers worldwide. Especially, the implementation of modern techniques and tools such as the finite element method (FEM) during the research of drilling is a topic with increasing interest.

Lotfi et al. [1] developed a 3D finite element (FE) model to investigate the heat and wear on the drill faces during drilling process of AISI 1045 steel. Moreover, authors verified the

* Corresponding author: pkyratsis@uowm.gr

simulated model with experiments and evaluated the surface roughness and built-up edge. Dou et al. [2] developed a new constitutive model to improve a model for prediction purposes of generated thrust force and torque in drilling. Additionally, a 3D FE model for simulating the drilling process of SiCp/Al6063 was established by implementing the new constitutive model. Finally, authors compared the numerical results with the experimental observations in thrust force, torque, and chip morphology. Similarly, Xiang et al. [3] analyzed in their work the micro mechanisms underlying the wear of macroscale tools during diamond machining of SiCp/Al6063 composites. Moreover, they developed the mechanism-based diamond wear model in relation to the dominant wear behaviors. Majeed et al. [4] investigated the effect of drill bit temperature on tool life of twist drills during drilling of AISI 4340 high-strength steel both numerically and experimentally. The 3D FE modelling of the hard drilling process was performed with DEFORM3D™ software. Oezkaya et al. [5] presented a 3D FEM simulation model to predict the performance of modified flow drilling tools. In addition, to validate the simulation model, authors investigated the flow drilling of AlSi10Mg with a non-pre-heated and pre-heated tool as well. Zhu et al. [6] conducted a new 3D schematic to fully understand the chip formation and morphology in dry drilling of Ti6Al4V alloys in order to supply a better understanding of chip formation theory and metallurgical properties in chips.

Even though many researchers in the past few years utilized 3D FE modelling for investigation purposes of machining, 3D analysis of alloy drilling is a topic with multiple aspects that require thorough examination. In this paper, 3D FEA is implemented in order to study the behavior of Al7075-T6 alloy during conventional drilling process with coated solid carbide tools. In particular, the thrust force, the tool tip – workpiece temperature distribution and the chip cone radius were determined.

2 Methodology

2.1 Experimental process setup

A set of nine drilling experiments were carried out with the utilization of HAAS VF1 CNC machining center. A plate of Al7075-T6 alloy with dimensions 150mm x 130mm x 15mm was used as the workpiece. A 10mm solid carbide drill from Kennametal with catalogue number B041A1000CPG (KC7325 grade), was used to perform the experiments. The drills of this series are double coated; they have a multilayer of TiN/TiAlN with 3.5 micron thickness and an outer layer of TiN with 15 micron thickness. Figure 1 illustrates some of the most important parameters of the tool. Finally, in order to measure the produced thrust forces, a Kistler type 9257B 3-component dynamometer was configured.

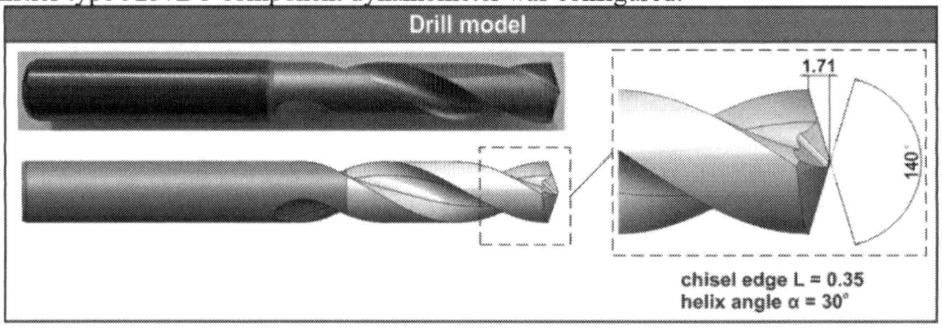

Fig. 1. Basic geometric characteristics of the B041A1000CPG drill.

2.2 Simulation setup

The 3D simulations of the nine drilling experiments were performed with DEFORM3D™ ver. 11.3 simulation software by maintaining the same execution order as the experiments. Moreover, the simulations were stopped when the thrust force achieved steady state for time saving purposes. The model of the workpiece was designed to be circular and thin, with a thickness equal to the radius of the drill. To further reduce the simulation time, a center drill spot based on the point angle of the tool was created on the workpiece (Figure 2a).

The workpiece was modelled as plastic and meshed with tetrahedral elements. The minimum element size of the mesh was set to 50% of the feed [7]. Furthermore, the mesh at the center of the workpiece was refined with a ratio of 10:1 (Figure 2a), so that the contact interface between the tool tip and the workpiece surface can be optimized. Due to the fact that the area located at the center of the workpiece is where high temperatures, large strains and deformations occur, an adaptive remeshing method was employed. In this way, the simulation time was improved as well as the formation of the chip.

To simplify the tool modelling, only the tip of the tool was used. It was modelled as rigid and meshed with approximately 25000 tetrahedral elements. A finer mesh with a 4:1 ratio was applied closer to the tip (Figure 2b). Finally, both the model of the workpiece and the drill were converted to STL file type prior to importing to DEFORM3D™.

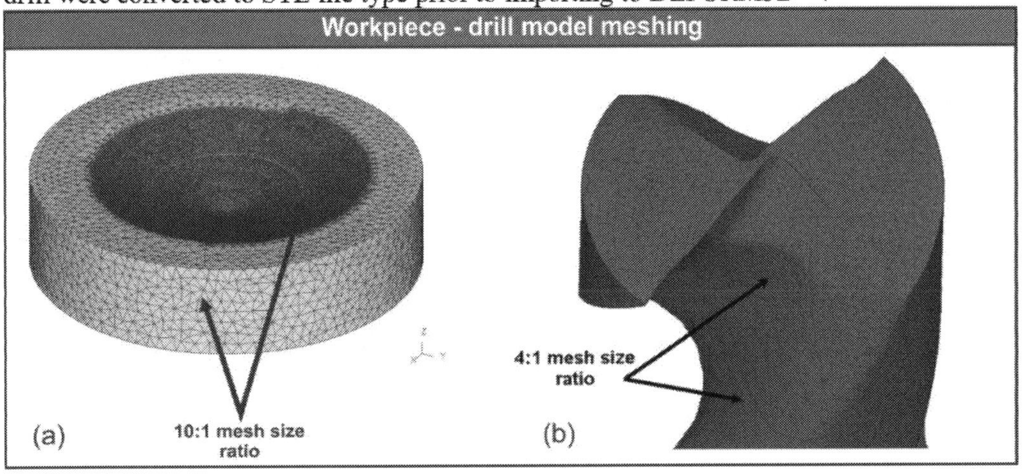

Fig. 2. The workpiece – drill model mesh setup.

In addition to the normal meshing of the tool, an extra layer of mesh was applied, so that the double coating of the drill can be taken into account to the numerical results. The basic properties of Al7075-T6 alloy, TiN and TiAlN coatings are presented in Table 1.

Table 1. Basic properties of workpiece material [8] and tool coatings.

Material	Al7075-T6	TiN	TiAlN
Young's modulus (GPa)	71.7	60	44
Density (kg/m^3)	2810	5220	5220
Poisson's ratio	0.33	0.25	0.23
Thermal expansion (μm/m°C)	2.2x10^{-5}	9.4x10^{-6}	9.2x10^{-6}
Thermal conductivity @ 20°C (W/mK)	130	25	12

The generalized Johnson - Cook model was selected for the simulation of the flow stress of Al7075-T6 during drilling since in such operations deformation is high. This model takes into consideration the effect of three parameters: the effect of strain, strain rate and temperature. The model can be described by equation 1.

$$\sigma = \left(A + B\varepsilon^{n}\right)\left(1 + C\ln\frac{\dot{\varepsilon}}{\dot{\varepsilon}_0}\right)\left[1 - \left(\frac{T - T_0}{T_m - T_0}\right)^{m}\right] \quad (1)$$

The parameters that form equation 1 are: A (initial yield stress), B (strain hardening modulus), C (strain rate dependence coefficient), ε (plastic strain), n (strain hardening exponent), m (thermal softening coefficient), $\dot{\varepsilon}$ (plastic strain rate), $\dot{\varepsilon}_0$ (reference plastic strain rate), T (reference temperature), T_0 (bulk temperature of workpiece material) and T_m (melting temperature of workpiece material). Table 2 contains the values of the parameters and constants for Al7075-T6. In this study, a reference strain rate of 1/s was used.

Table 2. The constitutive model parameters and constants for Al7075-T6 [9].

A (MPa)	B (MPa)	C	n	m	T_0 (°C)	T_m (°C)
546	678	0.024	0.71	1.56	20	635

To simulate the phenomenon of material separation, the damage model of Cockcroft and Latham was employed [9]. In addition, to define the tool - chip interface, a hybrid model (Shear – Coulomb) [10] was used for friction. This model approximates the developed frictional stresses around the tool tip and at the same time describes the friction across the sliding zone. For this study, the constant shear friction coefficient is m = 0.7 and the constant Coulomb friction coefficient is $\mu = 0.6$.

The nodes at the side of the workpiece were fixed in all three axes (X, Y, Z) so that velocity is zero and thus the workpiece is treated as stationary. On the other hand, a constant rotation around Z axis and a constant feed towards –Z axis were assigned to define the rotational and translational movement of the tool. The boundary conditions that relate to the heat exchange with the environment were defined and in addition, heat transfer through convection between the workpiece and the tool tip surface was defined. The default convection coefficient for dry cutting was used with a value of 0.02 N/sec/mm/°C. Moreover, to approximate the heat transfer through conduction in the tool - workpiece interface, the default coefficient was used with a value of 45 N/sec/mm/°C.

3 Experimental and simulated results

3.1 Assessment of thrust force

The nine simulations were performed on a desktop PC with dual-core CPU, 16GB RAM and SSD hard drive. The produced values of thrust force for each set were exported to Excel™ for processing and comparison with the experimental results. Figure 3 depicts a sample thrust force diagram (lower left diagram) at cutting conditions of V_c = 50m/min and f = 0.15mm/rev. It is concluded that thrust force increases rapidly and reaches a steady state quickly. In this case, the time before thrust force stops to increase and steady state begins is approximately 0.0171sec. Similar trend was observed in all simulations. The values of thrust force varied between 370 and 605N for the aforementioned cutting conditions, whereas the mean value was calculated roughly 500N.

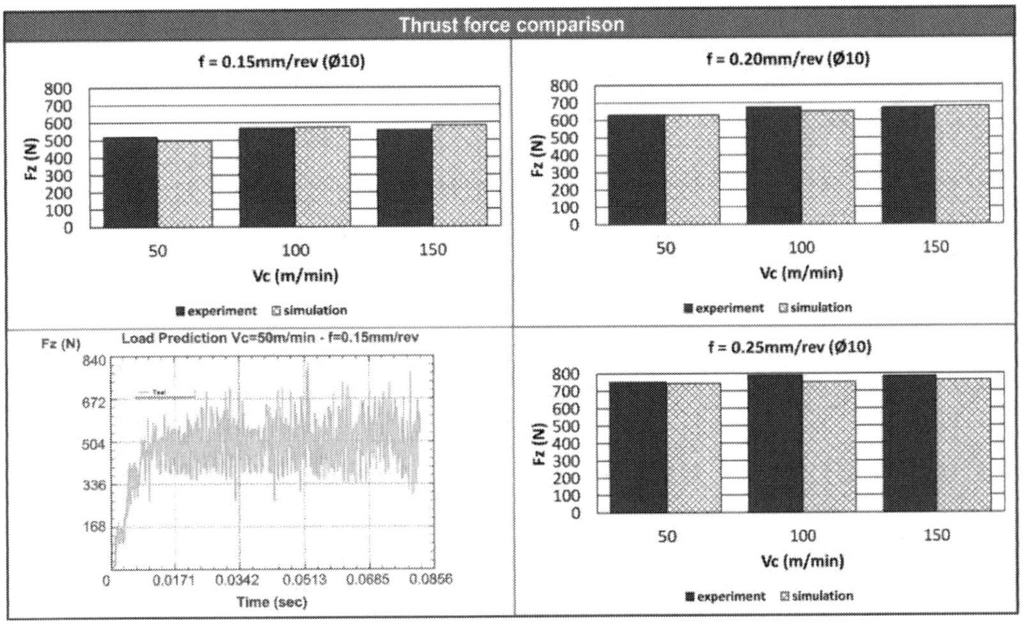

Fig. 3. Comparison between the experimental and the simulated thrust forces.

With the process of the simulated thrust force results and the acquisition of the experimental ones, a comparison between them was made. In addition, to the sample thrust force diagram, Figure 3 includes three bar charts that display this comparison for each of the three feed values.

In general, the results of thrust force are in high accordance with the experimental values. The agreement percentages vary between 96 and 102%, whereas in the cases with V_c = 50m/min and f = 0.20 and 0.25mm/rev the percentage reaches 99%. The maximum deviation between the experimental and the numerical results was found to be in the simulation with V_c = 100m/min and f = 0.25mm/rev. The default first order exponential smoothing was applied to the results and the standard deviation was calculated between 20 and 75N. Specifically, the minimum value of standard deviation was found on the simulation with the next conditions: V_c = 50m/min and f = 0.15m/min, whereas the simulation test with cutting speed of V_c = 50m/min and feed of f = 0.25mm/rev exhibited the maximum value of standard deviation. Furthermore, the mean value of standard deviation for all nine simulation tests was determined 52N.

3.2 Assessment of chip formation and temperature distribution

The formation of chip in drilling of Al7075-T6 follow a similar pattern for both the experimental and the simulated procedure. Figure 4a depicts the experimentally generated chip at cutting speed of V_c = 50m/min and feed of f = 0.15mm/rev. It is noticeable that the chip has a conical shape which was maintained on all experiments despite the change in feed and cutting speed. Additionally, the same chip morphology was observed for all simulation tests (Figure 4b). Moreover, the diameter of the chip's curling was found to be in good agreement between the experimental and the simulated one. As depicted in Figure 4b, it was determined approximately 5.33mm for the simulation and 6.05mm for the experiment, for the same cutting conditions and after approximately 4500 steps (this is when a full curling of the chip formed). The agreement percentage for this case is 86.5%.

Lastly, temperature distribution on the tool tip surface was introduced, due to the heat transfer between the tool and the workpiece. Figure 4c illustrates the temperature distribution on the 10mm tool tip (B041A1000CPG) after about 8182 steps (≈ 0.1mm penetration depth). It is obvious that the maximum temperature of around 67.9°C is located on the cutting lips. A slight increase in maximum temperature was observed as higher values of cutting speed and especially feed were chosen. The simulations were intentionally stopped after a number of steps in order to avoid very long completion times. Before a simulation was halted, it was allowed to reach steady state and ensured that it was maintained for a reasonable period.

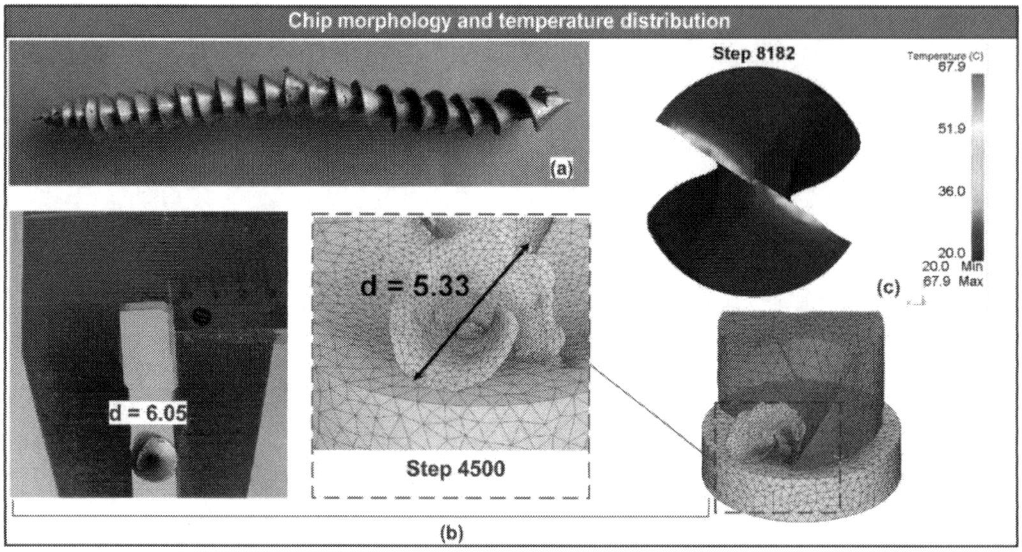

Fig. 4. Chip morphology comparison and temperature distribution on the tool tip.

4 Conclusions

Nine 3D simulations of Al7075-T6 drilling process with a 10mm solid carbide tool were performed under different cutting conditions; cutting speed between 50 and 150m/min and feed between 0.15 and 0.25mm/rev. The simulated results were validated with the implementation of experimental testing and the findings demonstrated an increased correlation between the simulated and the experimental results (between 96% and 102%). Moreover, the morphology and the dimensions of the chip during formation further confirm the validity of the 3D FE model presented in this paper. Subsequently, the findings of this study can lead to the following conclusions:

- An increase in cutting speed can increase thrust force by a small value.
- On the other hand, an increase in feed have significant impact in the produced thrust force.
- The generated chip morphology retains a typical conical shape which do not alter with a change on the cutting conditions.
- The maximum developed temperature on the drill is located on the cutting lips.

References

1. M. Lotfi, S. Amini, and I. Y. Al-Awady, Adv. Manuf., **6**, 2, pp. 204–214, (2018)
2. T. Dou, H. Fu, Z. Li, X. Ji, and S. S. Bi, Int. J. Adv. Manuf. Technol., **103**, 1–4, (2019)
3. J. Xiang, S. Pang, L. Xie, F. Gao, X. Hu, J. Yi and F. Hu, Materials (Basel), **11**, 2, (2018)
4. A. Majeed, A. Iqbal, and J. Lv, Int. J. Adv. Manuf. Technol., **95**, 5–8, (2018)
5. E. Oezkaya, S. Hannich, and D. Biermann, Int. J. Mater. Form., **12**, 3, (2019)
6. Z. Zhu, K. Gao, J. Sun, J. Li, Y. Liu, L. Chen and Y. Zheng, J. Manuf. Process., **34**, pp. 531–539, (2018)
7. SFTC, "DEFORM V11.3 (PC) Documentation." (2016)
8. P. Kyratsis, A. Markopoulos, N. Efkolidis, V. Maliagkas, and K. Kakoulis, Machines, **6**, 2, (2018)
9. M. G. Cockcroft and D. J. Latham, J. Institue Met., **96**, pp. 33-39, (1968)
10. İ. Ucun, J. Mech. Sci. Technol., **30**, 4, (2016)

MATEC Web of Conferences **318**, 01039 (2020) https://doi.org/10.1051/matecconf/202031801039
ICMMEN 2020

Design of 3D printed smart material compatible hand prosthesis

Abdalla M. Omar[1,*], and *Mohamed* Hassan[2]

[1]Department of Mechanical, Aerospace and Civil Engineering, University of Manchester. abdalla.omar@manchester.ac.uk
[2]Department of Mechanical, Aerospace and Civil Engineering, University of Manchester. mohamed.hassan@manchester.ac.uk

Abstract. Every year there are about 3500-5200 people suffering from upper limb amputations, most of which are wrist disarticulation and transcarpal. This paper investigates current upper limb prostheses and presents the disadvantages of current prostheses, including limited degrees of freedom (DOF), limited range of motion, weight, customizability, and appearance. The proposed design is the first stage of a series of papers that proposes designs that are compatible with shape morphing materials. The use of these materials as actuators allows the development and design of more advanced upper limb prostheses. Therefore, the prosthesis is modelled as needed for patients with transcarpal/wrist disarticulation amputations. The proposed model has 27 degrees of freedom (DOF), reduced weight, low cost, improved appearance, and is printable to fit.

1 Introduction

Orthopaedic prosthetic devices are external medical devices designed to substitute missing skeletal parts of the body. According to the World Health Organisation (WHO), more than 40 million people need a prosthetic device, but only 5-15% of them will have access to one [8]. Globally, the orthopaedic prosthetic devices market is expected to grow at a rate of 5% till 2027 and it is expected to represent 3 billion dollars by 2027. Among the different orthopaedic prosthetic devices, hand prostheses are the most common type of devices for work-related amputations worldwide.

The design and manufacture of upper limb prosthetics depends on the amputation case. Figure 1 highlights the nomenclature used for different upper limb amputation cases. Two classes of amputation are considered. The first corresponds to amputations that take place through the length of the bone such as Transhumeral (b), Transradial (d) and Transcarpal (f). The second class corresponds to amputations at the ends of the bones such as Shoulder disarticulation (a), elbow disarticulation (c) and wrist disarticulation (e) [2]. A study reports that the upper limb amputations: 16% trans-humeral, 12% trans-radial, 2% forequarter, 3% shoulder disarticulation, 1% elbow disarticulation, 2% wrist disarticulation,

* Corresponding author: Abdalla.Omar@manchester.ac.uk

61% transcarpal, and 3% bilateral limb loss [2]. Therefore, the case of interest will be transcarpal and wrist disarticulations.

Functionality, cost, and lack of personalisation are the main problems associated to commercially available lower limb prosthesis. Additive manufacturing, also known as 3D Printing, emerged as the ideal technology to produce affordable personalised prostheses. This technology, which allows the fabrication of prosthesis based on the patient's anatomy, has been used to fabricate "passive" devices equipped with DC motors, making the final prosthesis heavy, costly and limited in terms of functionality [6].

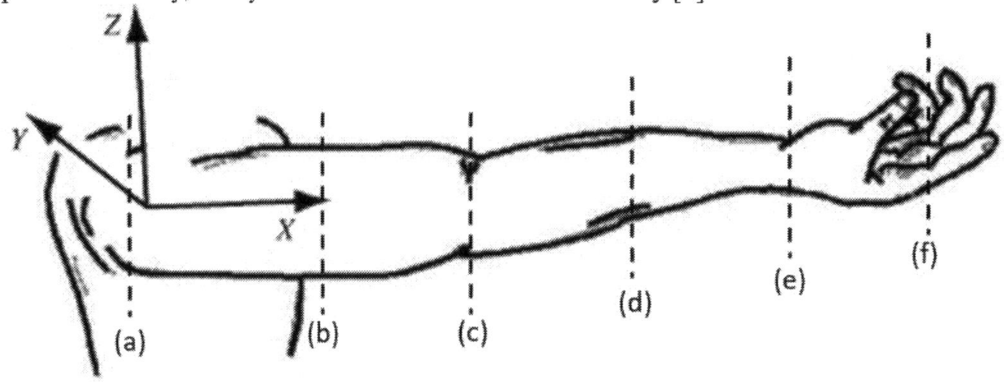

Figure 1. Model of a human hand with different amputation locations[3]

At the University of Manchester, we started a major project aiming to 4D print a full functional upper limb prosthesis. This paper presents preliminary results related to the design, materials, and degrees of freedom of the upper limb prosthesis that will be built on with the use of electro-active shape changing materials.

2 Methodology

2.1 Design

The wrist has 2 possible movements for the transverse axis, flexion, and extension, and in the longitudinal plane there is ulnar and radial flexion. All these movements depend on the carpal joints along the ulna and radius. The hand has also the same movements, and defined ranges of motion as shown in Figure 3. Based on these movements it is important to define an overall model of the hand's kinematics, critical to design a hand prosthesis. Different models considering different degrees of freedom (DOF) have been proposed [1,4,7,10]. Where the more degrees of freedom are considered the lower is the error when using specific grasps (e.g. circular, prismatic). Therefore, a model considering 24 DOF and providing very accurate dexterity and manipulation is commonly used [4,10]. Whereas, Morecki proposed a novel kinematic model of the hand based on 22 DOF, including the DOFs of the hand and 3 DOF for the wrist [1,7]. This is due to an assumption where any motion which is </5 degrees is a slack and that motion is disregarded. This is the model considered in this research work due to its simplicity and rigour.

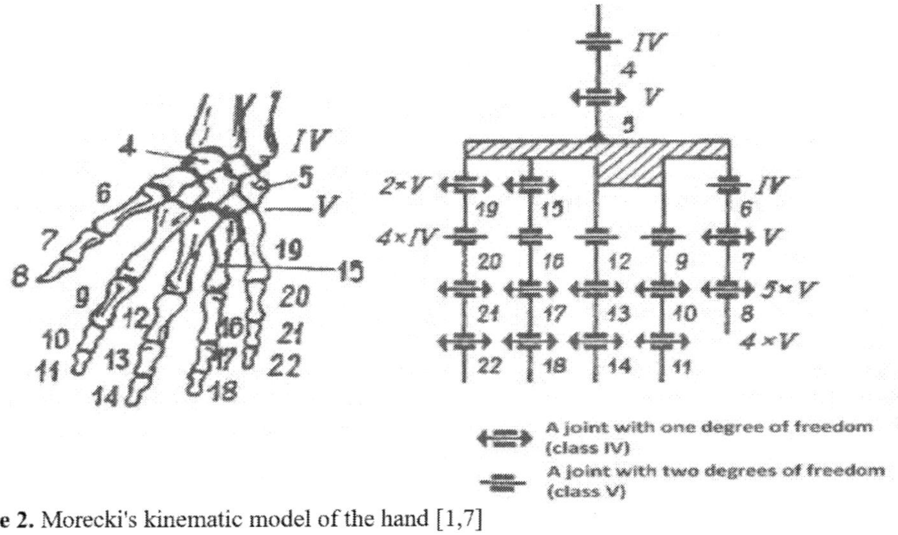

Figure 2. Morecki's kinematic model of the hand [1,7]

The hand prosthesis was designed in Dassault systems-Solidworks considering some anatomical measurements were used from literature and then scaled respectively to produce the hand. The hand-wrist connection has a ball-socket joint allowing the wrist to move in all directions. The metacarpophalangeal joints were modified to have 2 DOFs by considering a pivot joint but with a smaller accessory in the inside which allows abduction/adduction movements. The remaining finger proximal and distal joints are basic pivot joints with a small screw acting as a connector. The cad model of the designed hand model is presented in Figure 4.

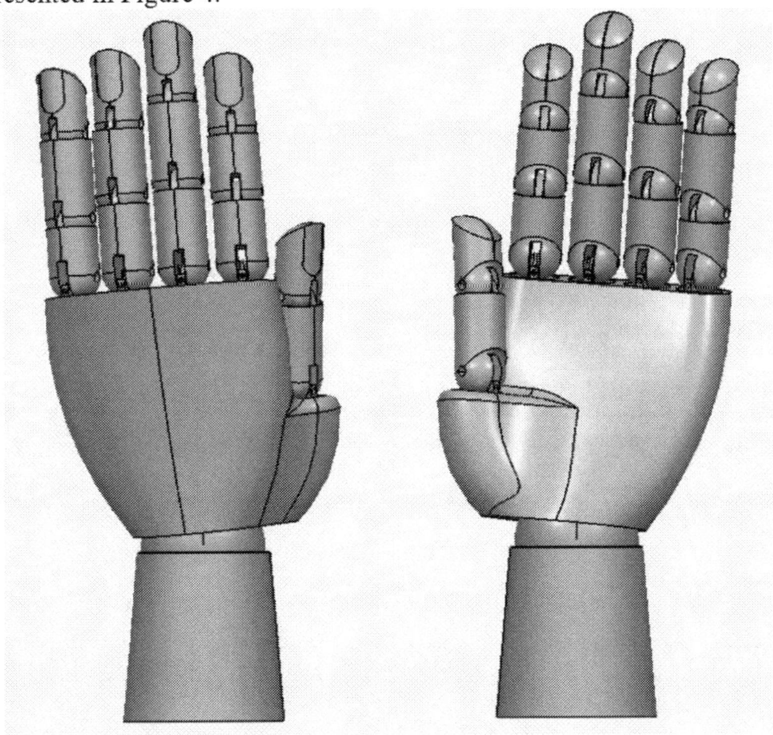

Figure 3. CAD model of the prosthesis; Dorsal view (left), Palmar view (right)

2.2 Fabrication

Moreover, SUNLU-PLA Plus dog bone structures were produced using Extrusion-based 3D printer. The Structures were designed using ASTM D638-14 standard. This standard is specific for testing the tensile properties of plastics. Structure was designed using Type V specimen dimensions, it was printed layer by layer with 100% infill. The same printing parameters were used to produce the hand; however, the infill percentage was reduced to 15% for the hand. The printing parameters are presented in Table 1 below:

Table 1. Printing parameters

Parameters	Values
Layer height (mm)	0.15
First layer height (mm)	0.27
Extrusion width (mm)	0.4 ± 0.05
Nozzle Temperature (℃)	205
Bed temperature (℃)	50
Printing speed (mm/s)	60
Travel speed (mm/s)	80
Infill (%)	100/15
Retraction length (mm)	1.5
Retraction speed (mm/s)	30

2.3 Mechanical tensile test

The mechanical properties of the PLA plus was evaluated using the uniaxial tension tests. The tests were carried out using an SAUTER TVL [Manual Crank Handle Test Stand] (Sauter GmbH, UK). The machine is fitted with a 500N load cell and two AC 01R Fine Point Clamps that are fitted to the platform and the loadcell. The 5 tensile parts were tested in a dry state to ~2% strain which is the yield point. The acquired data is presented in the form of a table alongside the respective average values.

3 Results and discussion

3.1 Design and fabrication

The hand model shown in Figure 4 has one ball and socket joint in the wrist with 3 DOF, five alternative pivot joint for all of the metacarpophalangeal joints each with 2 DOF, and 9 pivot joints for the remaining proximal and distal joints with 1 DOF. This results in a hand model with a total of 22 DOF. The length measurements of the hand are taken from the joints, and radius measurements are of all joints in the respective part. All of which are listed in Table 2. In addition, a sample of the printed index finger is presented in Figure 4.

Table 2. Length and radius measurements of the hand

Part	Length (mm)	Joint Radius (mm)
Thumb	70	9.1/9.7/9.7
Index finger	91.72	9.1/9.7/9.7
Middle finger	106.38	9.1/9.7/9.7
Ring finger	97.02	9.1/9.7/9.7

Little finger	91..61	9.1/9.7/9.7
Palm	99.43	33.61
Wrist	75.45	22.2
Overall	281.26	33.61

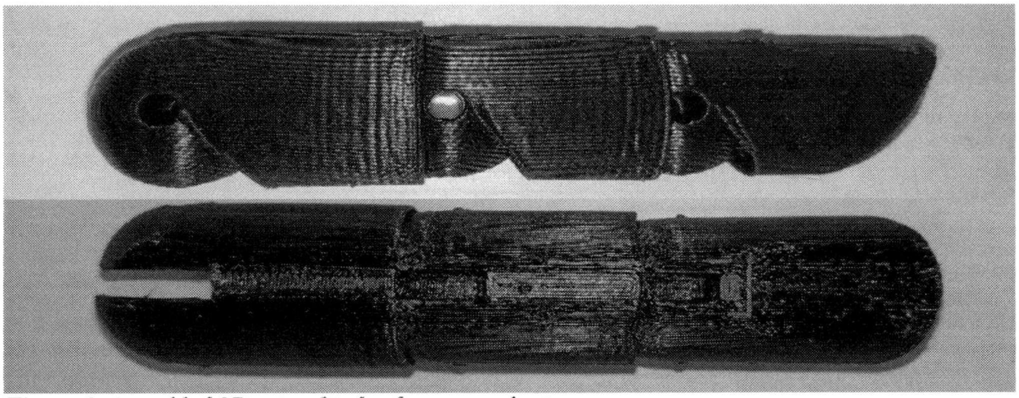

Figure 4. *Assembled 3D printed index finger prosthesis.*

Considering the use of PLA Plus as the 3D printer filament material, the weight of the full hand prosthesis is 220g including all the connections. The filament roll is 335meters and costs ~21£, and the length used to print the whole prosthesis is estimated to be 73.3 meters in length. Therefore, the material cost is approximately 4.60£. The operational costs can also be considered such as (e.g. printer cost, printing time, electricity, maintenance and more); However, there are no accurate models to do that, and so an approximation is made based on the following (printed object weight, printing time, electricity tariff, printing power, printer lifetime, time used) and an approximate value of 1.84£ is approximated during the printing of this prostheses.

3.2 Mechanical testing

Using the SAUTER TVL [Manual Crank Handle Test Stand] to test the dog bone specimens according to the ASTM D638-14 Type V standards were done. Through designing the specimen in the correct dimensions and were measured afterwards to make sure they did not have any defects or deviations from the designed model. Furthermore, the standard specifies the use of 5 different tests to acquire accurate and precise results as seen in Table 3. The testing stand was set to the peak setting to identify the maximum force when reaching the plastic region and was stopped at values of ~2% strain by calculating the displacement required beforehand.

Table 3. Tensile Mechanical properties of PLA Plus

Test	Force (N)	Displacement (mm)	Stress (MPa)	Strain (%)	Modulus (GPa)
Test 1	502.3	1.48	52.707	0.0233	2.261
Test 2	498.5	1.41	52.308	0.0222	2.355
Test 3	491.8	1.44	51.605	0.0226	2.275
Test 4	504.2	1.37	52.906	0.0215	2.452
Test 5	489.4	1.42	51.353	0.0223	2.296
Average	497.2	1.42	52.176	0.0224	2.328

However, there are multiple limitations in this approach as the use of a manual stand means that the strain rate is not fixed as when using an automated mechanical testing machine, which might result in some inaccuracies. Also, the point in which the tests were stopped are based on the hand calculations and approximations made to when necking might start in the small cross-section of the specimen.

The results mainly were used to evaluate the materials mechanical behavior and ensuring that a standard material like PLA can withstand forces up to 500N, as a maximum force of 538N must be withstood and the material used here starts to yield at that point [8]. Presenting a mostly safe elastic behavior in that range.

4 Conclusion

In conclusion, the work presented highlights the challenges and the desired design specifications by upper limb amputees for high quality prostheses. The significance of 4D printing is highlighted and how current DOF limitations can be ignored if such actuators are utilized. Then, the design, fabrication, and testing methodology are presented. The results highlight the 22 DOF achieved in this model using different joints, the cheap cost for a prosthesis skeleton of 6.44£, and a weight of 220g. Further work will build on this design by equipping Shape changing materials.

□

References

1. al., A.M.e., *Teoria mechanizmów i manipulatorów*. (2002)
2. A. Jette, C. Spicer, and J. Flaubert, (2017),
3. E. Pennestrì, R. Stefanelli, P. Valentini, and L. Vita, J. Biomech, **40**(6): p. 1350-1361, (2007).
4. F. Chen Chen, A.F., M. Mousavi, E.P. Ambrosio, S. Appendino, A. Battezzato, D. Manfredi, F. Pescarmona , B. Bona, ICES, (2011)
5. Flashforge, (2020)
6. L. Resnik, M. Meucci, S. Lieberman-Klinger, C. Fantini, D. Kelty, R. Disla, and N. Sasson, *Arch Physl Med Rehab*, **93**(4), pp.710-717, (2012)
7. L. Jaworski, and R. Karpiński, J. Technol. Exploit. Mech. Eng, **3**(1), pp.28-33, (2017)
8. M. Hichert, A. Vardy, and D. Plettenburg, *Prosthet and Orthot Int*, **42**(1), pp.84-92, (2017)
9. Organization, W.H., *S* (2015)
10. S. Cobos, M. Ferre, Uran Sanchez M.A, J. Ortego, and C. Pena, IROS, (2008)

The influence of chromium as a diffusive additive in the boronizing treatment of AISI 4140 steel on the corrosion resistance of the coating evaluated by Electrochemical Impedance Spectroscopy (EIS)

Dimitrios I. Zagkliveris[1,*], *Azarias* Mavropoulos[2], *Efstathios* Ntovinos[1], *Georgios* K. Triantafyllidis[1]

[1]Laboratory of Materials' Technology, Dept. of Chemical Engineering, Aristotle University of Thessaloniki, 541 24 Thessaloniki, Greece
[2]Physical Metallurgy Laboratory, Dept. of Mechanical Engineering, Aristotle University of Thessaloniki, 541 24 Thessaloniki, Greece

Abstract. A large variety of protective coating is being used in industrial applications to improve the resistance of the metallic substrates against corrosion. The pack-cementation method for boronizing and borochromizing is effective to produce extremely hard and corrosion resistant thick coatings and, additionally, is a low-cost and simple technique. In the present study, AISI 4140 steel specimens underwent boronizing and afterwards chromizing by the pack-cementation method using B_4C as boron source and Fe-Cr as chromium source, respectively. In both treatments the appropriate activators were used. After chromizing the boronized substrate, a mixed boride phase FeCrB was formed, as it was confirmed by X-ray Diffractometry (XRD). The boronized and the borochromized specimens were subjected to Electrochemical Impedance Spectroscopy (EIS). From the analysis of the frequency response of the coating systems (Bode and Nyquist display), the conclusion that the borochromized specimens were significantly more corrosion resistant was extracted. Finally, data of optical and electron microscopy contribute to the validity of the conclusions.

1 Introduction

A plenty of technical applications demand the utilization of materials that withstand intense wear and corrosion conditions such as frictional contact with other hard components and interaction with corrosive environments, probably in a wide temperature range.

One of the mostly effective ways for superficial protection of many metallic materials against the above referred destructive conditions is the boronizing treatment. Boronizing is a thermochemical diffusion-dominant process, in which boron atoms diffuse into the crystal

* Corresponding author: zagklive@cheng.auth.gr

lattice of the base metal, which is usually steel [1] or a titanium alloy [2]. As a result, a superficial layer of one or more boride phases of the base metal are formed. In steels as base metals, the boride phases are FeB and Fe_2B. Generally, the formation of a single-phase Fe_2B layer is desirable, because the presence of the FeB phase externally can lead to cracks due to the significantly lower thermal expansion coefficient and the internal stress agglomeration [3].

It is experimentally shown that the addition of another metallic element in the boride layer could result in an increase of the plasticity of the coating [4]. Another significant advantage, especially when chromium is used as additive, is the improvement of the resistance of the coating against corrosion [5].

The investigation of the borochromizing process as an optimized version of boronizing is an open field of research. In particular, experimental studies on chromizing a boronized steel (boronizing + chromizing) [6-8], boronizing a chromized steel (chromizing + boronizing) [9,10] and the simultaneous borochromizing [11-13] are referred in the literature. In some of these cases, the examined property of the borochromized materials was resistance against friction [6,10] and the corrosion resistance [12].

In the present study, the boronizing + chromizing technique was selected as the most efficient against corrosion, as it was shown from a series of preliminary experiments and as it is confirmed in [5]. The borochromized specimens were tested in Electrochemical Impedance Spectroscopy (EIS) using an aqueous solution of 3 wt. % NaCl as electrolyte. Their score in EIS method was compared with the simply boronized specimens. Both treatments were carried out by the pack-cementation technique. The last decade EIS became a valuable method for evaluating the corrosion resistance of boride [14] and other types of coatings on metallic materials [15-18].

2 Materials & methods

In the preliminary stage of the study, specimens of AISI 4140 steel were boronized and then half of them underwent chromizing. Both thermochemical treatments were carried out using the pack cementation technique. The chemical composition of the AISI 4140 steel is presented in **Table 1**.

Table 1: Chemical composition of AISI 4140 steel

Carbon	Silicon	Manganese	Chromium	Molybdenum	Phosphorus	Sulfur
0.38-0.43	0.15-0.35	0.75-1.00	0.80-1.10	0.15-0.25	≤0.035	≤0.040

The reacting powder which is used in boronizing treatment consisted of boron carbide (B_4C, 230 mesh, Sigma-Aldrich), potassium tetrafluoroborate (KBF_4, powder, Alpha Aesar, >99.5%) and silicon carbide (SiC, 400 grit, Struers). Moreover, the chromizing powder consisted of ferrochrome (Fe, 325 mesh, Acros Organics, 99% και Cr, powder, Acros Organics, >99%), ammonium chloride (NH_4Cl, powder, Merck, 99.8%) and aluminum oxide (Al_2O_3, powder, Panreac, >99.5%). The composition and purpose of use of each component are presented in **Table 2**.

For the thermochemical treatments, a conventional CARBOLITE CWF 1300 furnace was used and the experiments were carried out in an argon atmosphere. After boronizing or borochromizing treatment, the specimens were examined with X-ray Diffractometry (XRD) with a BRUKER D8 device, scanning electron microscopy (SEM) with energy-dispersive X-ray spectroscopy capability (EDS) and finally with electrochemical impedance spectroscopy (EIS).

Table 2: Experimental details of the boronizing and chromizing treatments

Boronizing (950°C for 8 h)			**Chromizing (1050°C for 6 h)**		
Powder composition					
Component	Content(wt.%)	Purpose of use	Component	Content (wt.%)	Purpose of use
SiC	80	Filling material	Al_2O_3	69	Filling material
B_4C	10	Boron source	Fe-Cr (50/50)	25	Chromium source
KBF_4	10	Activator	NH_4Cl	6	Activator

3 Results & discussion

3.1 Boronizing & borochromizing processes

During the boronizing treatment, a compact dense single-phase Fe_2B coating was formed, as XRD measurements revealed (**Figure 1**). Also, a metallographic image of the coating is depicted in **Figure 1a**. One can observe that the coating layer displays the common "tooth"-like morphology with the elongated crystallographically oriented (textured) Fe_2B monocrystallines.

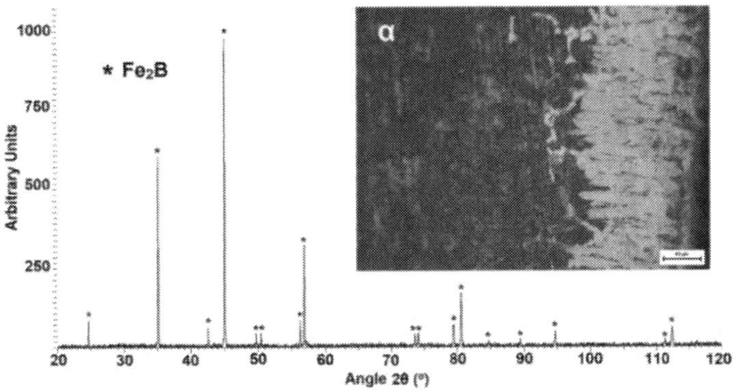

Figure 1: XRD spectrum and metallographic image of a boronized specimen

During the chromizing treatment, the previously formed Fe_2B phase was enriched with chromium up to half of its total thickness. As a result, an additional distinguishable borochromized layer appeared. Through XRD measurements the mixed boride phase FeCrB was recognized. Moreover, there is evidence of possible presence a chromium boride (Cr_5B_3) and traces of remaining elementary chromium (**Figure 2**).

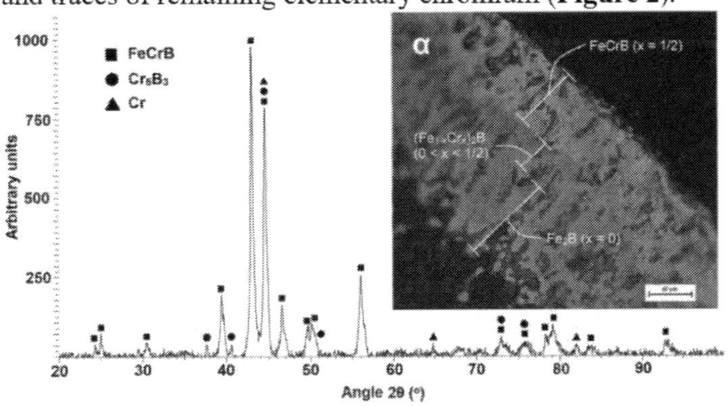

Figure 2: XRD spectrum and metallographic image of a borochromized specimen

The mixed FeCrB phase is the result of chromium substitutional diffusion into the Fe_2B crystal lattice. Cr atoms are able to replace Fe atoms with a diffusional mechanism, when the necessary chemical potential difference is present. As the superficial FeCrB phase indicates, the Cr→Fe substitution has an approximately 50% efficiency close to the surface. However, deeper into the coating layer, the substitution rate decreases till nihilism. The chromium substitution rate gradient across the depth of the coating was determined by SEM/EDS and is depicted in **Figure 3**.

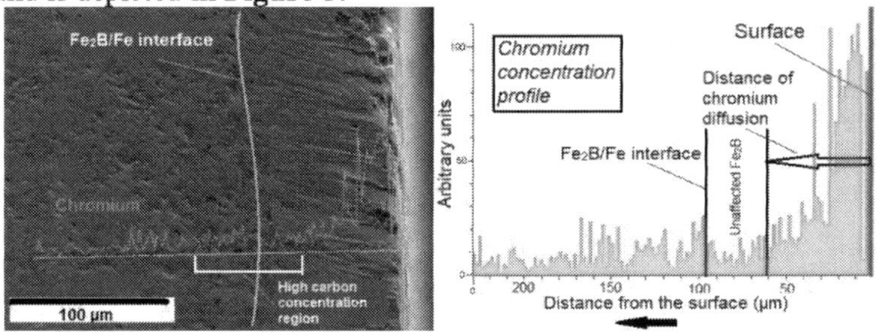

Figure 3: SEM/EDS analysis and chromium concentration profile of a borochromized specimen

The new chromium-modified Fe_2B coating can be formulated as: $(Fe_{1-x}Cr_x)_2B$ with $0 \leq x \leq ½$. The parameter x represents the chromium substitution rate, as it was described above. In accordance with this formula, when $x = ½$, the FeCrB phase results and, when $x = 0$, the formula is simplified to the initial Fe_2B phase (**Figure 2a**). Additionally, as it is shown mainly in **Figure 2a**, the "tooth"-like morphology is significantly modified during the chromizing process. This is probably due to local boron diffusion phenomena, because of the high temperature of the chromizing treatment. This local diffusional mobility could lead to the decrease of the interface energy at the Fe_2B/Fe interfaces. As an additional evidence of the local boron diffusion, the total thickness of the coating after chromizing is almost twice the thickness before (comparison between **Figures 1** and **2**), due to boron diffusion towards the metal substrate.

3.2 Electrochemical Impedance Spectroscopy

From the EIS experiments, data of real and imaginary part of impedance for each value of voltage frequency were collected. Bode and Nyquist diagrams were constructed, using these data (**Figure 4**).

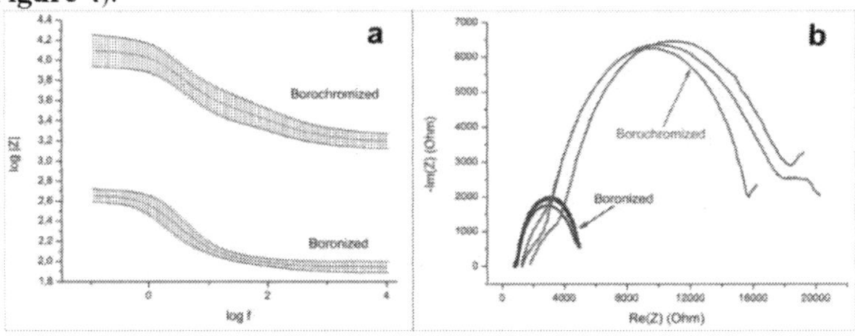

Figure 4: EIS analysis results. Bode (a) and Nyquist (b) diagrams

In **Figure 4a**, the depicted bands represent the statistical range of the measurements and the continuous line in the middle is the mean value. As it is obviously shown, the total impedance of the borochromized specimens (the |Z| in Bode diagram) is almost two orders of magnitude higher than the impedance of the boronized ones. This result leads to the safe conclusion that the chromizing treatment improves in a very high degree the impedance and, consequently, the corrosion resistance of the boride coating.

In full agreement with the Bode diagram, the significantly larger semicircular curves of the borochromized specimens in the Nyquist diagram (**Figure 4b**) indicate much higher ohmic resistance (real part of Z – x-axis) as well as higher capacitive resistance (reactance) (imaginary part of Z – y-axis) in comparison with the boronized specimens.

The EIS data underwent an equivalent circuit fitting by using appropriate software. In **Figure 5** the proposed equivalent circuits and the fitting results for two representative specimens are presented.

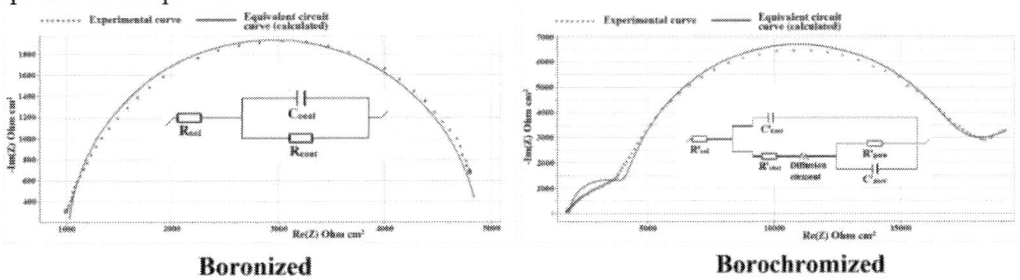

Figure 5: Equivalent circuits of the boronized and borochromized coatings

Both circuits consist of a main R-C loop, which corresponds to the coating's impedance (R_{coat}, C_{coat}). Another common element is the ohmic resistance of the electrolyte solution (R_{sol}). The experimental Nyquist diagram of the borochromized specimen (**Figure 5b**) consists of a small semicircle before the main one. This is attributed to the presence of pores with large tortuosity factor that can entrap corrosive substances in high concentration causing additional corrosion effect [14]. This pore system in the coating displays its own impedance and can be represented as a secondary R-C loop in the equivalent circuit (R'_{pore}, C'_{pore}). Moreover, a diffusion electric element is been added in the borochromizing equivalent circuit in order to simulate the increase of the right side of the curve.

The results of the comparative study between the boronized and the borochromized coatings via EIS are ambiguous. On the one hand, the borochromized coatings represent very higher coating impedance than the boronized ones, which means that they display higher initial corrosion resistance. On the other hand, the presumptive pore network and the diffusion phenomena, that the borochromized coatings display, constitute risk factors for the evolution of their resistance levels after a significant time period in a corrosive environment.

4 Conclusion

- The boronizing process at 950°C for 8 h on AISI 4140 steel results in the formation of a dense single-phase Fe_2B coating layer with approximately 80 μm thickness.
- When the boronized steel undergoes chromizing at 1050°C for 6 h, an external layer of the mixed boride phase is formed and the total coating doubles in thickness.
- The Electrochemical Impedance Spectroscopy showed that chromizing improves to a great extent the resistance of the boronized steel against corrosion.
- Although, after chromizing, a possible superficial pore network, which can facilitate intense local galvanic corrosion and ion diffusion, could lead to a decrease of the durability of the borochromized coating through time. These phenomena are absent from the simple boride coatings.

The authors would like to thank **prof. Eleni Pavlidou** (Dept. of Physics, A.U.Th.) for conducting SEM/EDS measurements and **Ms. Xanthi Dabou** for her contribution at XRD experiments.
This study was financially supported by **General Secretariat of Research and Technology (GSRT)** and **Hellenic Foundation for Research and Innovation (HFRI)** (Grant No: 530).

References

1. A.G. Matuschka, *Boronizing*, (Carl Hanser-Verlag, Munchen, 1978)
2. W. Fichtl, Mater. Des. **2** 276-286 (1981)
3. M. Rile, Met. Sci. Heat Treat. **16** 836-838 (1974)
4. J.R. Davis, *Surface Hardening of Steels: Understanding the Basics*, ed. J.R. Davis, (ASM International) 213-223.
5. P.A. Dearnley and T. Bell, Surf. Eng. **1** 203-217 (1985)
6. S. Sen and U. Sen, Ind. Lubr. Tribol. **6** 146-153 (2009)
7. S. V. Grachev, L. A. Maltseva, T. V. Maltseva, A. S. Kolpakov, and M. Yu. Dmitriev, Met. Sci. Heat Treat. **41** 465-468 (1999)
8. Yu. A. Balandin, Met. Sci. Heat Treat. **47** 103-106 (2005)
9. V. V. Shyrokov, Kh. B. Vasyliv, L. A. Arendar, and E. M. Rudkovskyi, Mater. Sci. **46** 440-447 (2011)
10. V. Samadi and A. Habibolahzade, Mater. Sci. Technol. **26** 41-46 (2010)
11. Suwattananont N. Solid State Sci. **14** 1669-1672 (2012)
12. N. Suwattananont and R. Petrova, Solid State Sci. **14** 1669-1672 (2012)
13. S. M. Chernega, Ya. V. Zaulichnyi, M. V. Karpets, and M. V. Belous, Powder Metall. Met. Ceram. **39** 594-598 (2000)
14. I. Campos, M. Palomar-Pardave, A. Amador, C. V. Velazquez, and J. Hadad, Appl. Surf. Sci. **253** 9061-9066 (2007)
15. B. R. Hinderliter, S. G. Croll, D. E. Tallman, Q. Su, and G. P. Bierwagen, Electrochim. Acta, **51** 4505-4515 (2006)
16. S. J. Cooper, A. Bertei, D. P. Finegan, and N. P. Brandon, Electrochim. Acta **251** 681-689 (2017)
17. S. J. Lee and S.I. Pyun, J. Solid State Electrochem., **11** 829-839 (2007)
18. S. Skale, V. Dolecek, and M. Slemnik, Corros. Sci. **49** 1045-1055 (2007)

Design of Multiple Mailboxes

Athena Baronos[1,*], *Odysseas* Manoliadis[2] *and Aristeidis* Pavlidis[1]

[1]University of Western Macedonia, Koila Kozanis 50100, Greece
[2]Democritus University of Thrace, Vas. Sofias 12 Xanthi 67100 Greece

Abstract. In today's world the design of multiple mailboxes comes to cover the evolution of logistics in delivering mail where the postman is not required to visit every user. In this research the 3D visualization is used for the design of multiple mailboxes for domestic use. It concerns the design of mailboxes in ergonomic building blocks and apartment complexes in 3D design so that they can be easily manufactured. Between the advantages of this design will be rapid production of ready-made products production of prototypes that enables testing at the design stage and reduces the time and the cost of production. The design when done with 3D CAD can be manufactured with modern machine tooling methods. In this paper after an extensive Literature Review the postal multiple mailboxes is used as a case study in the use of 3D CAD for 3D printing. A methodology is proposed that enables the examination of prototypes at the design stage according to specifications and allows the manufacturing department of a company to prepare the right tools and begin installing production lines. Conclusively this method gives the advantage of designing the product and supporting the production of scaffolds that can be functionally and ergonomically tested before finalizing the production.

1 Introduction

3D printing group can be utilized to rapidly create a scale model of a physical assembly utilizing three-dimensional Computer Aided Design (CAD) information. This is effectively done using computer aided design (CAD) for 3D printing [1]. 3D printing is an innovation that empowers people all over the world to plan custom items. Advertising effectiveness, customizability, and availability to reasonable generation, 3D printing is changing the flow of how we make things [2]. Nowadays, computer aided design (CAD) can design imaginative 3D printed items in a way that designers and engineers could previously only dream of producing [3].

The final thirty years 3D printing has been recognized as an effective and feasible innovation within the areas of advanced manufacturing. A recent research about of 3D CAD is conducted to advance 3D printing for a better performance of a companies' manufacturing and fabricating is presented in Figure 1 [4].

* Corresponding author: abaronos@gmail.com

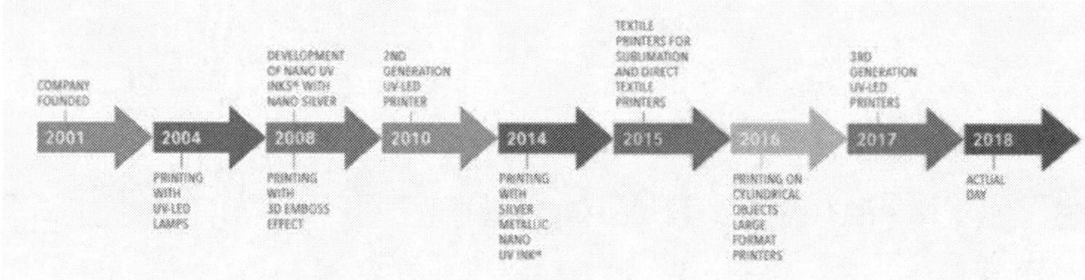

Fig. 1. Development of 3D printing for UV-LED Lamps[4]

This paper examines the application of 3D CAD to 3D printing in manufacturing. In order to understand the contribution of 3D CAD to 3D printing in the next paragraph few selected characteristics of 3D printing are described. Then the methodology used is described followed by a case study- A multiple mailboxes design to provide a comprehensive understanding of pairing 3D CAD to 3D printing.

2 3D Printing

At an essential level, 3D printing takes advanced representations of objects and makes them in physical shape by building up different layers of plastics, metals, powders, liquid resins, and other materials [5]. Individuals frequently allude to 3D printing as additive manufacturing to differentiate it with conventional "subtractive" fabricating methods, which include cutting, processing, or removing material to create an object [4].

A major portion of 3D printing's offer is that it permits for an exceptional degree of fast prototyping and mass customization [6]. Whereas media scope of 3D printing has regularly cantered on cheap plastic 3D printed products — such as smart phone cases— a few of the foremost critical progresses in 3D printing are taking put in trade applications 3D printing encourage openings for mass customization, perfection, and efficient flawlessness [7]. Already, major international breakthroughs in additive manufacturing are accelerating these trends leading to convergent applications.

Since late 1984 the first 3D printing appears. In 1988 it is available to the public after been patented in 1986 (Fig. 1). Stereo Lithography Apparatus (SLA) machine was developed and a sustainable resolution to reduce waste and production cycle has come around [8]. From the fabrication of complicated jet engine parts to simple cups, 3D printing is being regarded as an alternative and more efficient way to develop new product. Statistics show that the global market for 3D printing will be worth $16.2 billion by 2018, which is four times larger than that of 2014 [8].

Taken as a whole, 3D printing is transforming manufacturing and reconfiguring supply chains around the world. It is changing the way consumers get the products they need and shifting power to individuals with unique preferences [8].

3 Methodology

A framework for utilizing 3D CAD using a commercial computer-aided design (CAD) and drafting software application is proposed. The framework is a decision making tool presented in Figure 2. The basic steps of the procedure are:

Step 1 2D Sketches Considering Design Specifications
Step 2 Creation of Technical Drawings
Step 3 Creation of Technical Drawings Mesh & Files in AutoCAD
Step 4 Modelling 3D Objects
Step 5 3D Printing Process using either Polymer Ceramic or metal
Step 6 Production of Scaffolds
Step 7 Execute functional and ergonomic tests.
Step 8 If the tests are satisfactory proceed to the applications (**Step 9**). Otherwise go to step 4
Step 9 Application - Production

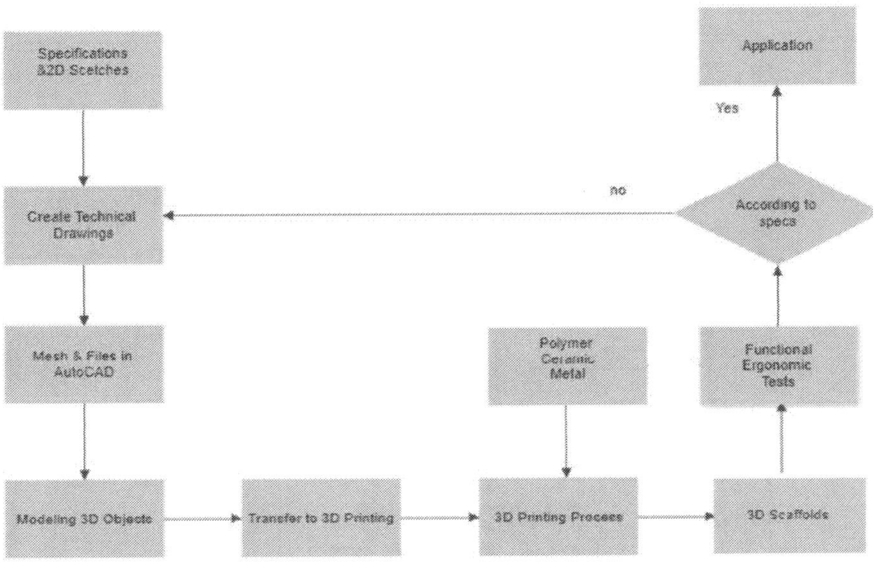

Fig. 2. CAD procedures in 3D printing for the case of multiple mailboxes.

4 Case Study - Multiple Mailboxes

Postal operators and nearly every organization that works in shipping, delivery, or other aspects of logistics should keep a very close eye on it. One of the uses of a multiple mailbox post relates to community housing, where there may be estates with multiple homes and there is no need to install individual mailbox posts for each house. Alternatively, one might own or lives in a property where one building is divided into several apartments or flats. In this case, multiple mailbox posts outside the building are ideal for ensuring that everybody can receive and access their own mail.

A multiple mailbox is installed into the ground and can withstand almost all weathers. On each side of the cast iron post are two mailboxes – securely fixed and with the respective neighbors' house numbers painted or identified in another way.

In urban areas the boxes are numerous but in rural areas the design works well if there are five to ten homes situated in a remote area or in a location that is difficult to access. Rather than forcing couriers and the mailman to come all the way to your front door, you can position a multiple mailbox post at the end of the road and affix several boxes to it.

The Postal Service has an unmatched last-mile conveyance arrange — no other organization covers as much domain as often as possible and frequently as the Postal

Service. It is regularly not fetched compelling for private conveyance firms to form partitioned stops to provide little, moderately cheap bundles, especially in rural zones.

However, the Postal Service visits these locations nearly every day. Accordingly, other delivery firms often use the mailboxes for final delivery of many packages. The Postal Service delivers nearly two-thirds of lightweight, commercial packages to their final destination [9]. Holding a competitive advantage in materials efficiency and less processing procedures, the technological and economic benefits of 3D printing has been widely reported and documented, and it became especially popular in the United States in the last decade Web of Science suggests that more than 12 papers have been published from 2005-2015 within the US, and it still presents a trend of fast increase in the basic theory exploration and technology development. Although several survey about 3D printing have been documented , almost all of them are written on behalf of its categories, processes, quality and cost benefits or focus on environmental impact of rapid prototyping technologies before 2006. The multiple mailbox concepts in delivering may be benefiting from the growth of 3D printing. Being different from traditional manufacturing processes, 3D printing going from designing to finishing, it saves the traditional processes of casting, forging and rough machining without reducing the final part quality. Difference of processes in 3D printing and traditional manufacturing is shown in Table 1.

Table 1. Comparison of 3d Printing and Traditional Manufacturing in Multiple Mailboxes

Items	3D printing	Traditional manufacturing
Manufacturing	Additive	Subtractive
Efficiency	higher	Lower
Production Waste	less	More
Built time	hours to days	Days to months
Cost	higher	Lower
Life cycle	Shorter	Longer

Besides the multiple mailboxes application of 3D printing is leading to dramatic changes in the fields of manufacturing and logistics of postal services, with major implications for postal operators and other transportation and delivery companies. In a July 2014 white paper, *If It Prints, It Ships: 3D Printing and the Postal Service* [10] it is described how the rise of 3D printing could significantly benefit the Postal. The Postal Service describes the developments in 3D printing and the logistics marketplace and examines what these changes could mean for the Postal Service that some parts used to assembly the final solution of a single mailbox. Multiple mailbox installations must meet the same criteria as single mailbox. The requirement precludes the use of heavy case and supporters that are based on the ground and consists the support system (Figure 3). In terms of the design of multiple mailboxes study results concerning details based on ergonomic study are as follows: The house number on a mailbox must be at least 3cm tall and positioned on the front or flag side of the mailbox (Figure 4). A lower section incorporates newspaper delivery box. The door or slot must be at the level of 41 to 45 inches from the ground. The final output of the design conducted by the authors of mailboxes is shown in Figure 3 together with a prototype of a commercial company for comparison (Figure 4).

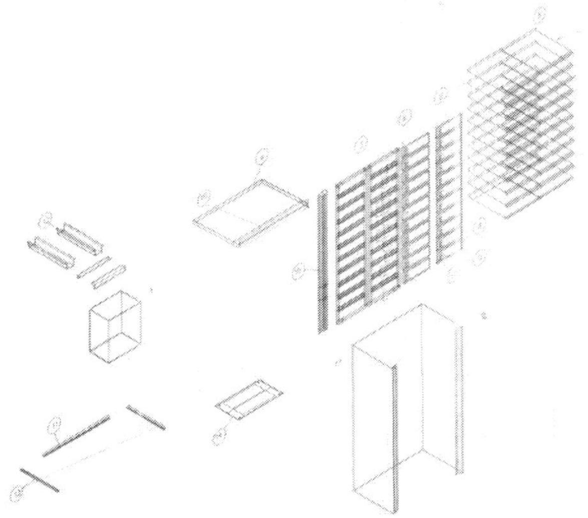

Fig. 3. Components of mailboxes produced designed in 3D CAD .

Fig. 4. Multiple mailboxes [9]

5 Conclusions

Construction of the part or assembly of multiple mailboxes is easily done using computer aided design (CAD) for 3D printing. It is used to create prototypes, models for plastics and metals. The information collected from comparison of tests and related references are at that point utilized to alter the model models and make strides the item or plan. CAD models are a significant instrument for quickly making cost-efficient plan adjustments some time before an item goes to fabricating where these changes will be greatly costly. Since the multiple mailboxes is state-of-the-art in keeping with usefulness and aesthetics and gotten from surveys custom-made to the necessities of inhabitants to coordinate single-family homes to expansive complexes of buildings 3D CAD gives the interesting advantage of creating autonomous geometry in terms of its versatility in utilize. The results have shown that 3D CAD is utilized to select variety of materials. The great advantage of this method is that checks requirement together with functionality and ergonomic tests as well as materials examined.

This paper is a by-product of the scientific program "Design of Multiple Mailboxes" of the University of Western Macedonia.

References

1. A. Bégin-Drolet, M.A. Dussault, and S.A. Fernandez, Addit. Manuf, **15,** 23-24 (2017).
2. T. Mikolajczyk, A. Borboni, X. Kong and A. Olaru, Int J Appl Mech Mater **772**, 477-481 (2015).
3. T. Mikolajczyk, H. Latos, T. Paczkowski, D. Y. Pimenov and T. Szynka, Procedia Manuf., **22,** 160-175 (2018).
4. D. Perrault, M. Stevanovic, and A. Ghiassi. Surgical applications of three-dimensional printing: a review of the current literature & how to get started https://www.ncbi.nlm.nih.gov/pmc/articles/PMC5220021 (2016).
5. CAD software: Geomagic (https://www.3dsystems.com/software) (2020).
6. S. Wang, Y. Liu and W. Shi., Mater J., **10**(9) (2017).
7. Campbell, T., Williams, C., Ivanova, O. and Garrett, B. Could 3D Printing Change the World? Technologies, Potential, and Implications of Additive Manufacturing, Atlantic Council https://www.atlanticcouncil.org/in-depth-research-reports/report/could- 3d-printing-change-the-world (2011).
8. Bloomberg Businessweek 3D printing, materials & services: global market size http://www.statista.com/statistics/315363/global-market- for-3d-priting-and-services/
9. An Update on 3D Printing and the Postal Service https://www.uspsoig.gov/document/update-3d-printing-and-postal-service (2015)
10. United States Postal Service, 3D Printing and the Postal Service, If It Prints, It Ships:https://www.uspsoig.gov/document/if-it-prints-it-ships (2014).

The design of mechatronic system for gel electrophoresis

Apostolos Tsagaris[1,*], *Konstantinos* Theodoridis[1], *Fotis* Stergiopoulos[1], *Dimitrios* Bechtsis[1], *Nikolaos* Nikolaidis[1], *Dimitrios* Triantafyllidis[1], *Asterios* Papaoikonomou[2] and *Anastasios* Filelis[2]

[1]International Hellenic University (IHU), Department of Industrial Engineering & Management, PO Box 141, Sindos, Thessaloniki, 57400, Greece
[2] Evresis S.A DA 12a Block:39b Industrial Area, 57400, Sindos, Thessaloniki

Abstract. The paper present the methodology to design and implement a mechatronic system for gel electrophoresis. It describes the steps towards the development of a prototype and presents its functions. The new automated system is characterized by high precision, small size and relatively low cost, thus achieving the introduction of a significant innovation in the area of medical and laboratory measurements.

1 Introduction

Electrophoresis is the phenomenon of the movement of chemical compounds due to the application of an electric field. The movement of substances within the electric field is directly related mainly to the applied electric charge and the molecular weight of the substances. Therefore, if an electric field E is applied to an electrophoresis device, the motion depends exclusively on the molecular weight of the substance, on a given substrate (pH, water content, temperature, etc.). For the electrophoresis process, it is necessary to use protocols for receiving and maintaining the sample, the characteristics of the electric field and the use of the appropriate substrate and the corresponding regulatory solutions.
The electrophoresis, as a process, is a simple technique, providing immediate results of high accuracy and low cost. It is used for the analysis of proteins in the blood serum but also in other biological fluids (e.g. urine, cerebrospinal fluid, saliva). Electrophoresis can be applied both as a research method and for clinical diagnosis of gammoglobinopathies, hemoglobinopathies, lipidemia, etc. As a diagnostic method, it provides clear quantitative (photometric evaluation of electrophoresis) and qualitative information for individual protein changes and contributes to the immediate and indirect diagnosis of pathological conditions, such as multiple myeloma, spinal cord disease, AIDS etc.
Nowadays the main technological methods for electrophoresis are two:
a) Electrophoresis with gel substrate such as agarose - natural polysaccharide (gel electrophoresis)
b) Capillary Electrophoresis

* Corresponding author: tsagaris@gmail.com

Specificity as well as sensitivity play an important role for correct and reliable results. According to the international literature, agarose substrate electrophoresis provides higher selectivity (98.9%) than capillary electrophoresis [1, 2]
The typical steps of an agarose substrate electrophoresis process can be summarized as follows:

1. Sample taking and placing on an appropriate agarose substrate
2. Application of the electric field (constant conditions)
3. Drying of the processed sample
4. Staining and unstaining of the processed sample
5. Drying of the sample
6. Scanning using a computer
7. Results analysis
8. Printing and presentation of the results

2 State of the art

Technological developments are gradually evolving the process of electrophoresis. Today the commercial devices that exist are distinguished into manual and semi-automatic. In manual devices the transition from one stage of processing to another is done with human intervention while the semi-automatic ones do not include the part of scanning and exporting results to the device.
In [3], , researchers have developed a fully automated system for two-dimensional (2D) electrophoresis procedures. It is designated for rapid and highly reproducible protein analysis. The system can automatically complete, apart from situ protein detection, also isoelectric focusing (IEF), on-part protein staining and sodium dodecyl sulfate polyacrylamide gel electrophoresis (SDS−PAGE). For the evaluation the resultscomparison with a commercial minigel system has been performed. The proposed system achieved high-resolution despite shorter focusing time and smaller part dimensions. The resulting reproducibility was better or comparable to the performance of the minigel system. Complete 2D separation was achieved within 1.5 h [3].
In another case, a new device is described for the casting of many PAGE gels with improved practicality and performance due to the fact that the system contains four detachable components. This produces multiple gels in less time and more easily. The system is characterized by ease of construction, versatility and low cost that makes it competitive in the market [4]. Development of the Capillary Electrophoresis (CE) led to more efficient systems, like the microchip electrophoresis [5]. The microchip system provides a fast and effective analysis as it contains various microchannels that are made of glass and/or fused silica substrates with optimal thermal characteristics. The advantages of CE provided the opportunity to researchers to develop reliable, fast and economic fully automated systems [6] or semi-automated portable devices [7].
In [8] a ready-to-use device is proposed, which runs on the Agilent 2200 TapeStation instrument and provides a Quality Control reproducible method for analyzing the integrity and quantity of genomic DNA in combination with the ease of an automated system. The preparation of the required samples is minimal and the loading is automated. The results are in digital format and can be presented as a gel image, a data table and an electro-application viewer. Expert software gives the system a lot of potential, greatly increasing its functionality and usability [8].
Finally, according to a market survey, the company Sebia (www.sebia.com) holds a dominant position in the international electrophoresis market, while Beckman Coulter (www.beckmancoulter.com) also holds a significant share. However, there are other smaller companies in some countries that are mainly active in the automation of specific

stages of electrophoresis, or in the manufacturing of devices for semi-automated processes, or finally in the production of consumables (film and gel). However, it is emphasized that even today, even the so-called "automatic" devices are not fully in the sense that they do not include a scanner and a computer for processing the film, and the final processing stage is always done in a separate computer with specialised software.

3 The Design of mechatronic system

As mentioned before, the aim of the paper is to present the design and manufacturing of a prototype of a new fully automated agarose gel electrophoresis device with significant reliability, small size, portability, and low cost advantages.
The steps to be followed for an automated electrophoresis process are summarized below (Fig. 1)

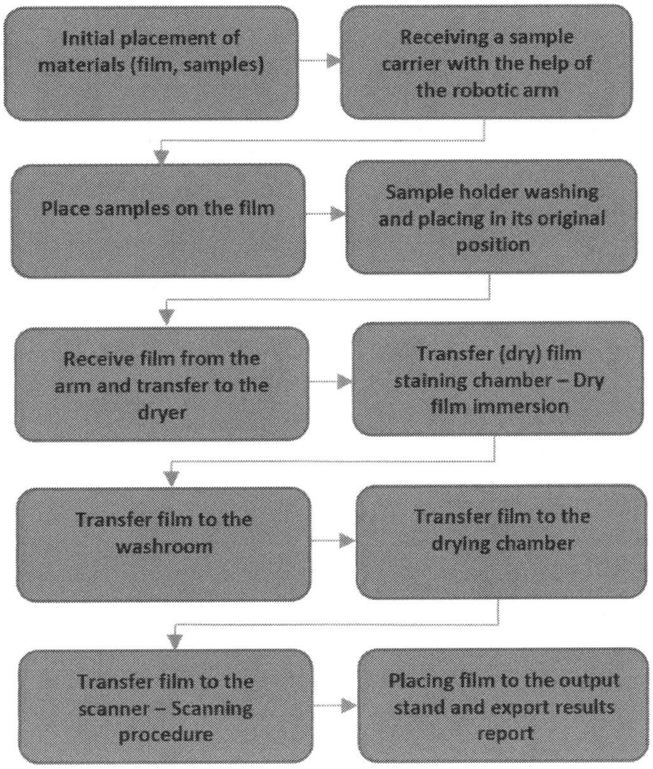

Fig. 1 – Steps for automated electrophoresis process

The construction methodology of the device performs the above steps and consists of four (4) basic phases:
1. In the first phase, the primary research is carried out for the analysis of the requirements, the 3D design of the individual parts and the analysis of the technical specifications for the electrical, electronic and mechanical parts of the device. The results of the preliminary study that have already been carried out and are used as the starting point.
2. In the second phase, the individual parts of the device are manufactured, as prototypes, for the implementation of mechanical components, arm and electronic and electrical parts, as well as firmware and software. The second phase can give feedback to the first phase to fine tune the technical specifications.

3. In the third phase, pilot tests are performed, with the interconnection of the individual sections developed in the second phase. Pilot tests are divided into two stages:

I. **Stage of creation of the original prototype**, where the initial assembly and control of both the individual parts and a first version of the device are performed. In addition, the operation of the individual parts is studied and issues of functionality and needs of redesign / improvements of parts of the device are highlighted.

II. **Stage of creation of the final prototype**, after receiving the feedback from the original prototype. Specifically, optimization and redesign will be performed in order to reduce the volume of the device and to adjust external covers. In the second stage, a substantial redesign of the original is carried out and all the problems presented during the first stage of construction are analyzed. Finally, all assembly procedures and synchronized hardware and software tests are performed to build the final and fully functional prototype.

The third phase may re-adjust the individual implementations of the second phase if necessary.

4. In the fourth phase, the study of the exploitation of the research results takes place. This phase is implemented immediately after the implementation of the standard and includes an exploitation study of the results, the investigation of possible patent results, the creation of publications and the development of a market promotion plan.

The following figure (fig. 2) shows the methodology.

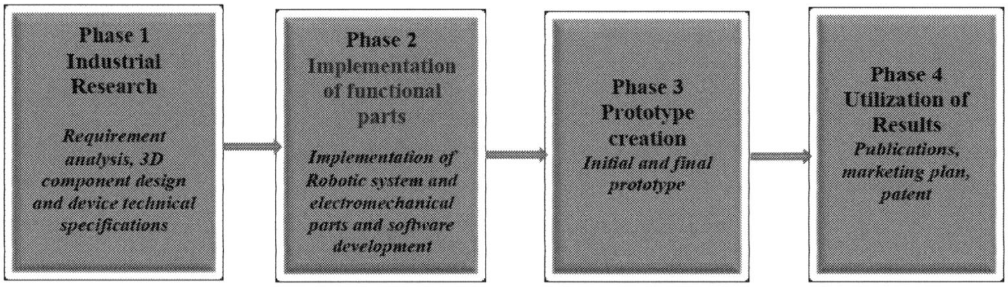

Fig. 2 – Implementation methodology

With the proposed methodology, the conditions for success will be created, as in each phase, possible errors will be highlighted and corrected in the next stage.

4 System Description

The electrophoresis mechatronic system consists of different subsystems which contribute to the necessary actions to complete the electrophoresis process. Emphasis was given on the design of the special parts and especially on the moving parts trying to incorporate many elements of intelligence into the mechatronic system. In the following figure (fig. 3) the main parts of the mechatronic system are shown which include: (a) a tube drawer system, (b) a bar code scanner, (c) a wash tank, (d) a sample carrier, (e) an electrophoresis chamber, (f) a robotic arm/gripper, (g) a staining-unstaining chamber, (h) a drying chamber, (i) a frame/film manipulator, (j) a camera analysis chamber.

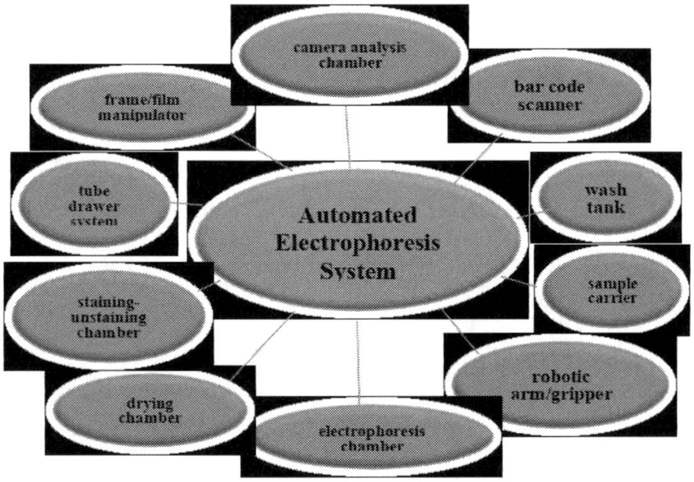

Fig. 3 – System configuration

The main part of the automatic electrophoresis machine is the mechatronic system (Fig. 4) that performs the sampling and movement routines of the film frame at the various stages of the electrophoresis process. It has the ability to move on two axes of motion in XX and ZZ. It has a properly designed clamp that has the ability to open and close in order to grab and hold components like a robotic end effector. The Closing of the end effector is ensured with a suitable sensor while the opening is performed with time calculated after the closing. The movement on the axes is limited by appropriate start and stop sensors.

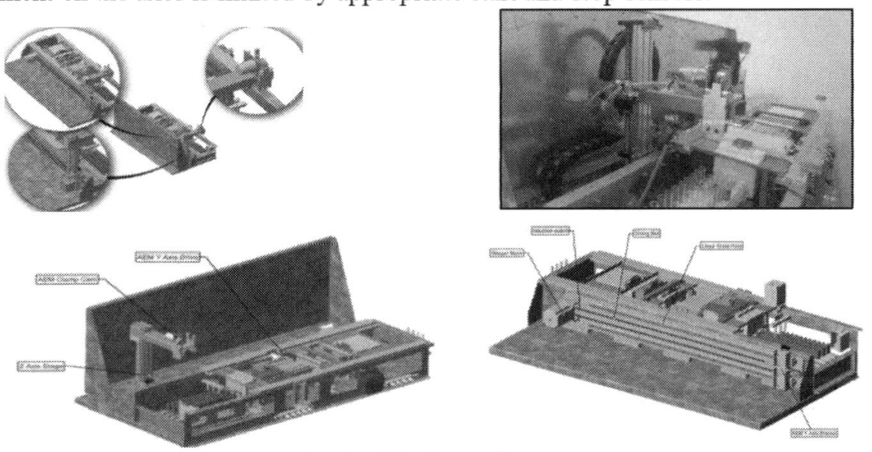

Fig. 4 – Mechatronic system

The XX axis of movement have a movement accuracy of 0.1mm and is able to accelerate the system with 0.3 m / sec2 up to a maximum speed of 0.7 m / sec. Two high-precision linear guides were used to reduce oscillations during acceleration and deceleration. The motor drivers and individual components such as position sensors were mounted on specially treated aluminum plate made of 6063T6 material to achieve the necessary level of tolerance and parallelism thus minimizing losses and reducing the system's rolling resistance. The vertical axis of movement ZZ has an accuracy of 0.05mm and accelerates to 0.1m / sec2. The high accuracy is crucial since the samples must be deposited in the agarose film without the screw of the sample carrier entering the substrate. Due to the increased required accuracy, the solution of the drive screw has been selected, through which high accuracy can be achieved by limiting the maximum speed.

5 Conclusion

An innovative fully automatic electrophoresis mechatronic system has been designed and manufactured that can perform electrophoresis processes on precast gel agarose films. The operation of the system was evaluated as well as the robustness of the machine by performing a complete electrophoresis process that includes protein analysis. The results showed high accuracy in operation and efficiency in completing the different phases of the electrophoresis process. An additional feature is the speed of execution of the process that helps the productivity of the system.

This research has been co-financed by the European Regional Development Fund of the European Union and Greek national funds through the Operational Program Competitiveness, Entrepreneurship and Innovation, under the call RESEARCH – CREATE – INNOVATE (project code: T1EDK-02403)

References

1. B. Lissoir, P. Wallemacq, D.Maisin, Ann Biol Clin (Paris).**61**, 557-562. (2003)
2. Ch. R. McCudden, S. P. Mathews, Sh. A. Hainsworth, J. F. Chapman, C. A. Hammett-Stabler, M. S. Willis and D. G. Grenache, Am J Clin Pathol, 129:451-458 458 (2008)
3. A. Hiratsuka, H. Kinoshita, Y. Maruo, K Takahashi, S. Akutsu, Ch. Hayashida, K. Sakairi, K. Usui, K. Shiseki, H. Inamochi, Y. Nakada, K. Yodoya, I. Namatame, Y. Unuma, M. Nakamura, K. Ueyama, Y. Ishii, K. Yano, K. Yokoyama, **79 (15)**, 5730-5739 (2007)
4. P. Maurye, A. Basu, T. K. Bandyopadhyay, J. K. Biswas, B. P. Mohanty, ELECTROPHORESIS 2017 **38 (16)**, 2060-2068 (2017)
5. B. C. Durney, C. L. Crihfield, L. A. Holland, Analytical and Bioanalytical Chemistry, **407(23)**, 6923-6938 (2015)
6. J. Bodnar, L. Hajba, A. Guttman, ELECTROPHORESIS, **37(23-24)**, 3154-3159 (2016)
7. T. A. Nguyen, T. N. Pham, T. T. Doan, T. T. Ta, J. Saiz, T. Q. Nguyen, T. D. Mai, J Chromatogr A, **1360**, 305-311 (2014)
8. D. Rabiller, D. McDade-Walker, A. Padmanaban, R. Salowsky, M. Gassmann, J Biomol Tech **24(Suppl):** S55 (2013)

A Space-Time POD Basis Interpolation on Grassmann Manifolds for Parametric Simulations of Rigid-Viscoplastic FEM

Orestis Friderikos[1,2,*], *Marc* Olive[1,**], *Emmanuel* Baranger[1], *Dimitrios* Sagris[2], and *Constantine N.* David[2]

[1]Université Paris-Saclay, ENS Paris-Saclay, CNRS, LMT - Laboratoire de Mécanique et Technologie, 94235, Cachan, France.
[2]Mechanical Engineering Department, Laboratory of Manufacturing Technology & Machine Tools, International Hellenic University, GR-62124 Serres Campus, Greece.

Abstract. Parametric simulations of thermomechanical metal forming processes still remain computational costly and difficult due to inherent strong non-linearities. To this end, Reduced Order Models (ROMs) are introduced to decrease the computational time in large scale models and provide near-optimal solutions in acceptable times. ROMs based on the Proper Orthogonal Decomposition (POD) are usually capable of accurately reproducing the dynamics of high-fidelity FEM simulations and offer the potential for near real-time analysis. However, ROMs are not robust with respect to parameter changes and must often be rebuilt for each parameter variation. This work aims to interpolate ROM POD basis associated with a limited number of training points on Grassmann manifolds, so as to obtain a robust ROM corresponding to a target parameter. A novel Space-Time (ST) POD basis interpolation, where the reduced spatial and time basis are separately interpolated on Grassmann manifolds, is proposed. Good correlations of the ROM ST models with respect to their associated high-fidelity FEM counterpart simulations are found. Hence, application of the ROM adaptation method for near real-time metal forming simulations using off-line computed ROM POD databases can be possible.

1 Introduction

Computational metal forming has been widely used in a variety of applications in academia laboratories and manufacturing industry over the last decades. Simulations of such problems are characterized by multiple sources of strong non-linearities, thereby are often prohibitively expensive for an exhaustive design analysis of large scale models. Moreover, several core issues are still of significant importance and development, among others, parallel computing, adaptive meshing and remeshing, modeling of micro-structure evolution, multiscale modeling, optimization of processes or parameter identification, stochastic approaches and machine learning (see a thorough review in [1, 2]). One of the key feature challenges mentioned in [1], is the model order reduction methods to decrease computational time. Indeed, simulation of complex configurations can be intractable since the computational cost can highly increase

*email : ofriderikos@ihu.gr
**email : marc.olive@math.cnrs.fr

for a parametric analysis. ROMs via the POD have been chosen to reduce the problem dimensionality while maintaining solution accuracy. For solving a parametric problem, the method starts by a *training stage* during which the problem is solved for a limited number of parameters. The full-order field 'snapshots' are then compressed using the POD to generate a ROM that is expected to reproduce the most characteristic dynamics of its high-fidelity counterpart solution. For a new parameter value, an interpolation method is proposed using the underlying spanned subspaces of the POD basis [3]. Since these matrices are obtained via the Singular Value Decomposition (SVD), they describe the optimal projection operators, which reduce the problem from the full space to a low dimensional subspace. POD basis interpolation is performed using local maps on Grassmann manifolds by evaluating the geodesic paths between the subspaces on this manifold, all this being done in the framework of *Riemannian geometry*, which is a specific matter of differential geometry. In this context, a novel non-intrusive *Space-Time* POD interpolation is proposed here allowing near-real time predictions since no new ROM FEM models have to be solved, as required by the standard POD approach.

The coupled thermomechanical analysis is based on the Rigid Visco-Plastic (RVP) formulation using an implicit solution strategy. In RVP, the elasticity effects are completely neglected due to the fact that elastic components of strain remain small as compared with irreversible strains. Therefore, the RVP formulation turns out to be very similar of fluid flow problems and it is also called as *flow formulation*. Although it is not possible to calculate the residual stresses and the spring-back effect, the flow formulation presents outstanding advantages. Unlike the elasto-plastic FEM, the RVP formulation, even though more approximate, is simpler to be implemented in computer codes, more stable and can use relatively larger time increments, thus improving the computational efficiency. A thorough overview of the foundation of the theory can be found in [4]. In this work, RVP simulations are performed using a mechanical FEM solver [5] and FEM solver for the thermal analysis in Matlab. A staggered approach for the thermomechanical coupling is used.

2 Proper Orthogonal Decomposition and Grassmann Manifolds

Let us consider here some *snapshot matrix* obtained from a FEM solution of a RVP problem

$$\mathbf{S}_{jk}, \quad 1 \leq j \leq 3N_s, \quad 1 \leq k \leq N_t$$

with N_s the number of spatial points and N_t the number of time steps. Such matrix encodes in fact N_t vectors $\mathbf{u}_k := u(\cdot, t_k) \in \mathcal{H}_{\text{spatial}}$, and we write

$$\mathbf{S} := [\mathbf{u}_1, \ldots, \mathbf{u}_{N_t}]$$

As a classical result [6], POD of mode p can be obtained from any singular value decomposition so to define a *projection matrix*. Taking $\phi_1, \ldots, \phi_k$ ($k = \min(n, N_t)$) to be the spatial vectors associated to the singular values $\sigma_1 \geq \sigma_2 \geq \ldots \geq \sigma_k$ of $\mathbf{S}$, POD of mode p is obtained using

$$\mathbf{S}_p := \mathbf{\Pi}_p \mathbf{S}, \quad \mathbf{\Pi}_p := \mathbf{\Phi}_p \mathbf{\Phi}_p^T, \quad \mathbf{\Phi}_p := \left[\phi_1, \ldots, \phi_p\right] \in \mathrm{Mat}_{n,p}(\mathbb{R}).$$

so that the projection matrix $\mathbf{\Pi}_p$ is defined using the matrix issued from vectors ϕ_i. A main observation is that a projection matrix can be defined using *any* orthonormal basis $\mathbf{v_1}, \ldots, \mathbf{v_p}$ of the vector space $\mathcal{V}_p$ generated by $\phi_1, \ldots, \phi_p$.

Finally, POD of mode p is thus characterized by the p dimension subspace $\mathcal{V}_p$ of $\mathbb{R}^n$ and not by the spatial vectors $\phi_1, \ldots, \phi_p$. Any interpolation made on POD translate into interpolation made on p dimensional subspaces.

Mathematically, the set of all p dimensional subspaces of $\mathbb{R}^n$ is known as a Grassmann manifold $\mathcal{G}(p,n)$ which is also a Riemannian manifold. Thus, a POD of mode p made on a snapshot matrix $\mathbf{S}$ leads to some *point* $\mathbf{m}$ in $\mathcal{G}(p,n)$, and any interpolation has to be done on this *curved* manifold. This means that computations on $\mathcal{G}(p,n)$ can only be done using *local charts*, and the Riemannian structure allows us to use *normal coordinates* obtained from an *exponential map* defined using geodesics.

3 Space-Time Interpolation on Grassmann Manifolds

Consider a set of snapshot matrices $\mathbf{S}^{(1)},\ldots,\mathbf{S}^{(N)}$ corresponding to parameters $\lambda_1,\ldots,\lambda_N$. Any POD of mode p gives rise to points $\mathbf{m}_1,\ldots,\mathbf{m}_N$ (resp. $\mathbf{m}'_1,\ldots,\mathbf{m}'_N$) in $\mathcal{G}(p,n)$ defined from the spatial part (resp. the time part).

To address the question of interpolation for a target value $\widetilde{\lambda}$, we have to make interpolation on the Grassmannian manifold $\mathcal{G}(p,n)$. A strategy introduced in [3] make use of normal coordinates, and thus the exponential and the logarithm map. This leads to the following, for the spatial part (while the same is done from the time part):

- *Step 1:* Choose a reference point $\mathbf{m}_0 \in \{\mathbf{m}_1,\ldots,\mathbf{m}_N\} \subset \mathcal{G}(p,n)$. For each *vector space* $\mathbf{m}_i$, take $\mathbf{Y}_i$ to be the matrix associated to one of its orthonormal basis.
- *Step 2 (the logarithm step):* First compute thin SVD

$$\mathbf{Y}_i\left(\mathbf{Y}_0^T\mathbf{Y}_i\right)^{-1} - \mathbf{Y}_0 = \mathbf{U}_i\mathbf{\Sigma}_i\mathbf{V}_i^T$$

  and define a tangent vector $\mathbf{Z}_i := \mathbf{U}_i\tan^{-1}\mathbf{\Sigma}_i\mathbf{V}_i^T$.
- *Step 3:* Make Lagrangian interpolation

$$\widetilde{\mathbf{Z}} := \sum_{i=1}^{N}\prod_{i\neq j}\frac{\widetilde{\lambda}-\lambda_j}{\lambda_i-\lambda_j}\mathbf{Z}_i$$

- *Step 4 (the exponential step):* Take a thin SVD of $\widetilde{\mathbf{Z}}$ given by $\widetilde{\mathbf{Z}} = \widetilde{\mathbf{U}}\widetilde{\mathbf{\Sigma}}\widetilde{\mathbf{V}}^T$ and define

$$\widetilde{\mathbf{Y}} = \left[\mathbf{Y}_0\widetilde{\mathbf{V}}\cos\left(\widetilde{\mathbf{\Sigma}}\right) + \widetilde{\mathbf{U}}\sin\left(\widetilde{\mathbf{\Sigma}}\right)\right]\widetilde{\mathbf{V}}^T \tag{1}$$

A *Space-Time* (ST) interpolation is finally some ROM snapshot matrix $\widetilde{\mathbf{S}}$ obtained from a spatial part $\widetilde{\mathbf{\Phi}}$, a time part $\widetilde{\mathbf{\Psi}}$ (both obtained from previous algorithm), and a *mixed part* $\widetilde{\mathbf{S}}^*$ defined to be some square matrix of size $p \times p$, so that

$$\widetilde{\mathbf{S}} := \widetilde{\mathbf{\Phi}}\,\widetilde{\mathbf{S}}^*\,\widetilde{\mathbf{\Psi}}^T.$$

Remark 3.1. ST interpolation offers the advantage of reconstructing a ROM snapshot matrix without relaunching ROM-FEM computations. Thus, it results into near-real time solutions due to direct matrix multiplications.

4 Rigid-Viscoplastic FEM Formulation

Classical RVP problems consider the deformation of an isotropic body occupying a domain $\Omega \subset \mathbb{R}^3$ and its associated boundary $\vartheta\Omega$, which represents the current configuration according to the *Updated Lagrangian* formulation. A weak form is applied to obtain a solution that satisfies the governing equations and the boundary conditions. The essence of the variational formulation is to calculate the total potential of a scalar quantity Π (functional) of the system, and to invoke the *stationarity* of Π, i.e. $\delta\Pi = 0$ with respect to arbitrary changes of the state variables. The functional Π is defined by an integral form in accordance with the virtual work-rate principle

$$\Pi(v) = \int_{\Omega} \bar{\sigma}\dot{\bar{\varepsilon}} dV - \int_{\vartheta\Omega_F} F_i v_i dS \tag{2}$$

where $\bar{\sigma}$ is the effective stress, $\dot{\bar{\varepsilon}}$ is the effective strain rate and F_i denotes prescribed surface tractions on the boundary surface Ω_F. The first term in eq. (2) represents the internal deformation work-rate, whereas the second term represents the work-rate done by the external forces. The real velocity field gives to the functional Π a stationary value, which means that the first order variation of Π vanishes. In order to meet the incompressibility constraint condition $\dot{\varepsilon}_v = \dot{\varepsilon}_{kk} = 0$ on an admissible velocity field, a penalized form of incompressibility is used

$$\delta\Pi = \int_{\Omega} \bar{\sigma}\delta\dot{\bar{\varepsilon}} dV + \frac{1}{2}\int_{\Omega} K\dot{\varepsilon}_v \delta\dot{\varepsilon}_v dV - \int_{\vartheta\Omega_F} F_i \delta v_i dS = 0 \tag{3}$$

where K is a large positive constant which penalizes the dilatational strain-rate.

For the thermal problem, a thermodynamically sound derivation is adopted based on the conservation of energy

$$-\rho c \frac{\vartheta T}{\vartheta t} + k\nabla^2 T + \xi\bar{\sigma}\dot{\bar{\varepsilon}} = 0 \tag{4}$$

where ρc is the volume specific heat of the material, $\xi\bar{\sigma}\dot{\bar{\varepsilon}}$ represents the work heat rate per unit volume due to plastic deformation, k is the thermal conductivity, T is the temperature and ξ is a coefficient that presents the fraction of the deformation energy dissipated into heat. Using the weak form and after the application of the divergence theorem, the energy balance eq. (4) can be written

$$-\int_{\Omega} \xi\bar{\sigma}\dot{\bar{\varepsilon}}\delta T dV + \int_{\Omega} k\nabla T\delta(\nabla T) dV + \int_{\Omega} \rho c \frac{\vartheta T}{\vartheta t}\delta T dV - \int_{\vartheta\Omega} q_n \delta T dS = 0 \tag{5}$$

where $q_n = k\vartheta T/\vartheta n$ is the heat flux across the boundary $\vartheta\Omega$ and n denotes the unit normal vector to the boundary surface $\vartheta\Omega$. The mechanical and thermal analysis can be *loosely coupled* in such a way that plastic work and friction are considered as heat source in the thermal analysis while the updated temperature field is used to determine the flow stress behavior during the deformation analysis.

5 Numerical Investigations

To evaluate the performance of ST POD interpolation, a rectangular cross section bar is compressed between two parallel flat dies under the conditions of constant shear friction factor m at the die/workpiece interface. The initial workpiece has dimensions $h = 20$ mm in height and $w = 20$ mm in width. Plane strain conditions are considered. Due to the symmetry of the problem, only one quarter of the cross section is analyzed (Figure 1). The lower die is stationary while the upper die velocity is set to $v = 1$ mm/s. The workpiece is compressed till a 35% reduction in height is achieved. The final state is accomplished in 7 time steps with a constant time increment $\Delta t = 0.5$ s. A 4-node quadrilateral element with bilinear shape functions and four point integration is used. However, a reduced integration scheme which imposes the volume constancy over the linear element (one point integration) is used for the dilatational term (second integral of the functional in eq. 3). A total number of 100 elements with 121 nodes are used in the simulations. Considering the friction coefficient m as the investigated parame-

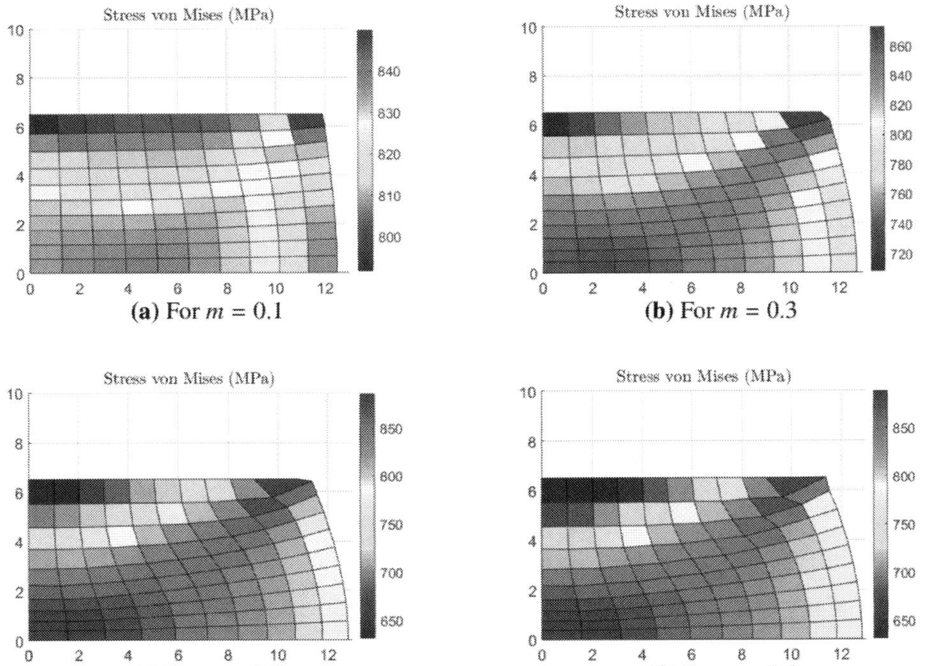

(a) For $m = 0.1$ **(b)** For $m = 0.3$

(c) For $m = 0.5$ **(d)** For $m = 0.9$

Fig. 1. Deformation patterns of the benchmark metal forming example using different values for the shear friction coefficient m represented by the variable λ.

ter, the following training points are selected, denoted by the variable $\lambda \in \Lambda_{tr} = \{0.1, 0.5, 0.9\}$. A simple strategy for sampling the parametric space is used by equi-distribute the points over the associated range. The target point is set to $\tilde{\lambda} = 0.3$. For each parametric simulation, a sequence of snapshots uniformly distributed over time using an increment of $\Delta t = 0.5$ s is extracted for all nodes of the workpiece. The space-time snapshot matrices $\mathbf{S}(\lambda_i) \in \mathbb{R}^{n \times N_t}$ of size 242×7 are associated to nodal velocity fields. Using the POD/SVD, the ROM spatial basis $\{\mathbf{m}_i\}_{i=1}^{N}$ and time basis $\{\mathbf{m}'_i\}_{i=1}^{N}$ are constructed for the training points λ_i. Then, the set of the low-dimensional POD basis is interpolated on Grassmann manifolds using the algorithm presented in section 3. ST interpolation is compared against the high-fidelity FEM solution. Considering the interpolated and the HF-FEM snapshot matrices $\tilde{\mathbf{S}}$ and $\mathbf{S}^{\text{FEM}}$, respectively, the following error measure is defined

$$e_{L_2}(\tilde{\mathbf{S}}) = \frac{\|\tilde{\mathbf{u}}_i - \mathbf{u}_i^{\text{FEM}}\|_{L_2}}{\|\mathbf{u}_i^{\text{FEM}})\|_{L_2}}, \quad i = 1, \ldots, N_t \tag{6}$$

where $\mathbf{u}_i$ denotes the i-th column of the snapshot matrix $\mathbf{S}$. The eigenvalue spectrum of snapshot matrices $\mathbf{S}_i$ corresponding to training points $\lambda_i \in \Lambda_{tr}$ is shown in a log scale in Figure 2. We can observe that the distance between the first and the last eigenvalue is of the order of 10^5-10^6. The relative L_2-error norm $e_{L_2}(\tilde{\mathbf{S}})$ between the interpolated and the HF-FEM solution for various POD modes is shown in Figure 3. The relative error for all POD modes lies within a range of 0.0175 up to 0.038. For this particular case, the interpolation error is minimum with $p = 2$ and $p = 3$ POD modes. In general, it can be observed that the interpolated ST POD solution delivers good accuracy and is reliable enough to predict the velocity field for the investigated target point. Due to the fact that the online stage of ST

POD interpolation consists solely of algorithmic matrix operations, it outperforms standard POD methods which require a FEM solution for the new ROM. Thus, in ST POD, the computational time can be several orders of magnitude lower than standard POD approaches, and even more against the high-fidelity FEM counterpart solution making it specially tailored for near real-time computations.

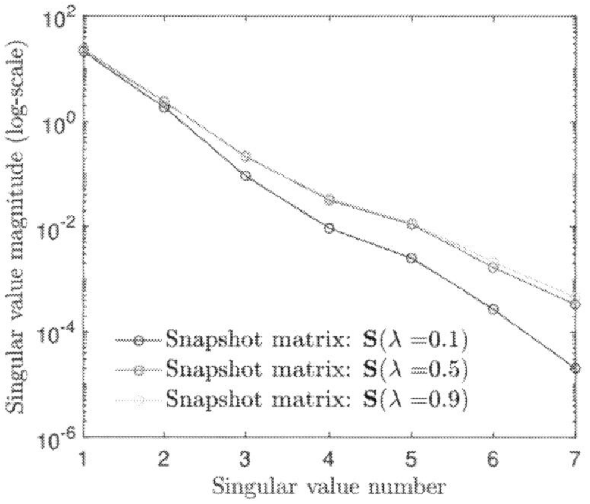

Fig. 2. The eigenvalue spectrum of snapshot matrices $\mathbf{S}_i$ corresponding to training points $\lambda \in \Lambda_{tr} = \{0.1, 0.5, 0.9\}$.

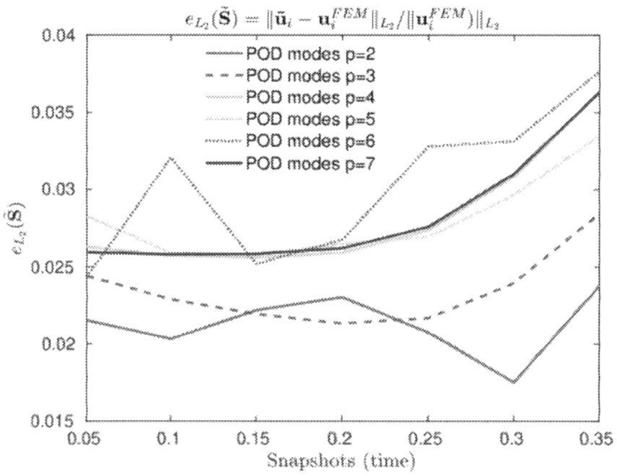

Fig. 3. Relative L_2-error norm $e_{L_2}(\tilde{\mathbf{S}})$ for various POD modes; training points: $\mathbf{m}_0(\lambda = 0.5)$ (reference point); $\mathbf{m}_1(\lambda = 0.1)$; $\mathbf{m}_2(\lambda = 0.9)$; target point: $\tilde{\mathbf{m}}(\lambda = 0.3)$.

References

[1] S. Andrietti, J.L. Chenot, M. Bernacki, P.O. Bouchard, L. Fourment, E. Hachem, E. Perchat, CMMS, AKAPIT, **15**, 265-293, (2015).
[2] Z. Gronostajski, Z. Pater, L. Madej et al., Archives CME, **19**, 898-941, (2019).
[3] D. Amsallem, J. Cortial, K. Carlberg, C. Farhat, IJNME, **80**, 9, 1241-125, (2009).
[4] S. Kobayashi, SI Oh, T. Altan, *Metal Forming and the Finite Element Method*, Oxford University Press, (1989).
[5] O. Friderikos, 4th ICMMEN, 3-5 October, Thessaloniki, Greece, (2011).
[6] R. Mosquera, A. Hamdouni, A.El Hamidi, C. Alleryet, DCDS-S, **12**, 6, 1743-1759, (2018).

Modeling and simulation of the nanosecond pulsed laser engraving process

Evangelos Nikolidakis[1], *Aristomenis* Antoniadis[1,*]

[1]School of Production Engineering & Management, Micromachining & Manufacturing Modeling Lab, University Campus Kounoupidiana, Technical University of Crete, 73100 Chania, Greece

Abstract. In this paper a 3D finite element simulation model of the nanosecond pulsed laser engraving process will be presented. With this model simulations of laser engraving process will be performed for some widely used materials using a wide range of process parameters in order to estimate the removed material layer thickness at each laser scan over the surface of the workpiece. Determining the removed material layer thickness is an important task because the machine must receive this value as input from the operator to calculate how many passes-layers need to be made in order to achieve the desired final depth of engraving. Since there is no simulation tool for this purpose at this time, the removed material layer thickness is determined through an experimental procedure. However, this procedure is time consuming as it has to be carried out each time separately depending on the process parameters, the material used, etc.

1 Introduction

Laser engraving is a machining process that achieves a high degree of precision compared to other traditional methods such as micro-machining with cutting tools. During laser engraving process, a laser machine generates laser beam pulses which impinge the target material surface. Each laser pulse ablates a very small amount of material. In order to produce the desired 3D geometry, the laser beam is scanned across the surface of the workpiece, sending the pulses at predefined positions. The quality of the result depends on the laser beam type, the workpiece material, the scanning strategy, the process parameters selection, etc. In order to correctly select the machining parameters and predict the machining result or final workpiece, the need to develop simulation tools was created.

Modeling and numerical simulation of laser engraving process is a complex task. Therefore, many researchers have dealt with modeling and numerical simulation of the laser ablation process of the materials. Wang et al. [1] proposed a two-dimensional (2D) modeling and simulation of the pulsed laser ablation process of aluminium caused by a single pulse using Abaqus Finite Element Analysis (FEA) software. R V Davydov and V I Antonov [2] presented a mathematical model for femtosecond laser ablation of metals. Roth et al. [3] presented a simulation model of femtosecond laser ablation in aluminium using double

* Corresponding author: antoniadis@dpem.tuc.gr

pulses. Tani et al. [4] developed a simulation package for micro-manufacturing processes using nanosecond pulsed lasers. L Dasallas and W Garcia [5] generated a numerical model of laser ablation process of copper caused by a femtosecond laser pulse at an oblique angle of incidence. T Hurtony [6] created a simulation tool for laser ablation caused by a single pulse shot in Matlab numerical computing environment. Aghaei et al. [7] presented a one-dimensional (1D) mathematical thermal model for a nanosecond pulsed laser ablation of copper in helium ambient gas.

In this work an experimentally confirmed 3D finite element simulation model of the nanosecond pulsed laser engraving process will be presented. Using this model the final 3d engraved geometry will be predicted taking into account the workpiece material properties, the laser machine process parameters, the interaction between the laser beam pulses and the irradiated target material, the ablation process mechanism, the material vapour-plasma formation and the laser beam shielding. Furthermore, the removed material layer thickness will be calculated using the predicted 3d final engraved geometry. The prediction of the removed material layer thickness is a major challenge because the laser machine must receive this value as input from the operator to calculate how many passes-layers n need to be made in order to achieve the desired final depth of engraving. Since there is no simulation tool for this purpose until now, the removed material layer thickness is determined through an experimental procedure. However, this procedure is time consuming as it has to be carried out each time separately depending on the process parameters, the material used, etc. and requires a remnant of the workpiece material for the experiments.

2 Modeling

2.1 Thermal model

For the thermal model of the laser beam a 2D surface Gaussian heat flux model was adopted [8]. The reason a surface heat flux model was chosen is that for nanosecond laser pulses on metals the absorption-penetration depth is very small as compared to the lateral heat diffusion dimensions of laser beam [9].

For the engraving of a 3D geometry laser beam pulses scan the surface of the workpiece moving on straight tracks and on multiple layers in a predetermined way (laser map). The laser map includes the coordinates of the focal point of the beam as well as the beam power P at each time point of the laser beam processing. As far as the laser power P as function of time, it was assumed that repeated pulses are generated with power equal to peak power [10,11] using a period step function. In order to generate the laser map the scanning strategy being used is cross hatching with hatching distance H_d to be equal with the track distance T_d and calculated by dividing the is the scanning speed V with the repetition rate F [12].

In order to calculate the energy absorbed by the target material coming from the incident laser beam, three cases are distinguished depending on the phase of the material as shown in Fig. 1.a. The first case shown on Fig. 1.a.1, the laser beam irradiates the target material and it's temperature is still below the melting temperature and the material is in solid phase. At this case, only a smart portion of the laser energy is absorbed by the surface of the material and the rest is reflected to the environment. The reflectivity of the laser beam on the metals is an optical property of the material and depends on the laser wavelength [13], the temperature and phase of the material (solid or liquid) [14] and is quite high for the most materials. Aluminium and brass reflects almost 95% of the radiation from Nd:YAG 1064nm laser beam at its solid phase. The second case shown on Fig. 1.a.2, the irradiated target material temperature has exceeded it's melting temperature but has not reached its evaporating temperature. That means that the material which absorbs the incident laser beam is in liquid phase. The reflectivity of aluminium, copper and other reflective materials drops sharply once the material melts [15,16].

The third case shown on Fig. 1.a.3 the laser ablation has begun. The irradiated target material has reached it's ablation temperature which means that the target material is in liquid phase and simultaneously material vapour is generated [17] from the ablated material. As the generated material vapour above the surface interacts with the laser beam it absorbs a part of energy (laser beam shielding) [18,19] causing it's ionization and a high temperature and high electron density plasma is formed [20]. The radiation absorbed by plasma during shielding increases the thermal energy of the plasma and this results in plasma emitting irradiance itself, a part of which returns to the workpiece and is absorbed by the material.

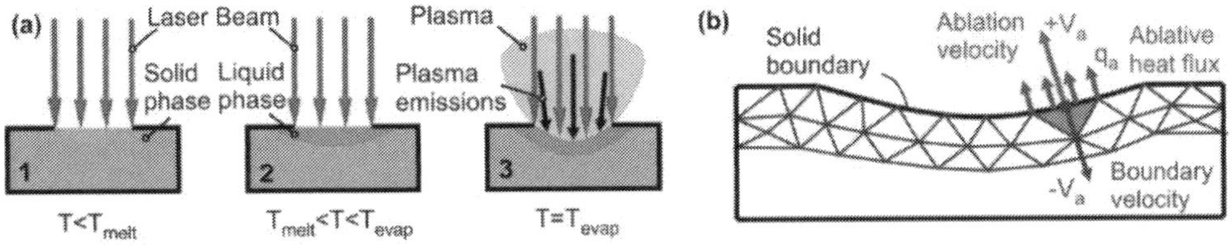

Fig. 1. (a) Ablation process thermal model **(b)** Laser ablation and material removal modeling

2.2 Material removal

Through the laser beam, the target material is being heated, leading to an increase in its temperature locally until it reaches the ablation temperature. When the surface material is at its ablation temperature there is mass loss from the surface and simultaneously a heat loss occurs due to the removal of the material vapor. The outward heat flux duo to material removal is governed by the mass flux of the ablated material and the heat of sublimation as follows:

$$q_a = M_a \, H_s \, , M_a = \rho \, v_a \tag{1}$$

where q_a is the ablative heat flux, M_a is the ablated material mass flux, H_s is the heat of sublimation, ρ is the material density and v_a is the material ablation velocity which has equal magnitude and opposite direction to the velocity of the solid boundary as shown in Fig. 1.b. Aiming at calculating the ablative heat flux at each iteration during finite elements analysis the heat-transfer version of Newton's law is used. The ablative heat flux is expressed as:

$$q_a = h_a(T_a - T) \tag{2}$$

where T_a is the ablation temperature and T is the temperature at each FEA iteration. A fictional temperature dependent heat transfer coefficient h_a is introduced that is zero for $T < T_a$ and increases linearly with a very steep slope for $T > T_a$, in order to ensure that the temperature of the solid cannot exceed the ablation temperature. By combining the two above equations, at each iteration of the finite element analysis solving, the velocity at which the boundary surface of the grid must be moved is calculated. So, by calculating the mesh displacement of the moving front boundary at each iteration, the final geometry of the workpiece for each time step is obtained.

2.3 Initial geometry, material, initial-boundary conditions and meshing

For the simulations, a rectangular block was used as initial geometry whose dimensions were adjusted so that there is enough space to be engraved the rectangular pocket. The materials for which the model was built and tested as presented in this work are yellow bronze C26000 and aluminium 7075. The initial temperature T_0 of the geometry was set to be equal to the ambient temperature $T_0 = T_{amb}$. As far as the boundary conditions, the top surface that is heated by the laser beam was set to be a diffuse surface (only surface to ambient radiation). All the other surfaces were considered to have prescribed temperature $T = T_{amb}$ (Dirichlet condition). For meshing the 3D geometry tetrahedral elements were used. In order to achieve mesh independent solutions, for the simulation tests a mesh with maximum element size of 5μm with denser distribution in the direction of depth was chosen by performing a mesh independence study.

3 Simulation and experimental tests

Using the model presented above, a large number of simulations were performed to verify its proper operation and accuracy. As initial workpiece a rectangular block made by yellow brass C26000 or aluminium 7075 was used and the requested geometry was a 150x150um rectangular pocket made by four-layer laser beam scans-passes. Regarding the process parameters, cases were examined for average power P=8, 12, 16W, for repetition rate F=20, 30, 40kHz and scanning speed V=200, 300, 400, 500 mm/s. The simulation's output-result was the 3D engraved final workpiece geometry at all intermediate stages of processing, as would have occurred if the actual process carried out. From the exported 3D geometry it can be easily calculated the removed material layer thickness D_z by dividing the whole depth of the pocket D_{zn} by the number of the layers n. The removed material layer thickness D_z is the average thickness of the material removed on each laser beam scan-pass.

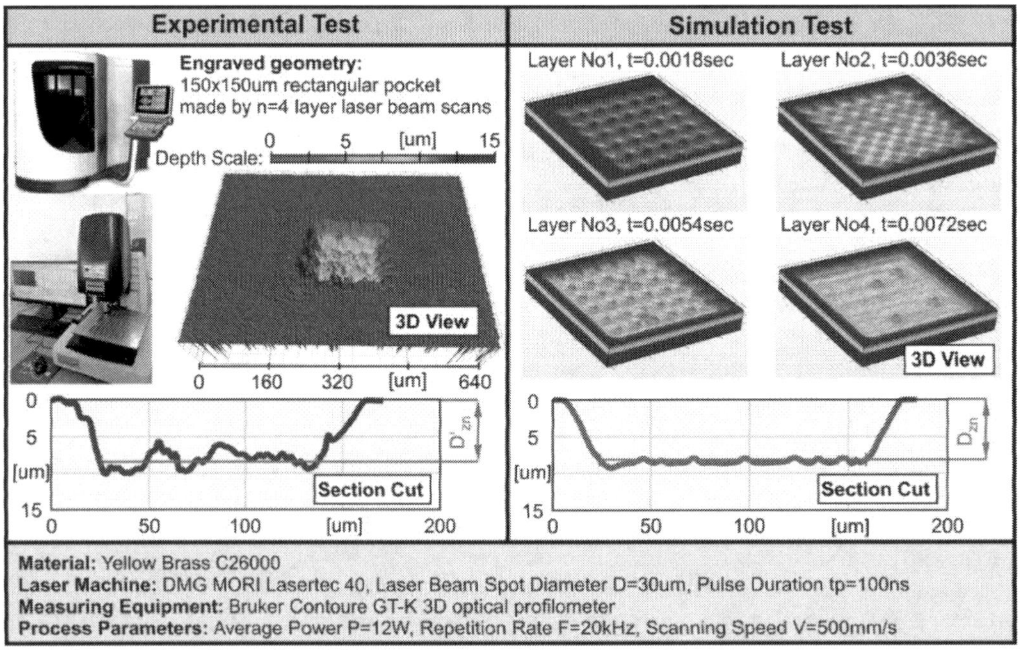

Material: Yellow Brass C26000
Laser Machine: DMG MORI Lasertec 40, Laser Beam Spot Diameter D=30um, Pulse Duration tp=100ns
Measuring Equipment: Bruker Contoure GT-K 3D optical profilometer
Process Parameters: Average Power P=12W, Repetition Rate F=20kHz, Scanning Speed V=500mm/s

Fig. 2. Presentation of experimental test compared to simulation test

A set of experiments was performed for the purpose of comparing the experimental with the simulation results. The laser engraving experimental samples were conducted using the DMG MORI Lasertec 40 machine with spot diameter D=30um and pulse duration t_{pulse}=100ns in 5mm thickness yellow brass C26000 and aluminium 7075 plates. The process was performed in exactly the same way as in the simulations. The sample geometry was a 4x4mm square pocket engraved by a constant number of 50 layers for brass and 30 layers for aluminium (number of layers selected to prevent engraving depth reach the whole plate thickness). The whole depth of each pocket measured using the Bruker Contoure GT-K 3D optical profilometer and the laser machine probing system was divided by the number of layers n to calculate the removed material layer thickness. In Fig. 2 is presented an example of a simulation test and the corresponding experimental test. The whole depth of the simulated pocket was predicted throw the simulation to be D_{zn} =8.072um and the corresponding one of the engraved pocket was measured by the optical profilometer to be D'_{zn}=8.210um.

The removed material layer thickness from the experimental and the corresponding simulation tests are presented in Fig. 3 for brass and aluminium. By observing the diagrams in Fig.3 it becomes apparent that the numerical values are in agreement with the experimental ones. Not only are the numerical points close to the experimental ones, but the curves have the same tendency. In addition, from the diagrams some basic conclusions can be drawn about the effect of process parameters on removed material layer thickness: Power as a process parameter has a great bearing on the results of machining as it is observed that a slight increase in power P results in a significant increase in depth D_z. An equally important parameter can be considered speed as a small increase in scanning speed V results in a significant decrease in removed material layer thickness D_z. As far as the repetition rate F is concerned, the diagrams show that its effect on the removed material layer thickness D_z is not very significant as it does not cause any noticeable change in it, which may be due to the limited number of experiments.

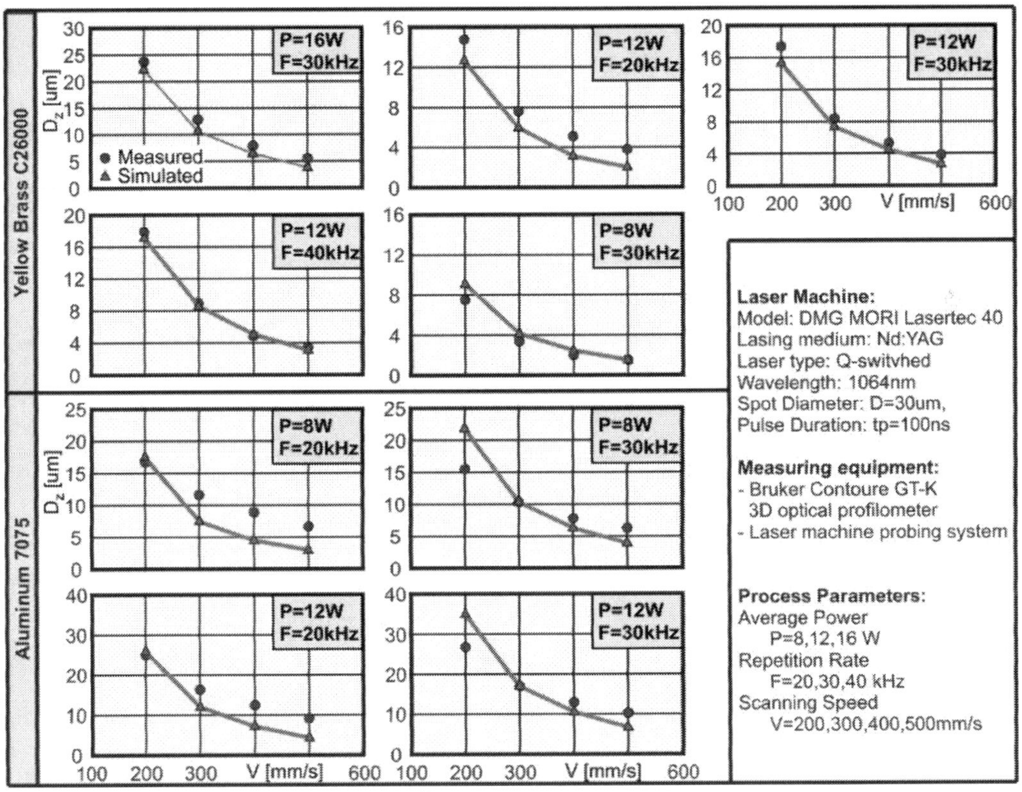

Fig. 3 Removed material layer thickness D_z

4 Conclusions

In this paper, a 3D simulation model for laser engraving process in nanosecond fields using finite element method (FEM) was presented. At first, it was developed the laser map of the laser engraving process which represents the timing and the positioning points of the pulses to be generated in order to be engraved the desired geometry. Then, the incident laser beam which provides the energy to cause the material ablation was modelled using a surface Gaussian distributed heat source. During the ablation process, the interaction between the laser beam and the material vapour-plasma was taken into account. The geometry shape change by ablated material removal was modeled by coupling heat conduction problem equations with the moving front boundary equations. A large number of simulations were performed for the case of engraving a rectangular pocket in yellow brass C26000 and aluminium 7075 for various values of process parameters: average power, repetition rate and scanning speed. The simulation's output-result was the 3D final workpiece geometry with the engraved pocket on it and the removed material layer thickness during the laser processing. Experimental tests were also performed using DMG MORI Lasertec 40 laser machine. The experimental results were compared with the numerical ones showing great agreement between them.

This research is co-financed by Greece and the European Union (European Social Fund- ESF) through the Operational Programme «Human Resources Development, Education and Lifelong Learning» in the context of the project "Strengthening Human Resources Research Potential via Doctorate Research – 2nd Cycle" (MIS-5000432), implemented by the State Scholarships Foundation (IKY).

References

1. Y. Wang, N. Shen, G.K. Befekadu, C.L. Pasiliao, Int. J. Heat Mass Transf. **113**, 1246 (2017).
2. R. V. Davydov, V.I. Antonov, J. Phys. Conf. Ser. **769**, 1 (2016)
3. J. Roth, A. Krauß, J. Lotze, H.R. Trebin, Appl. Phys. A Mater. Sci. Process. **117**, 2207 (2014).
4. G. Tani, A. Fortunato, L. Orazi, G. Cuccolini, Int. J. Nanomanuf. **3**, 279 (2009)
5. L.L. Dasallas, W.O. Garcia, Mater. Res. Express. **5**, 1 (2018)
6. T. Hurtony, 31st Int. Spring Semin. Electron. Technol. Reliab. Life-Time Predict. ISSE 2008. (2008)
7. M. Aghaei, S. Mehrabian, S.H. Tavassoli, J. Appl. Phys. **104**, 5 (2008).
8. A. Maleki, M. Kavosh Tehrani, H. Saghafifar, M.H. Moghtader Dindarlu, Chinese Phys. B. **25**, 3 (2016).
9. J.G. Lunney, R. Jordan, Appl. Surf. Sci. **127–129**, 941 (1998)
10. M.J. Kim, J. Zhang, Appl. Math. Model. **25**, 203-220 (2001).
11. K.A. Ghany, H.A. Rafea, M. Newishy, Int. J. Adv. Manuf. Technol. **28**, 1111 (2006).
12. D. Teixidor, I. Ferrer, J. Ciurana, T. Özel, Robot. Comput. Integr. Manuf. **29**, 209 (2013).
13. M.M. Quazi, M.A. Fazal, A.S.M.A. Haseeb, F. Yusof, H.H. Masjuki, A. Arslan, Crit. Rev. Solid State Mater. Sci. **41** (2016).
14. J. Xie, A. Kar, Weld. J. (Miami, Fla). **78** (1999).
15. O. Benavides, O. Lebedeva, V. Golikov, Opt. Express. **19**, 21842 (2011).
16. O. Benavides, L. de la C. May, A.F. Gil, Opt. Express. **21**, 13068 (2013).
17. M. Aden, E. Beyer, G. Herziger, H. Kunze, J. Phys. D. Appl. Phys. **25**, 1 (1992).
18. N.M. Bulgakova, A. V. Bulgakov, Appl. Phys. A Mater. Sci. Process. **73**, 199 (2001).
19. R.K. Singh, O.W. Holland, J. Narayan, J. Appl. Phys. **68**, 233 (1990).
20. J.M. Vadillo, J.M. Fernández Romero, C. Rodríguez, J.J. Surf. Interface Anal. **27** (1999).

MATEC Web of Conferences **318**, 01045 (2020) https://doi.org/10.1051/matecconf/202031801045
ICMMEN 2020

Computer modelling and simulation of a novel printing head for complex tissue engineering constructs

Gokhan Ates[1,*]

[1]School of Mechanical, Aerospace and Civil Engineering, The University of Manchester, Manchester, UK

Abstract. In tissue engineering, three-dimensional functional scaffolds with tailored biological properties are needed to be able to mimic the hierarchical structure of biological tissues. Recent developments in additive biomanufacturing allow to extrude multiple materials enabling the fabrication of more sophisticated tissue constructs. These multi-material biomanufacturing systems comprise multiple printing heads through which individual materials are sequentially printed. Nevertheless, as more printing heads are added the fabrication process significantly decreases, since it requires mechanical switching among the physically separated printheads to enable printing multiple materials. In addition, this approach is not able to create biomimetic tissue constructs with property gradients. To address these limitations, this paper presents a novel static mixing extrusion printing head to enable the fabrication of multi-material, functionally graded structures using a single nozzle. Computational fluid dynamics (CFD) was used to numerically analyze the influence of Reynolds number on the flow pattern of biomaterials and mixing efficiency considering different miscible materials.

1 Introduction

Natural tissues are complex structures composed of a hierarchical multi-functional and multi-material extracellular matrix and cells. Biomanufacturing, the combined use of additive manufacturing (3D Printing), biocompatible and biodegradable materials, cells and biomolecular signals (growth factors), emerged as a suitable approach to create tissue constructs [1]. Two main strategies have been considered: scaffold-based and cell-laden. In the case of a scaffold-based approach, additive manufacturing is used to produce three-dimensional (3D) biodegradable and biocompatible porous structures (scaffolds) that provide the physical substrate for cells to attach, differentiate and proliferate. After fabrication, scaffolds are usually seeded with cells and pre-cultured in a bioreactor before implantation. In the case of a cell-laden approach, additive manufacturing is used to print bioinks (hydrogels containing cells and growth factors) allowing the fabrication of soft tissues [2]. Among the different additive manufacturing techniques currently available,

* Corresponding author: gokhan.ates@manchester.ac.uk

MATEC Web of Conferences **318**, 01045 (2020) https://doi.org/10.1051/matecconf/202031801045
ICMMEN 2020

extrusion-based processes are the most suitable for both scaffold-based and cell-laden approaches. A wide range of printing heads (e.g. screw, pressure, or pneumatic-assisted) are being used allowing to process a wide range of materials but limited to the fabrication of single-material scaffolds or cell-laden constructs. In order to address this limitation multi-printhead systems have been developed to produce multi-material structures. Co-axial printing heads have been also developed enabling the fabrication of core-shell or hollow fibres. However, none of these systems are able to create functionally graded structures enabling the fabrication of structures based on a specific number of materials but with gradients of properties in a tailored way. In order to address this, our group is investigating the use of a Kenics mixer printing head. In order to design this printing head, it is important to understand fluid flow conditions and the mixing process in order to guarantee a good mixture of at least two materials. This paper presents preliminary results of a simplified model of the printing head considering two biocompatible and biodegradable materials. In order to create functionally graded tissue-engineered scaffolds understanding the mixing characteristic of different biomaterials is vital. CFD analysis was used to evaluated critical parameters such as mixing index, pressure drop and velocity.

2 Modelling and simulation

2.1 Static mixer printing head and simulation conditions

In this research, Kenics mixer was used as a part of printing head. It consists of a series of blades alternating clockwise and counter-clockwise twist which are arranged axially within a bifurcated pipe so that the leading edge of a blade is at right angles to the following edge of the previous blade. Blades were designed considering a twist angle of 180° and a rotation angle of 90° relative to the previous element. The aspect ratio of the Kenics blades (Ar), which corresponds to the ratio between the length l and width w of the element, is 1.5.

The commercial CFD software ANSYS CFX 19.2 (ANSYS Inc., USA), was used to numerically analyse the flow and mixing behaviour inside printing head. Multicomponent, the steady-state, laminar, incompressible, and isothermal flows were taken into account in all simulations. The domain discretization was carried out with unstructured tetrahedral elements. Simulations were performed considering two biocompatible and biodegradable natural polymers, alginate and gelatin, commonly used in tissue engineering. Details of these two materials are shown in Table 1.

Table 1. Simulation conditions and physical properties of the two considered fluids

	Alginate (Inlet 1)	**Gelatin (Inlet 2)**
ρ: density (kg/m^3)	1255	1255
μ: viscosity (N.s.m^{-2})	0.00925	0.01
Inlet velocity range (m/s)	737×10^{-5}-1474×10^{-3}	796×10^{-5}-1593×10^{-3}
Operating temperature (^{0}C)	40	40

2.2 Governing equations

The mixing process of two miscible fluids in a modified Kenics static mixer was analysed by solving the Navier-Stokes and continuity equations in the laminar flow region. In the case of incompressible flow, the mass conservation is simplified to a volume continuity equation as follows:

$$\nabla . \mathrm{u} = 0 \tag{1}$$

where u is the fluid velocity (m/s).

The momentum conservation equation in the motionless mixer with helical segments is given by following equation:

$$\nabla . (\rho u u) = -\nabla \mathrm{p} + \nu \nabla^2 u \tag{2}$$

where ρ is the fluid density, v is the kinematic viscosity. In the mixing process, the mass transport occurs through both diffusion and advection. Hence, the scalar transport equation of advection-diffusion type is provided as follows:

$$(u.\nabla)C = \alpha \nabla^2 C \tag{3}$$

where α is the kinematic diffusivity coefficient and C is the concentration of gelatin and alginate solutions in the static mixer.

2.3 Boundary conditions

In order to solve the governing equations of a laminar flow presented in the previous section, the following boundary conditions were imposed. Velocity boundary condition was employed for mixer inlets and the velocities are fixed for each fluids (alginate and gelatin) corresponding to the Reynolds number ranging from 10 to 2000. The Reynolds number for both entrances was defined based on the diameter of the bifurcated pipe:

$$Re = \frac{\rho V D_h}{\mu} \tag{4}$$

where ρ is density, V is the average velocity, D_h is the diameter of the pipe and μ is the dynamic viscosity of the fluid. A uniform velocity profile was assumed at each inlet. The static pressure was set to zero at the outlet and the no-slip condition was applied at the walls and on the surface of the mixing elements. A low diffusivity coefficient of 1×10^{-10} $(m^2 . s^{-1})$ [3] was assumed for gelatin so that the molecular diffusion can be neglected. In order to discretize the advection terms a high-resolution scheme of second-order approximation was employed. A converge criterion of root-mean-square (RMS) residual value of 1.0×10^{-6} was established and the number of iteration was set to 5000.

2.4 Mixing index

The distributive mixing capacity of the Kenics mixer was analysed for different Re numbers. Distributive mixing represents the spatial distribution of the components across the fluid domain and compared to dispersive mixing it is rather easier to achieve. As the distributive mixing capacity of a mixer cannot be judged depending on the visual contour plots only, it is essential to quantitatively evaluate the mixing performance. Therefore, mixing efficiency of the static mixer was analysed utilizing a statistical measurement method based on the concept of intensity of segregation. As previously reported, the mixing processes can be quantified using the mixing index (M.I) [4] at a cross-sectional plane perpendicular to flow direction, according to the following equation [1–6]:

$$M.I = \sqrt{\frac{\sigma^2}{\sigma_{max}^2}} \tag{5}$$

where σ^2 refers to actual variance, σ_{max}^2 denotes maximum possible variance at a cut-plane normal to the flow direction and σ^2 is defined as:

$$\sigma^2 = \frac{1}{n}\sum_{i=1}^{n}(c_i - \bar{c})^2 \tag{6}$$

where c_i is the mass fraction at ith sampling point, $\bar{c}$ is the optimal mixing mass fraction and n is the number of sampling points on the associated plane. The optimal mass fraction ($\bar{c}$) at the cut plane is equal to 0.5 for symmetrical boundary conditions. The maximum variance (σ_{max}^2) is determined from the following equation:

$$\sigma_{max}^2 = \bar{c}(1 - \bar{c}) \tag{7}$$

As the optimal maximum mass fraction $\bar{c}$ is equal to 0.5, in the case of equal flow of the fluid streams the value of σ_{max}^2 is considered to be equal to 0.25. The mixing index varies from 0 (unmixed state) to 1 (completely mixed fluid). As the mixing index decreases from 1 to 0, a lesser amount of mixing is achieved. Conversely, the higher mixing index represents higher mixing performance. However, it has been reported that the numerical results might overestimate the mixing quality due to numerical diffusion [5].

3 Results and Discussion

The influence of Reynolds number on the mass fraction distribution of both the gelatin and alginate was evaluated by plotting mass fraction contours at successive cross-sectional planes across the mixing zone. The mass fraction at the inlets was considered as reference for calculations and a step function was observed through the mixing channel width, with half of the bifurcated pipe filled with gelatin solution (red colour) and the other with alginate (blue colour) as shown in Fig.2. As the fluids flow into the mixing zone a transverse dispersion is observable at the end of each mixing element due to the stretching and folding process. The interfacial area became progressively blurred with two materials mingled together through stretching, splitting and recombining of the fluid-streams toward the complete mixing at the end of the mixer. For the lowest Reynolds number (Re=10), the thickness of the interfacial area at the cross-section before the first mixing blade is wider than those at Re=100, 500, 1000, 1500, and 2000. This can be attributed to fact that the lowest velocity results in the longest residential time of the fluids that provides the longest time for diffusion [5,7]. In order to quantitatively assess the degree of mixing, mixing index calculations were conducted across the mixing channel. Fig. 2a shows a comparison of predicted mixing indexes at the outlet of the mixer at different Re numbers. As observed, there is an inverse relationship between the Reynolds number and the mixing quality. The highest mixing index of 0.99 was achieved at Re=10, while the lowest value of 0.96 was obtained at the highest Re number. Overall, the proposed Kenics mixer for 3D Printing shows good mixing performance regardless of the Reynolds number owing to the presence of chaotic advection.

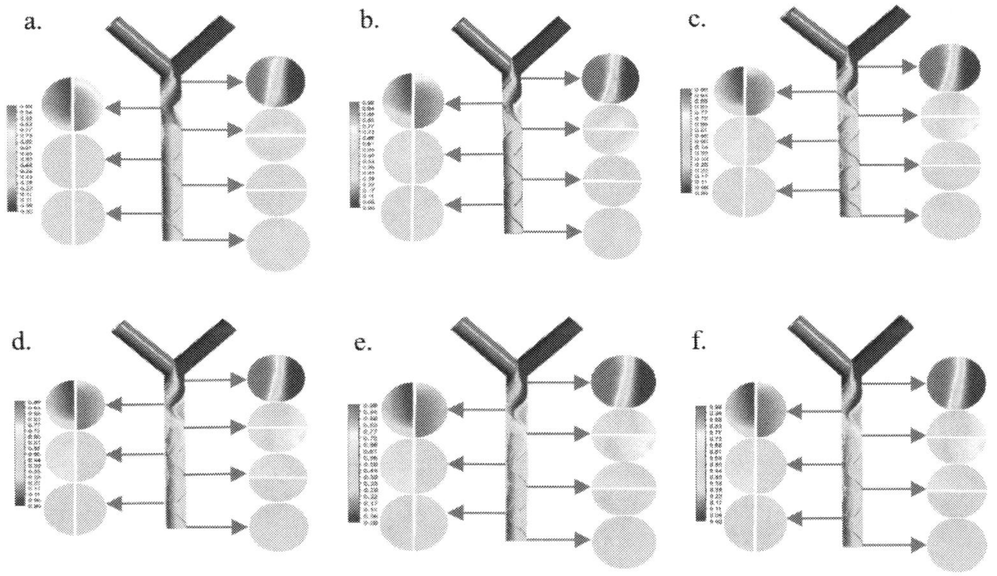

Fig. 1. Gelatin mass fraction (red colour) distributions at seven successive cross-sections and across the mixer for various Re numbers: (a) Re=10, (b) Re=100, (c) Re=500, (d) Re=1000, (e) Re=1500, and (f) Re=2000.

As the number of mixing units increases, the enhancement of the mixing index is generally smaller (Fig. 2b). After two mixing elements, the mixing index rapidly increases, and its value is higher than 0.7 over the all range of Reynolds numbers. The lowest mixing was observed at the cross-sectional plane where no mixing element is present. As the fluids proceed to the outlet the mixing index significantly increases. From Fig. 2c. it is also possible to observe that after the fourth mixing element (near to the outlet) the mixing index is almost the same for all the considered cases. This phenomenon, defined as tailpipe or downstream effects, is associated to the fact that in the case of laminar flow no further mixing takes place in this region, while in the case of turbulent flow the mixing process proceeds as the extra turbulence produced by the mixing units dies out [11].

a b. c

Fig. 2. (a) Effect of Re number on the mixing index at the outlet; the influence of the number of mixing elements on mixing performance: (b) Mixing index vs Re number, (c) Mixing index vs number of mixing units.

4 Conclusion

The flow behaviour of two miscible biomaterials in a bifurcated shape Kenics mixer was studied using a numerical method and considering different Re numbers. It was observed that splitting and recombining the fluids in a continuous manner played a significant role to achieve high levels of mixing. The highest velocity magnitude was obtained at Re=2000 with a corresponding peak pressure drop of around 100 kPa. The results show that the proposed mixer design can be used to achieve high mixing quality even at the low Reynolds numbers.

References

1. C. Mota, D. Puppi, F. Chiellini, and E. Chiellini, J. Tissue Eng. Regen. Med. **9**, 3 (2015)
2. J. Malda, J. Visser, F. P. Mechels, T. Jüngst, W. E. Hennink, W. J. A. Dhert, J. Groll, D. W. Hutmacher, Adv. Mater. **25**, 36 (2013)
3. A. Afzal and K. Y. Kim, Sensors Actuators, B Chem. **211**, 198–205 (2015)
4. M. Rafeie, M. Welleweerd, A. Hassanzadeh-Barforoushi, M. Asadnia, W. Olthuis, and M. E. Warkiani, Biomicrofluidics, **11**, 1 (2017)
5. N. Ait Mouheb, D. Malsch, A. Montillet, C. Solliec, and T. Henkel, Chem. Eng. Sci. **68**, 1 (2012)
6. I. Shah, S. W. Kim, K. Kim, Y. H. Doh, and K. H. Choi, Chem. Eng. J. **358**, 691–706 (2019)
7. J. Zhang and X. Luo, Micromachines, **9,** 5 (2018)
8. S. Sarkar, K. K. Singh, V. Shankar, and K. T. Shenoy, Chem. Eng. Process. Process Intensif. **85**, 227–240 (2014)
9. H. S. Santana, J. L. Silva, and O. P. Taranto, Chem. Eng. Sci. **132**, 159–168 (2015)
10. W. Raza, S. Hossain, and K. Y. Kim, Sensors Actuators, B Chem. **258**, 381–392 (2018)
11. E. L. Paul, V. a Atiemo-obeng, and S. M. Kresta, *Handbook of industrial mixing: Science and practice* (Hoboken: John Wiley & Sons, Incorporated, 2003)

MATEC Web of Conferences **318**, 01046 (2020) https://doi.org/10.1051/matecconf/202031801046
ICMMEN 2020

Starch Sandstones in Building Bio-materials

Gabriel Mansour[1], *Maria* Zoumaki[1,*], *and Dimitrios* Tzetzis[2]

[1]Department of Mechanical Engineering, Laboratory of Machine Tools and Manufacturing Engineering, Aristotle University of Thessaloniki, Greece
[2]International Hellenic University, School of Science and Technology, Greece*

Abstract. A review of the recent literature shows that the use of more sustainable, eco-friendly recycled waste materials made from natural biopolymers is an important step of the planning process to reduce the environmental impacts of traditional building materials such as cement and concrete products. This study introduces the application of maize starch in the production of a novel biodegradable construction material. The samples prepared in this investigation were formed by heating a mixture of different proportions of starch, water and sand. The structural properties, morphology and chemical composition of materials were investigated by scanning electron microscopy (SEM) coupled with thermal gravimetric analysis (TGA). The structural characteristics and morphology of the study material to a certain extent resemble natural sandstones, the most common type of sedimentary rocks. Based on the uniaxial compressive strength classification schemes, comparing with the brittle deformation behavior of natural rocks, it can be considered that this material behaves as a polymer - matrix composite with a ductile - thermoplastic mechanical behavior.

1 Introduction

Global warming has become an almost inevitable threat to human life and requires climate change adaptation and mitigation strategies with rapid reduction in global greenhouse gas emissions [1]. The construction sector produces a considerable amount of waste and greenhouse gas emissions that drive global warming. The recent literature on construction research indicates that biodegradable materials made from renewable sources are sustainable concrete alternatives that use natural polymers (such as starch and cellulose) instead of traditional cement [2-5].

The product developed by this project was obtained by heating a mixture of starch, water and sand. The strength and durability performances of such artificial starch – sandstone were also investigated as a function of the composition materials. Based on the uniaxial compressive strength classification schemes, the mechanical behavior of the product can be characterized as a polymer - matrix composite with a more ductile fracture response compared to the brittle mechanical behavior of natural rock materials. In applied soil and rock mechanics, materials can be divided into natural building materials like sand, clay and natural rocks, and the artificial construction materials like cement, concrete and

*Corresponding author: mariazita687@gmail.com

bricks. The materials can be also divided into two classes based on their relative behavior under stress conditions (when they fracture): The brittle solid materials under external loading exhibit little elastic deformation and negligible plastic deformation, while ductile materials under external loading exhibit negligible elastic deformation and substantial plastic deformation. Most building materials and natural rocks under stress (at laboratory temperature) behave as brittle materials with a tendency to break into smaller pieces [6]. The classification schemes of materials [7, 8] are generally based on three important engineering properties of rocks (uniaxial compressive strength, Young's modulus of elasticity and tensile strength) involving different types of compressive stress-strain curves, such as Type I: Elastic (e.g., basalts, quartzite, diabase, dolomites and strong limestones), Type II: Elastic – plastic (e.g., softer limestones, siltstone and tuff; type I or type II are usually typical of most traditional building materials, such as cement mortars, concrete and cement bricks), Type III: Plastic-elastic (e.g., sandstone, granites and schist cored parallel to the foliation), Type IV: Plastic-elastic-plastic (e.g., marbles and gneiss), Type V: Plastic-elastic-sigmoid plastic (e.g., schist cored perpendicular to the foliation), Type VI: Elastic – plastic – creep (e.g., rock salts and other evaporite rocks) and Type VII: Ductile - thermoplastic mechanical behavior (e.g., polymeric materials). Natural rocks based on their geological and engineering properties may be also classified into strong, moderate strong, medium and weak materials bounded by the uniaxial compressive strengths of 100, 50, 25 and 15 MPa, respectively, with a ratio (R) of tensile strength to compressive strength of approximately 0.23 – 0.41 and a compressive modulus of elasticity of about 5 – 100 GPa [9, 10]. For example, moderate strong mudstone, moderate strong limestone, medium strong sandstone, strong limestone, strong sandstone and strong ironstone were found to have a compressive strength of about 50, 42, 85, 143, 168 and 190 MPa, respectively, with a ratio R of about 0.70, 0.43, 0.29, 0.26, 0.39 and 0.23, respectively, with typical values of uniaxial compressive modulus of elasticity for natural rocks (such as basalt, marble, granite, dolomite, limestone, sandstone and evaporite rocks) ranging from 5 – 100 GPa [8- 11].

This study introduces the application of maize starch in the production of a novel biodegradable construction material formed by heating a mixture of starch with sand and water with good cost-effective and environmentally friendly properties and satisfactory compressive strength comparable to that of common red clay bricks. The strength and durability performances of the study material were also investigated as a function of the water to starch ratio. For each starch to sand ratio the maximum compressive strength corresponds to specific water content. Based on modulus of elasticity classification schemes for natural rocks, comparing with the brittle deformation behavior of natural rocks, it can be considered that this starch-sand composite behaves as a polymer - matrix composite with a ductile - thermoplastic mechanical behavior. One of the main objectives of this work is to validate the feasibility and accuracy of a novel preparation technique of this starch - sand material so that it can be easily transferred from the laboratory to an industrial scale.

2 Materials and Methods

In this study the starch – sandstone artificial material was prepared by heating a mixture of sand with corn starch and water (with constant sand to starch ratio of 5:1) according to two different heating techniques, using a microwave oven and a convection oven [5]. The cubic and cylindrical specimens of the starch-based material and natural rocks were illustrated in Figures 1a and 1b, respectively. The bio-composites based on thermoplastic starch and quartz sand were obtained by casting the above mixture in silicone rubber moulds (Figure 1c) and heated in a microwave oven (Daewoo, output power 800 W, model KOR-6C17) for 5min, first at thaw mode for 1.5 min and then at full power mode for 3.5 min or heated at a temperature of 105 °C for 24 h in an industrial in an industrial convection oven. Uniaxial compressive strength test (Figure 1d) and Brazilian test (Figure 1e) were performed on cubic and cylindrical samples (at room temperature of about 20°C) in a servo-controlled (M500-50AT Testometric tensile test machine), according to ISRM [12] suggested methods. The compressive Young's modulus of elasticity can be further determined as the slope of the tangent to the uniaxial compression stress-strain curve. It is generally taken at a stress level equal to 50 % of the compressive strength. Tensile strength was further determined by an indirect test method, the Brazilian test, in which a diametrical line load is applied over the curved surface area of a cylindrical specimen having a length (L) - to - diameter (D) ratio = 1/2 [6].

A disadvantage of the microwave heating procedure is the requirement to place a surcharge load (weight) above the samples in order to prevent deformation and the subsequent cracking and pop up of specimens while heating (Figure 1f). However, it was not considered necessary to place additional weight on the samples by the oven heating method [5]. This study introduces a novel preparation technique that will mitigate some of the problems above. The composites were placed in the oven by compressing the moulds between two flat Teflon plates, securely fastened together with Teflon screws (Figures 1g,1h and 1i) in order to prevent the cracks and pop up of specimens while heating by the microwave heating method (Figure 1j). The above method is suggested as an alternative technique designed to enable with insignificant cost the transition from the laboratory (Figure 1f) to production on an industrial scale (Figure 1j).

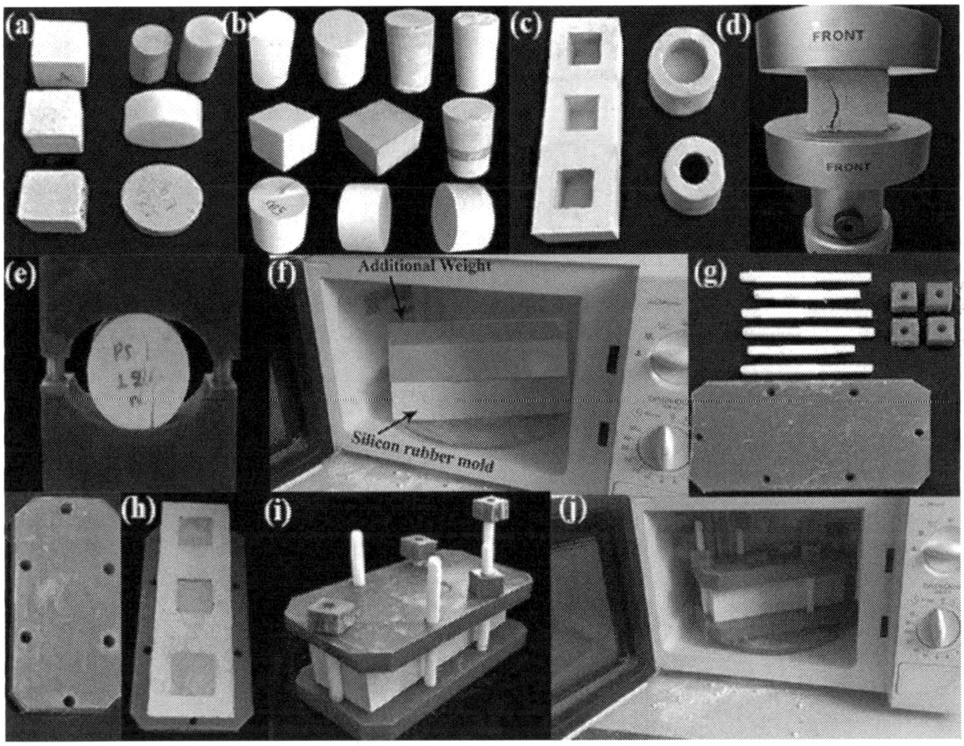

Fig. 1. (a),(b) Cubic and cylindrical specimens of the study material and natural rocks, (c) silicone rubber moulds, (d) uniaxial compressive strength test, (e) Brazilian test, (f) an additional weight over the samples while heating in a microwave and (g) - (j) compressing the moulds between two flat Teflon plates (securely fastened together with Teflon screws) in order to prevent cracks and pop up of specimens while heating in a microwave oven.

3 Results and discussion

This study deals with the use of maize starch as a bio-based construction material. The artificial sandstones prepared in this investigation were formed by heating a mixture of starch, water and sand in the microwave oven (Figure 1j). The uniaxial compressive strength of the prepared artificial solid material as a function of the composition of the mixture (with different proportions of water content and constant sand to starch ratio of 5:1) is presented in Figure 2. As shown in Figure 2, with an increase in water content, the compressive strength increases. At optimum water content of 12%, the maximum compressive strength of 30 MPa was achieved. However, a decrease in the compressive strength was observed above the optimum water content. A similar behaviour was observed for the stiffness of the testing specimens. The structural characteristics and chemical composition of materials were investigated by the scanning electron microscopy (SEM type JEOL JSM-840A, Japan).The mechanical properties and durability of the obtained solid starch-based materials were negatively affected by the highly porous structure morphology, as also confirmed by the SEM image (illustrated in Figure 3).

Thermogravimetric analysis (TGA) is a technique used for characterizing the thermal stability of a material [3]. The TGA curve in Figure 4 shows the mass loss percentage of a sample of this (starch – based) material as a function of increasing temperature (with constant heating rate of 5 °C/min, using a TGA analyzer TAQ50, TA Instruments, USA). The first mass change appears at around 100 °C, which is associated with the loss of the absorbed water (in the structure of the investigated material) at low temperatures in the range of 90 - 150 °C. With further heating, the removal of moisture becomes difficult because water molecules are trapped within the gelatinized starch matrix. The result clearly indicates that the greater mass change is associated with the loss of starch content. This mass loss coincides with the thermal degradation of starch matrix at temperatures above 200 °C with a maximum mass loss rate at around 290 °C (the burning temperature of native corn starch) and the degradation ended at a temperature of 450 °C. The technique of preparing this type of solid polymeric material uses gelatinized corn starch as a binder and sand as filler, and therefore it morphologically resembles an artificial sandstone material. Most building materials and natural rocks after compression tests tend to break into smaller pieces and therefore are classified as brittle materials (types I – V) in Figure 5a. However, as evident from Figure 5b, the evaporitic rock salt revealed the characteristic plastic – creeping deformation before failure (type VI) and the artificial starch – sandstone material exhibited the typical thermoplastic behaviour of polymeric materials (type VII).

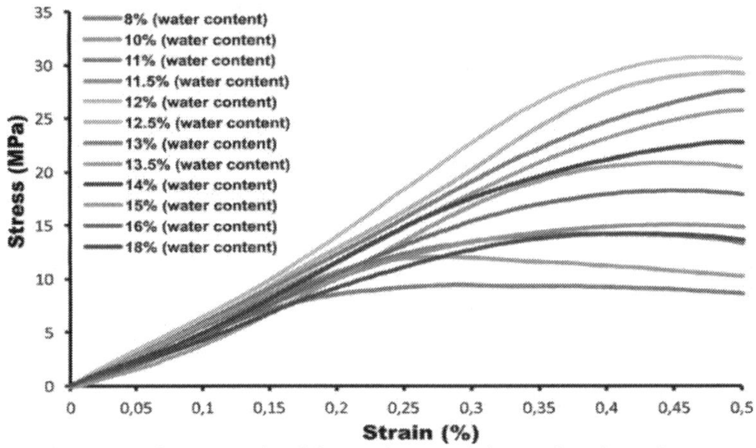

Fig. 2. The uniaxial compressive strength of the study material as a function of water content (with constant sand to starch ratio 5:1)

Fig. 3. SEM image of the structural characteristics of the obtained starch-based solid material

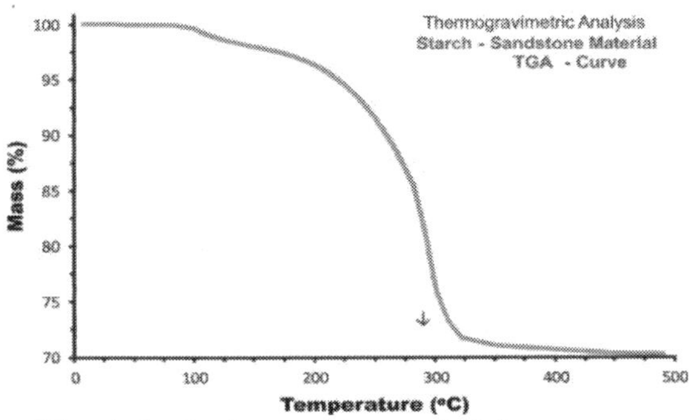

Fig. 4. The TGA curves of the study starch - sandstone material

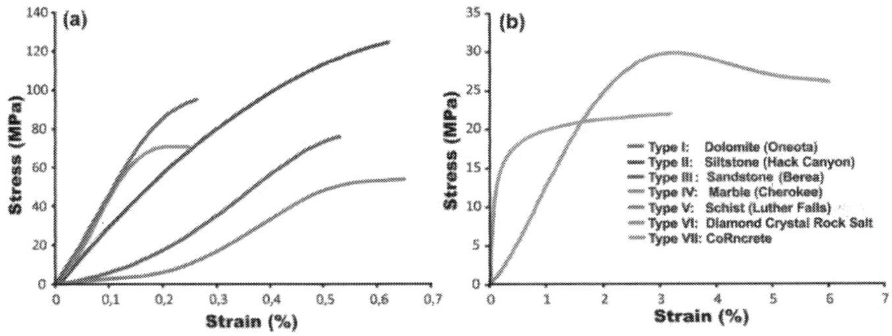

Fig. 5. Stress-strain curves for different types of materials in uniaxial compression strength test to failure.

Samples of the study material were tested and found to have a compressive strength of about 30 MPa with significantly low values of compressive modulus of elasticity (less than 2 GPa) and a ratio (R) of tensile strength to compressive strength of approximately 0.20. On the other hand, the most common types of building materials (such as concrete and cement mortars) and natural rocks, in general, tend to have a tensile strength of 1/10 of their compressive strengths [13, 14]. As can be seen in the representative examples illustrated in Figure 5, the uniaxial compressive strength ranges from about 20 to 130 MPa, and the compressive modulus of elasticity ranges from about 13 GPa to about 50 GPa, while the starch - sandstone is characterized by a very low value of elastic modulus of about 1.5 GPa.

4 Conclusions

The technique of preparing this type of solid polymeric material is based on the physical modification of starch and the formation of a gelatinized starch matrix that binds sand grains and forms a rapid hardening composite structure. The most important factors that were found to affect the physical and mechanical properties of this starch-based composition were the proportions of starch, water and sand, heating scheme, sand particle size distribution, maize starch gelatinization process and the bonding of starch matrix with sand grains. A maximum uniaxial compressive strength was obtained at the optimum water - starch ratio corresponding to the maximum dry density. Generally, this product exhibits similar physical properties with cement mortars as well as some common types of natural rocks and morphologically resembles an artificial sandstone material. The test results showed that this type of novel construction material exhibited the typical thermoplastic behaviour of polymer - matrix composite materials, however as the starch to sand ratio gets smaller, the plastic zone deformation reduces, and the failure becomes more brittle.

This research is co-financed by Greece and the European Union (European Social Fund- ESF) through the Operational Programme «Human Resources Development, Education and Lifelong Learning» in the context of the project "Strengthening Human Resources Research Potential via Doctorate Research" (MIS-5000432), implemented by the State Scholarships Foundation (IKY).

References

1. Daniel Oudin Åströml, D.O., Forsberg B., Ebi K.L., Rocklöv J., Nat. Clim. Change **3**, 1050–1054 (2013).
2. Zoumaki, M., Tzetzis, D., Mansour, G., IOP CONF. SER. MATER. SCI. ENG **564**, 012037 (2019).
3. Mansour, G., Zoumaki, M., Marinopoulou, A., Raphaelides, S., Tzetzisc, D., Zoumakis, N., Starke **72**, 1900166 (2020).
4. Le, A.T., Gacoin, A., Li, A., Mai, T.H., Rebay, M., Delmas, Y, Constr Build Mater. **61**, 106–113 (2014).
5. Kulshreshtha, Y., Schlangen, E., Jonkers, H.M., Vardon, P.J., LA Van Paassen, L.A., Constr Build Mater. **154**, 411-423 (2017).
6. Johnson, R.B., Degraff, J.V. *Principles of Engineering Geology* (John Wiley & Sons, New York 1988).
7. Miller, R.P., (*Engineering classification and index properties for intact rock, Ph.D., Thesis*, Univ. Illinois 1965).
8. Tatsuoka, F., Shibuya,S., *Proceedings of the 9th Asian Regional Conference on SMFE*, Bangkok, 101–170 (1992).
9. Agustawijaya, D.S., Civ. Eng. Dimens. **9**, 9-14 (2007).
10. Kanji, M.A., J. Rock Mech. Geotech. Eng. **6**, 186-195 (2014).
11. Dobereiner, L., De Freitas, M.H., Geotechnique **36**(1), 79-94 (1986).
12. ISRM Suggested Methods *Rock characterization testing and monitoring* (Editor: E.T. Brown, Published for the commission on testing methods, International Society for Rock Mechanics, Pergamon Press 1981).
13. Sari, M., Glob. j. Earth sci. eng. **5**, 1-25 (2018).
14. Stagg, K.G., Zienkiewicz, O.C., *Rock Mechanics in Engineering Practice* (John Willey & Sons, USA 1975)

Research Methods And Experiments of Piezoelectric Plates

Andrzej Wróbel[1,*]

[1]Department of Engineering Processes Automation and Integrated Manufacturing Systems, Faculty of Mechanical Engineering, Silesian University of Technology, Konarskiego 18a Street, 44-100 Gliwice, Poland

Abstract. The correct design of technical systems requires consideration of both geometrical and material parameters not only of the mechanical part, but also of the electrical part of the system. The correct assessment of characteristics and testing of intelligent systems should be carried out by both theoretical analyzes and laboratory tests. Usually this description is by means of mathematical equations. The ideal solution is in which it is possible to refine the object in such a way that the obtained model is the most accurate, and even the same as the physical model. Considering the complexity of the model, computational cost, computational possibilities and time-consuming calculations would practically prevent the use of this model. Intelligent materials are more often used for active vibration damping. They can be defined as a group of materials that have the ability to change their physical properties in an appropriate way as a result of external influences. One of the most popular materials in this group are piezoelectric materials. They are used where it is necessary to reduce vibrations with small amplitudes of μm and work in high frequency ranges. In the presented article, the author will attempt to refine the model of intelligent materials with particular emphasis on piezoelectric plates.

1 Introduction

The article attempts to present processing of analysis results of mathematical analyzes with modeling and testing of phenomena occurring in intelligent materials, in particular piezoelectric plates. The submitted studies are part of the conducted research at the Department of Automation of Engineering Processes and Integrated Manufacturing Systems of the Silesian University of Technology. Intelligent materials can be defined as a group of materials that are available to change properties under the influence of external factors. The most popular materials in this group include metal alloys with shape memory, magnetorheological fluids as well as piezoelectric materials. Due to the great possibilities of applications, these materials are often used in various fields of science and technology for several decades. Piezoelectrics are most often used in automatics, micromanipulation, measuring technology and medicine, used as actuators or sensors. One of the disadvantages

* Corresponding author: andrzej.wrobel@polsl.pl

of this kind of sensors is the small displacement, which can be increased by complex systems.

The correct identification and analysis of the piezoelectric systems should be carried out both by theoretical analysis and laboratory tests. Under certain conditions, physical examination of vibrating objects is difficult or impossible to implement. In cases where laboratory tests are expensive and their implementation is economically unjustified, a correspondingly similar mathematical model should be designed.

The description of the analyzed physical phenomenon is usually presented by means of mathematical equations. These are most often differential equations, which are a mathematical model of the studied phenomenon. This is usually a continuous or discrete model.

The ideal solution is the situation in which the designer will such elaborate a detailed object, so that the obtained model was the most accurate and even perfect with the physical model. However, considering the complexity of the model, computing capabilities, and time-consuming calculations would make it practically impossible to use this model. In the presented study, the author attempts to rather detail analyze the longitudinally vibrating piezoelectric plate and to examine the influence of external and internal phenomena occurring in it.

This article is a part of undertaken analysis and experimental research of piezoelectric plates by author on physical phenomena occurring in intelligent materials, working as subassemblies or assemblies in machine construction.

2 Model under examination

The basic model adopted for analysis by the author is the model of a single piezoelectric plate, as in Figure 1.

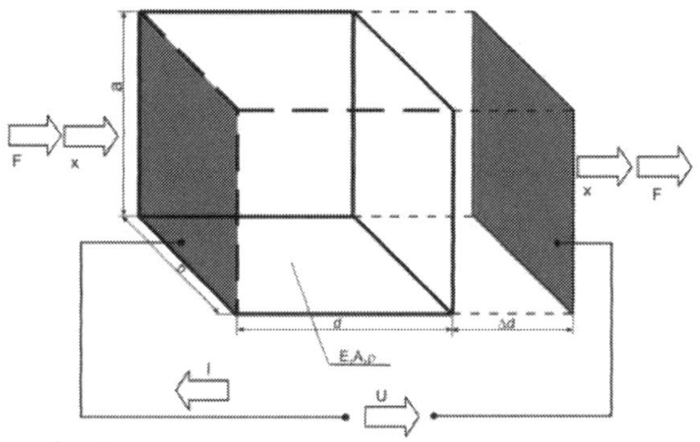

Fig. 1. Model under examination

The accepted model is characterised by the segment constant section. Parameters describing the piezoelement model are: face area, Young's modulus, mass density, power voltage applied to connectors and current. Dislocations (x) , and current (I) are caused by exciting forces (F). The exaggerated model is very convenient for mathematical analysis, especially for matrix analysis, in which case the system can be treated as the n port system. Linear dislocations are measured only on the x axis starting from the beginning of the co-ordinate system.

Assuming the right matrix [1, 3, 5] containing both mechanical and electrical elements presented in equation 1, it is difficult to apply the already known and previously used methods for the analysis of mechanical systems.

$$\begin{bmatrix} F_1 \\ U \\ F_2 \end{bmatrix} = \begin{bmatrix} \dfrac{Z}{\tan kd} & \dfrac{h}{\omega} & -\dfrac{Z}{\sin kd} \\ \dfrac{h}{\omega} & \dfrac{1}{\omega C_0} & -\dfrac{h}{\omega} \\ \dfrac{Z}{\sin kd} & \dfrac{h}{\omega} & -\dfrac{Z}{\tan kd} \end{bmatrix} \begin{bmatrix} u_1 \\ i \\ u_2 \end{bmatrix}. \quad (1)$$

The use of a matrix presented in equation 1 in a detailed analysis of the piezoelectrics may prove to be a time-consuming activity. Therefore, the authors idea was performed modeling of the system using a replacement system. In order to maintain the transparency of a representation, the characteristics are marked, in brackets, by Arabic numbers according to [2, 6, 7]. As a result of introduction of the substitute matrix, there was obtained matrix 2x2, which is the basis for further analysis by network methods.

In order to apply the method of structural numbers, the piezoelectric plate system should be analyzed in the simplest possible form. The main benefit of such a conversion is the reduction of the four-vertex graph (Fig 2.) being a representation of the first system, minimization of the number of branches and vertexes, with constant input-output relations.

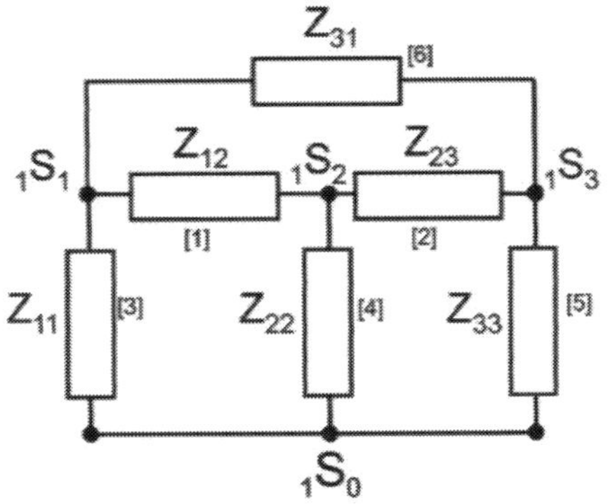

Fig. 2. The representation of the first system for analysis

As a result of the reduction system (Fig. 2) into the reduced system we expect to receive a system shown in Fig. 3.

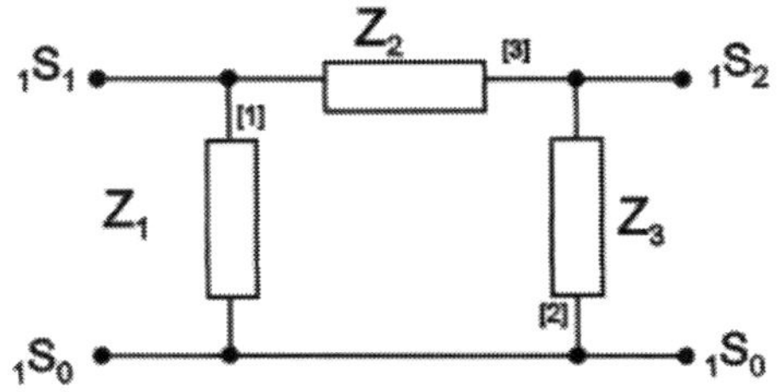

Fig. 3. The representation of the replacement circuit accepted for further analysis

The structural number of the oryginal system is:

$$A = [136][256][345]. \tag{2}$$

By multiplying the factors of the structural number (2), we obtained:

$$A = \begin{bmatrix} 1 & 1 & 1 & 1 & 1 & 1 & 1 & 1 & 3 & 3 & 3 & 3 & 3 & 6 & 6 & 6 & 6 & 6 \\ 2 & 2 & 2 & 5 & 5 & 6 & 6 & 6 & 2 & 2 & 5 & 6 & 6 & 2 & 2 & 2 & 5 & 5 \\ 3 & 4 & 5 & 3 & 4 & 3 & 4 & 5 & 4 & 5 & 4 & 4 & 5 & 3 & 4 & 5 & 3 & 4 \end{bmatrix}. \tag{3}$$

The edges of the replacement graph were calculated based on the formula:

$$Z_k = \frac{\det\limits_z \left(\dfrac{\partial A}{\partial D^{\mu_1 k}} \cap \dfrac{\partial A}{\partial D^{\mu_2 k}} \right)}{\det\limits_z \dfrac{\partial A}{\partial D}}. \tag{4}$$

Using the signs shown in Equation 4, there was written:

$$Z_{1'} = \frac{\det\limits_Z (A_{2'} \cap A_{3'})}{\det\limits_Z A_{2'3'}} = \frac{Z_1(Z_3 + Z_4) + Z_3(Z_2 + Z_4)}{Z_1 + Z_2 + Z_4}. \tag{5}$$

The replacement value $Z_{3'}$ of the graph was calculated in accordance with the written conjunction:

$$Z_{3'} = \frac{\det\limits_Z (A_{1'} \cap A_{3'})}{\det\limits_Z A_{1'3'}} = \frac{Z_1(Z_2 + Z_6) + Z_6(Z_2 + Z_4)}{Z_1 + Z_2 + Z_4}. \tag{6}$$

The replacement value $Z_{2'}$ of the system was calculated in accordance with the written:

$$Z_{2'} = \frac{\det\limits_Z (A_{1'} \cap A_{2'})}{\det\limits_Z A_{1'2'}} = \frac{Z_4(Z_2 + Z_5) + Z_5(Z_1 + Z_3)}{Z_1 + Z_2 + Z_4}. \tag{7}$$

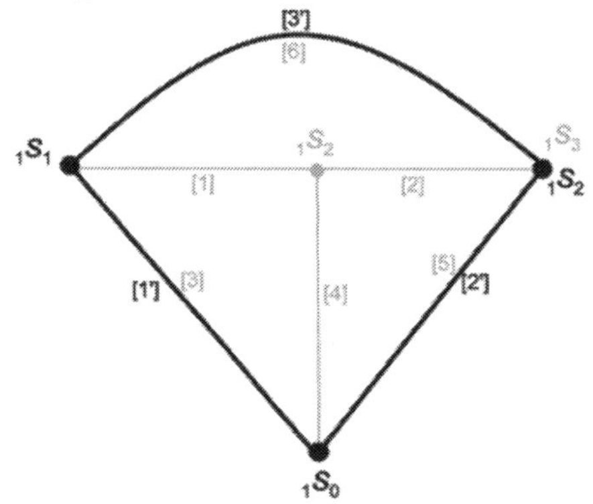

Fig. 4. A substitute system calculated by the structural number method after reduction

After calculating the substitute values, the correct replacement system was checked (Fig. 4). In the circuit simulation program, the circuits were created, and the branches marked [1]...[6] the impedance values were assigned. From the analysis of presented methods of reduction of the primary system to the replacement system by the method of transfigurations and the method of structural numbers, it follows that the calculated impedance values of individual branches of the replacement system are identical in both methods. This confirms the validity of the presented hypothesis that the presented methods can be used to study and analyze of the piezoelectric systems.

3 Experimental measurement

In this point theoretical considerations on the analysis of piezoelectric plates' vibrations will be validated with the conducted experiment. During the generation of characteristics determining the displacement of the piezoelectric stack in the frequency domain, there was used analysis, which main assumption is the distribution of the examined signal into elementary components and to definite the amplitudes of these components.

Fig. 5. View of piezoelectric plates used for experimental research

The tests were carried out using the laser vibrometer and the generated signal was recorded using an oscilloscope. The results are shown in the figure 6.

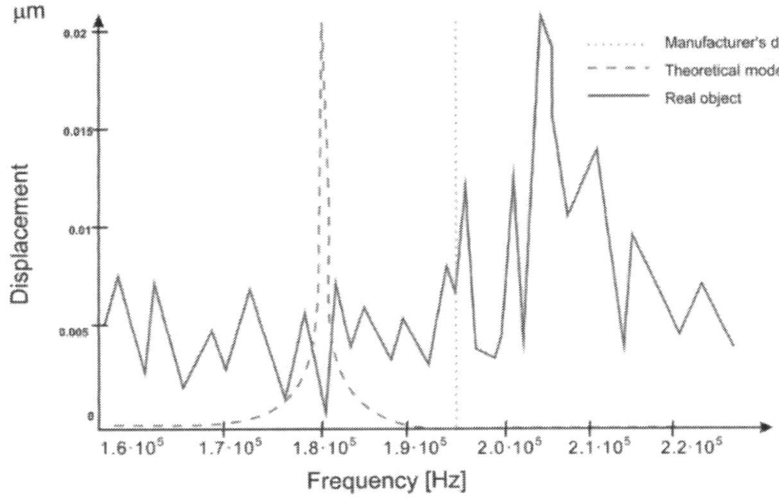

Fig. 6. Comparison of resonant frequencies of the algebraic method and research on the piezoelectric element

The analyzed courses originating from the laser vibrometer showed the vibration courses of the piezoelectric system, whereas for a readable data to be read, there was used a Fast

Fourier Transform. This analysis allowed to read data in the frequency domain and a readout of the amplitude with given frequency extortion.

4 Conclusions

The tests were performed on single piezoelectric elements and the piezoelectric stacks made of two plates. The test results are presented in the form of comparative diagrams. On the basis of the diagrams, there were compared the algebraic method and the research on real elements, but a slight error of both methods was noted. Analyzing the graphs presented in Fig. 6, it was found that the percentage error in relation to the research performed on the real element is not greater than 12%. In the next step, the results of experimental and mathematical models were compared with the data contained in the piezoelectric plates manufacturer's card. The cataloged values of the frequency of vibrations of the tested elements differ about 8% from the values obtained as a result of testing, and about 7% in reference to models calculated using the algebraic method. Analyzing the research results, it can be stated that the adopted mathematical model is derived correctly.

References

1. A. Wróbel A., A. Buchacz, M. Płaczek, *Selection of the geometric and materials parameters in piezoelectric sensors level.* Vibroengineering PROCEDIA, **vol. 10**, p. 277-281 (2016)
2. W.P. Mason, *Electromechanical Transducers and Wale Filters*. Van Nostrand, (1948)
3. A. Buchacz, M. Płaczek, A. Wróbel: *Modelling of passive vibration damping using piezoelectric transducers – the mathematical model.* Eksploatacja i Niezawodnosc – Maintenance and Reliability, **16 (2):** 301–306, (2014)
4. S. Bellert, *Prace wybrane*. Państwowe Wydawnictwo naukowe. Warszawa (1981)
5. W. Soluch (red.), *Wstęp do piezoelektroniki*. WKiŁ, Warszawa 1980,
6. A. Wróbel, M. Płaczek, A. Baier: *Calculations of flexibility module in measurements instruments*, IOP Conf. Ser.: Mater. Sci. Eng. 227 012137 (2017)
7. A. Wróbel, *Analysis of possibility of applying the PVDF foil in industrial vibration sensors*. Modern technologies in industrial engineering, IOP Conference Series; Materials Science and Engineering ; **vol. 95** 1757-8981, (2015)

Model-based definition capabilities and its impact on industrial productivity

Emmanouil Varitis [1,*], *Konstantinos* Rinos[1], and *Nikolaos* Kostis[2]

[1]University of Western Macedonia, Department of Mechanical Engineering, 50100 Kozani, Greece
[2] University of Patras, Department of Mechanical Engineering and Aeronautics, 26500 Patras, Greece

Abstract. The need for optimization of the product manufacturing cycle and data management resulted in the creation of the model-based definition technology. The beneficial impact in most processes (CAD, CAE, CAPP, CAM, etc.) of a product manufacturing cycle, made the MBD technology implemented in various industries. The following paper has the purpose of highlighting the benefits and limitations of the current MBD capabilities and makes its impact on the industrial collaboration known by summarizing the communication file formats, which support this technology. Special interest is attributed to lightweight collaboration files, especially 3D PDFs, which are capable to deliver information and data to the domains of a company inside or outside of the manufacturing cycle and support all the necessary downstream activities of the product.

1 Introduction

It has always been industry's intention to create economical, accurate, reliable, and time-saving methods of distributing product data across its sub-branches such as design, manufacturing, managing departments, etc. The first big step was the digitalization of information like 2D schematics, which enabled companies to fasten-up their designing processes by eliminating the need for hand sketching and therefore decreasing errors originating from human personnel [1]. However, its impact on data distribution and minimization of the necessity for supplementary documents is not exactly spot on. Thankfully a very promising candidate emerges in the name of model-based definition (MBD). The resourceful idea of implementing all the important information on a 3D model without the infringement of the principles of mechanical design has positive effects on the product's life-cycle management (PLM) and especially in the manufacturing processes [2]. The following research summarizes the evolution and the capabilities of MBD technology and points out the limitations of its current use. Furthermore, this paper divulges the impact of MBD technology in industrial collaboration and product data management (PDM), based on the interoperability of the existed exchange information formats.

* Corresponding author: evaritis@uowm.gr

2 MBD evolution and research overview

The Model-Based Definition (MBD) technology is considered as an innovative tool for the digital definition of a product, as well as a helpful guide for its subsequent manufacturing processes. Besides, this technology is not a current discovery but, it is originated in the 1990s when the aerospace industry and specifically the Boeing used it as a tool for the reduction of the aircraft production cycle [3]. As soon as possible (1997), the American Association of Mechanical Engineers (ASME), in cooperation with the Boeing Company, started its research for this new methodology and by July 2003 developed the standard "ASME Y14.41-2003, Digital Product Definition Data Practices" [2]. This standard destined for the use of the engineering team of the American Department of Defense (DoD) to fully define the requirements and the best practices for the manufacturing of its products (the latest version ASME Y14.41-2019). For the same reason, ASME launched in February 2009, the standard "ASME Y14.5-2009, Dimensioning and Tolerancing", which has been revised to its current version in 2018. This standard operates as guidance for the establishment of rules and requirements of the implementation of dimensions and tolerances. MBD technology uses both of these standards for the integrated definition of the manufacturing intent of a part or assembly. At the same time, European countries wanted to include this technology to their international standards, developed in 2006 the standard "ISO 16792-2006, Technical Product Documentation - Digital Product Definition Data Practices", to specify the operational requirements of the 3D model definition data, including the engineering data of 2D drawing (the latest version ISO 16792-2015). These standards assist in straddling the chasm between the distinguished use of MBD technology and its general use from a variety of industrial domains. Specifically, MBD is now fully adopted by the engineering software industry as the main idea for the development of a variety of software programs SOLIDWORKS MBD, CATIA MBD, PTC CRO MBD, etc.

Despite the current capabilities of MBD, there is a necessity for optimization and implementation in more industrial fields. Therefore, this technology has been the point of interest in many kinds of research over the last years. Quintana et al. achieved to define the limitations [4], which restrict the full adoption of MBD as the resource of information for the product life cycle. Based on their research, there are still many industrial requirements that should be overcome, but according to their benefits, MBD has the potentials to replace the 2D drawings in the future. Furthermore, this technology is considered a useful tool for inspection and manufacturing processes. As a consequence, Neng Wan et al. introduced two different methods (forward and reversed) for the creation of a 3D machining process based on the MBD technology [1]. These methods enhance the efficiency of the process planning with the combination of machining knowledge and model information in the same document. Nevertheless, MBD documents are great resources for commerce information (resources, marketing, management, finance, etc.) of a product and they can improve the collaboration between interior (branches inside the company) and exterior (partners, customers, etc.) teams. For that reason, Hartman et al. proposed a framework for the evaluation of different collaborative MBD product representations based on four different data formats [5]. This framework includes the necessary criteria of the workflows, which included in a product life cycle insight a Model-Based Enterprise (MBE) [7].

3 MBD technology and capabilities

3.1 Concept and applications

The Model-Based Definition (MBD) is a revitalizing technology that utilizes 3D models as the principal carrier of all vital information of the overall product life cycle. The main idea is to establish 3D models as the main source of information for processes such as design, production, technical documentation, management, etc. Specifically, MBD data sets are composed of 3D geometry, dimensions, tolerances, and manufacturing annotations [6]. A demonstration of dimensions viewed in a 3D model is shown in figure 1. Additional information regarding the bill of materials (BOM), technical information, and production properties can be implemented. Moreover, it provides the fluency of automation regarding repetitive tasks by standardizing data [8]. The enrichment of 3D models with drawing annotations and geometrical-topological information optimize their implementation in CAM machining processes, mitigating this time-consuming procedure [2]. Besides that, the avoidance of the error-prone process of converting 3D data to 2D drawings is of great importance. From a marketing perspective, the 3D visualization of the final product is useful for promoting and communicating with customers [4]. Consequently, MBD can have a central and pivotal role in managing and distributing data across many brunches avoiding excess, supplementary, and duplicated documentation. As a sequence time and cost of engineering procedures are decreased.

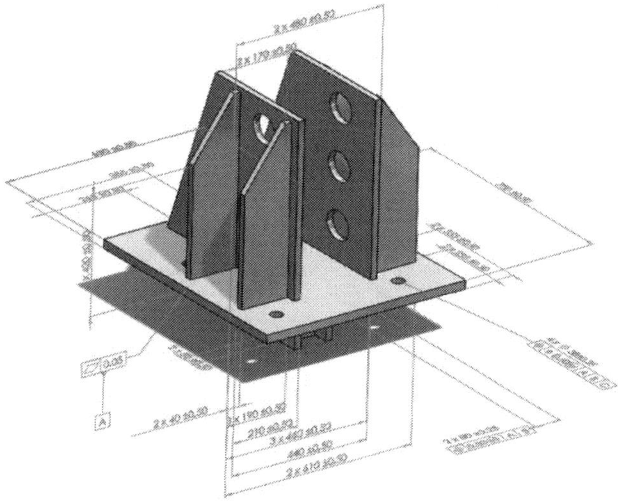

Fig. 1. MBD dimensions and tolerances implementation in 3D model.

3.2 Benefits

The beneficial impact of MBD technology does not pertain only to its capabilities per se. Firstly, the ease of manipulation of the model in cooperation with the well-structured CAD software environment, lead in the deep understanding of even the most complex geometries and optimize the interaction between the user and the model. Moreover, modern MBD-CAD software provides automation in dimensioning, positioning, tolerancing, etc. enabling the user to use commands like auto-dimensioning or classify groups of features, which have the same properties. Besides that, the use of some automation in complex geometries or assemblies may cause many problems, but its implementation in simple parts can improve the productivity of the designer as well as the efficiency of the model. Furthermore, MBD technology aims the understating of the design intention of the model and improves the interpretation of non-designer users [4]. Considering this, the attached information on the model operates as a supplementary factor for the subsequence processes of the 3D model. For instance, the feature-recognition process is a fundamental step for the manipulation of the 3D models in neutral files form and exists in every process (CFD, CAPP, CAM, CMM, etc.) of the product manufacturing cycle after the design [1,9]. Therefore, MBD technology aims this process by providing important information, like the existence of a number of holes or chamfers with the same dimensions, and prevents the user from doing it manually. Another advantage of the cooperation between MBD technology and modern CAD software is the automatic recognition of standardized features [10]. Standards like ASME Y14.41, ASME Y14.5, and ISO 16792 are already installed in the CAD software, and therefore, the MBD users do not need to spend valuable time for dimensioning standardized features (holes, metric threads, etc.) [2]. Last but not least, the combination of 2D drawings and 3D models in a single document, as well as the continuous updating of this document for changes, resulting in the improvement of the PDM process. Consequently, the easy in managing and storing has a great impact on its later use from departments inside or outside of the product manufacturing cycle (sales, technical support, marketing, etc.).

3.3 Limitations and fields of improvement

Despite all the privileges, MBD has to offer there are still handicaps to overcome in order to take a form that will make feasible its smooth immersion in industry. Considering the benefits that occur from the model-based definition method, its adaption from the industry seems to be highly possible. Namely, access to CAD software must be secured for diverse users across the conceptualization, design, manufacturing, quality control, and promotion phase of a product. Printing MBD files is not suggested, due to its nature, it is only readable in 3D environments. Consequently, not only a need for new hardware and software acquirement might arise but also a portion of personnel will have to get acquainted with 3D modeling manipulation. Moreover, highly important is the fact that different portions of the design process have personalized data requirements. To this point, this is arranged by the creation of different types of 2D drawings (assembly, detail, layouts, manufacturing, etc.). Therefore, MBD has to be able to deliver the necessary data to each subgroup while allowing them to use it for their specific tasks without actually modifying it [4]. Taking into account the design of highly complex solid parts, the establishment of international standards will guarantee compliance under strict rules for implementing information to the 3D document properly and universally reducing the risk of misconceptions. For instance, in cases of complex assemblies, and surfaces even experienced users might have to deal with the highly demanding process of defining dimensions as well as extensive and dense engineering data correctly in order to accomplish a successful and efficient inspection. Nevertheless, the existing automation for dimensioning can facilitate this process, but there is still a necessity for additional processing from the user, such as the positioning of the dimensions in proper views for the better interpretation of the model. However, MBD is still a more efficient and simple way for dimensioning, than the traditional 2D drawings, which require technics like model decomposition in its sub-bodies.

4 Efficiency in collaboration

The process of product development depends on various factors such as the complexity of manufacturing, the material, the size, the existence of electronic parts, etc. As a result, the final form of the product originates from the cooperation of different branches from the same or separate company, which uses a variety of software and types of data. Speaking about a model-based enterprise (MBE), the MBD technology accompanies the product form the design process until its deposition for future rehashing. Considering that, this technology is supported by various file formats, which include different information or representations of the product for a variety of uses. Native MBD files can only be shared between software (CAD, CAM, PDM, etc.), which use the same methods for data management and storage, but they allow the user to make changes to the model directly from its design history [5]. Furthermore, neutral MBD files like STEP AP 242 "Managed Model-Based 3D Engineering" include data for inspection, manufacturing, engineering visualization, supplier information, etc., which enables the collaboration between different software (SOLIDWORKS, CATIA, AUTOCAD, etc.), with the cost of possible data loses (design tree, mates, etc.) [11].

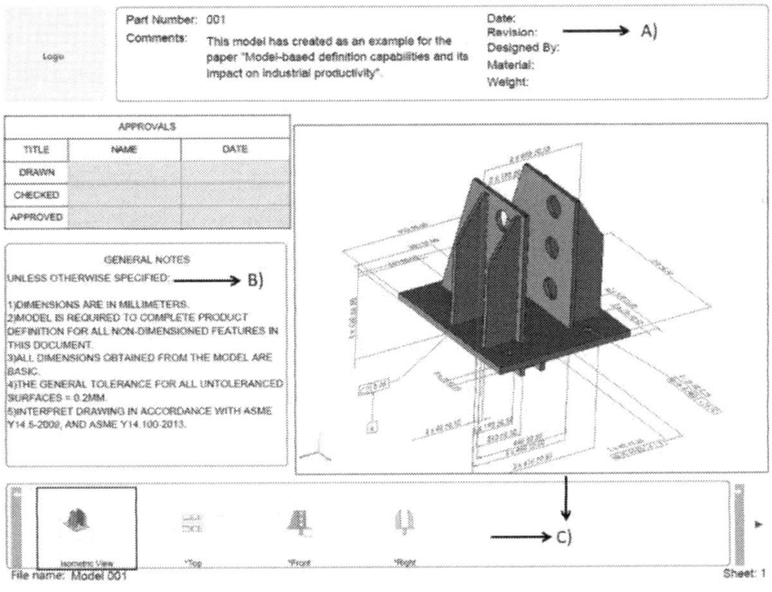

Fig. 2. Detailed MBD data exchange 3D PDF format.

Lightweight formats (3D PDF, 3D XML, JT, etc.) are another category of files, which have as purpose to share information and data of a 3D model, without the typical cost of advanced hardware and software [5]. The main difference between these files and the other two categories is that the Lightweight files contain information for the visualization of the model and are destined to enhance the collaboration between the labor force and not as a linking tool between CAD software. One of the most useful files of this category is the 3D

PDF format. These files enable the user to create 3D views of the MBD model or assembly and export them, including important information for the downstream activities of the model. As an example, figure 2 demonstrates a 3D PDF, which operates as an animated environment for the visualization of a 3D MBD model and its properties. This file includes three different categories of information. The first category (A) includes general information about the creation and revision of the part, as well as some physical properties. Category (B) is a section where the user can add important complementary information regarding various aspects of design. For example, in the case of assembly, notes which address mechanical joints and the way different parts are connected, can ease and guide the process. Moreover, category (C) enables users to have an integer visualization and 3D manipulation of the model. Also, 3D PDF provides useful tools such as section view, measurement, changing transparency etc. that assist the user to achieve better model inspection. Last but not least, these types of formats have the option of including different pages with various information like the bill of material for the supplier chain, manufacturing notes, etc.

5 Conclusions

MBD technology's capabilities have been increased drastically in recent years. They have evolved in a form capable of optimizing and supporting product life cycle management (PLM). With the current interoperability of data formats that support this technology, it is expected that MBD will become the main source of product information not only in the manufacturing process but also in other divisions of the industry. Moreover, this research enlightens the advances in collaboration through formats like 3D PDF files based on information (dimensioning, bill of materials) that can be distributed by them. Even though there are still aspects to improve, this technology has the credentials for further research due to its potential for improving manufacturing and industrial collaboration.

References

1. N. Wan, R. Mo, L. Liu, J. Li, Comp.In.Indu,**65**,537-549,(2014)
2. W. ZHU, M. BRICOGNE, A. DURUPT, S. REMY, B.LI, B.EYNARD, IFAC. Pap.On.Li, **49**,485-490,(2016)
3. B. Friendland, R. Malone, J. Herrold, INCOSE. Int. Sym, **26**, 386-398,(2016)
4. V. Quintana, L.Rivest, R. Pellerin, F.Venne , F. Kheddouci, Comp.In.Indu, **61**, 497-508,(2010)
5. N. Hartman, P. Rosche, K. Fischer, Int.Con.Prod.Lif.Rep.In.Prod.Lif.Wor, **388**,424-434,(2012)
6. Y. W. Huang, IOP Conf. Ser.: Mater. Sci. Eng. 616, (2019)
7. S. Frechette, CIRP.Int.Con.on.Man.Sys, (2011)
8. M. Alemanni, F. Destefanis, E. Vezzetti, Int.Jour.Adv.Manuf.Tec, **52**,1-14,(2011)
9. L. Li, Y.Zhengd, M. Yange, J.Lengf, Z.Chengd, Y. Xieg, P. Jiange, Y. Mad, Rob.Com.Int.Man, **61**, (2020)
10. W.D Li, S.K. Ong, A.Y.C. Nee, Com.Aid.Des, **34**,849-868, (2002)
11. A.Sarkara, D. Šormaza, Int. Con. Flex. Aut. Int. Man, **38**,513-523,(2019)

MATEC Web of Conferences **318**, 01049 (2020) https://doi.org/10.1051/matecconf/202031801049
ICMMEN 2020

The mechanical performance of 3D printed hierarchical honeycombs using carbon fiber and carbon nanotube reinforced acrylonitrile butadiene styrene filaments

Michel Theodor Mansour[1,*], *Konstantinos* Tsongas[2], and *Dimitris* Tzetzis[3]

[1]Technical University of Crete, School of Production Engineering and Management, Chania, Greece
[2]Aristotle University of Thessaloniki, Department of Mechanical Engineering, Thessaloniki, Greece
[3]International Hellenic University, School of Science and Technology, Thermi, Greece

Abstract. The aim of this paper is to design hierarchical honeycombs as well as manufacturing such structures with a commercial 3D Printer using Fused Filament Fabrication (FFF) technique. The materials under study are commercial filaments such as acrylonitrile butadiene styrene (ABS), acrylonitrile butadiene styrene/carbon fibers (ABS/CF) and acrylonitrile butadiene styrene/carbon nanotubes (ABS/CNTs). The fabricated hierarchical honeycombs were examined by compression tests in order to evaluate the mechanical behaviour of such honeycomb 3D printed structures. The compression behaviour of the hierarchical honeycombs was assessed also with finite element analysis (FEA) and at the end there was a comparison with the experimental findings. The results reveal that the 2nd order hierarchy presented an increase both in stiffness and strength in comparison with the 0th and the 1st hierarchies which make such designs a suitable for structures require such properties. Also, the results reveal that ABS/carbon fiber constructs outperform the other materials under study.

1 Introduction

In the last few years, 3D Printing had a tremendous impact on the manufacturing industry globally. This satisfactory outcome could be ascribed mainly to its exceptional capability to directly fabricate complex parts without special tools and at the same time to the decrease of material waste, time and cost of fabrication [1, 2]. It is widely acknowledged that there is a shortage of printable materials on the market nowadays for special applications due to their inferior mechanical performance, so it is very essential to develop printable materials with superior mechanical behaviour [1-3]. For that reason, composite filaments have been regarded as promising 3D-printing feedstocks due to their compelling combination properties of mechanical robustness, lightweight and versatility. Combining such materials with

* Corresponding author: mmansour@isc.tuc.gr

honeycomb structures, it is possible to make a lighter and at the same time stronger structures possibly with high-energy absorption capability and good crashworthiness behaviour [4].

In the current work, the aim was to present the design and fabrication of hierarchical honeycombs structures with three different printing materials but with the same polymer matrix. The materials used were an acrylonitrile butadiene styrene (ABS) plastic filament as well as composites such as acrylonitrile butadiene styrene with carbon fibers (ABS/CF) and acrylonitrile butadiene styrene with carbon nanotubes (ABS/CNTs). A computer generated algorithm was developed in order to create parametrically each hierarchical honeycomb structure, giving as input dimensions of the honeycomb such as width, number of hexagons horizontally and vertically and the hierarchy of the structure. Furthermore, the mechanical behaviour of these hierarchical honeycomb structures was evaluated by uniaxial compression tests. Finally, the experimental results were compared with predictions from finite element analysis (FEA), in order to understand these honeycomb structures and utilizing them for real life structures and components.

2 Design of hierarchical honeycombs

In the current work the vertices of a regular hexagonal lattice structure are replaced with smaller hexagons to achieve a shape with one level of hierarchy. Simultaneously the wall thickness is diminished so as to maintain the overall density. It is anticipated that higher hierarchies will demonstrate stiffness superior to that of its regular hexagonal counterpart of equal relative density. Fractal appearing honeycombs can be achieved by such approach with higher orders of structural hierarchy, if this replacement procedure for three-edge vertices is repeated. Figure 1 (a) shows the regular, 1st and 2nd hierarchies of the hexagonal honeycomb cells. According to A. Ajdari et. al. [5] for each level of hierarchy the structural organization of the honeycomb can be defined as the ratio of the introduced hexagonal edge length (b for the 1st order hierarchy and c for the 2nd order hierarchy), to the original hexagon's edge length, (a), as described in Figure 1 (a) i.e., γ_1= b/a and γ_2= c/a. For a honeycomb with 1st order hierarchy, $0\leq b\leq a/2$ and thus, $0\leq\gamma_1\leq0.5$, where γ_1= 0 denotes the regular honeycomb structure. For a honeycomb with 2nd order hierarchy, there are two geometrical constraints, $0\leq c \leq b$ and $c \leq a/2 - b$. In terms of the ratio parameters, the constraints are $0\leq \gamma_2\leq \gamma_1$ if $\gamma_1\geq 0.25$ and $0 \leq \gamma_2 \leq (0.5 - \gamma_1)$ if $0.25 \leq \gamma_1 \leq 0.5$. The dimensionless relative density, can be given in terms of t/a:

$$\rho = \frac{2}{\sqrt{3}} \times (1 + 2\gamma_1 + 6\gamma_2) \times \frac{t}{a} \quad (1)$$

where t is the thickness of the cell walls, from which the special cases of $\gamma_2, \gamma_1 = 0$ can be read off immediately. For regular honeycomb, $\rho = \frac{2}{\sqrt{3}} \times \frac{t}{a}$ and for honeycomb with first order hierarchy, $\rho = \frac{2}{\sqrt{3}} \times (1 + 2\gamma_1) \times \frac{t}{a}$ and finally for honeycomb with second order hierarchy, $\rho = \frac{2}{\sqrt{3}} \times (1 + 2\gamma_1 + 6\gamma_2) \times \frac{t}{a}$. This relation clearly shows that t/a must be reduced to maintain fixed relative density as γ_1, γ_2 are increased.

3 Experiments

3.1 Manufacturing of 3D printed hierarchical honeycombs

The hierarchical honeycomb structures were fabricated using a commercial open source Ultimaker 2+ 3D Printer through FFF printing method with a 3mm extrusion nozzle and also using Cura software for fine-tuning the honeycomb specimens. Moreover, a nozzle extrusion temperature of 240°C was used with a build-plate temperature of 100°C and also a printing speed of 35 mm/s. These are the very basic printing parameters that were used on the hierarchical honeycomb structures under investigation. The deposition speed was kept constant in order to avoid any variabilities in the 3D printed constructs. Additionally, all specimens were 3D printed in room temperature conditions without humidity control. Commercial Acrylonitrile Butadiene Styrene (ABS), acrylonitrile butadiene styrene reinforced with 15% carbon fibers (ABS/CF) and finally acrylonitrile butadiene styrene reinforced with carbon nanotubes (ABS/CNTs) were selected as the printing materials for the fabrication of the hierarchical honeycomb structures in this paper due to their good mechanical behaviour. The weight percentage of the CNTs was proprietary figure, not disclosed from the manufacturer. In addition, all the specimens had relative density of ρ=0.12. The original hexagon's edge length was a=20mm and also the thickness of the cell wall was measured as t=2mm. Furthermore, the hierarchical honeycomb with 1st order hierarchy had γ_1=0.3 and t=1.25mm and at the same way the hierarchical honeycomb with 2nd order hierarchy had γ_1=0.3, γ_2=0.12 and t=0.86mm. Also, there was a reduction in the cell wall thickness for honeycombs with the increase in the hierarchy, so as to maintain the overall relative density constant. This was kept similar to the FEA calculations.

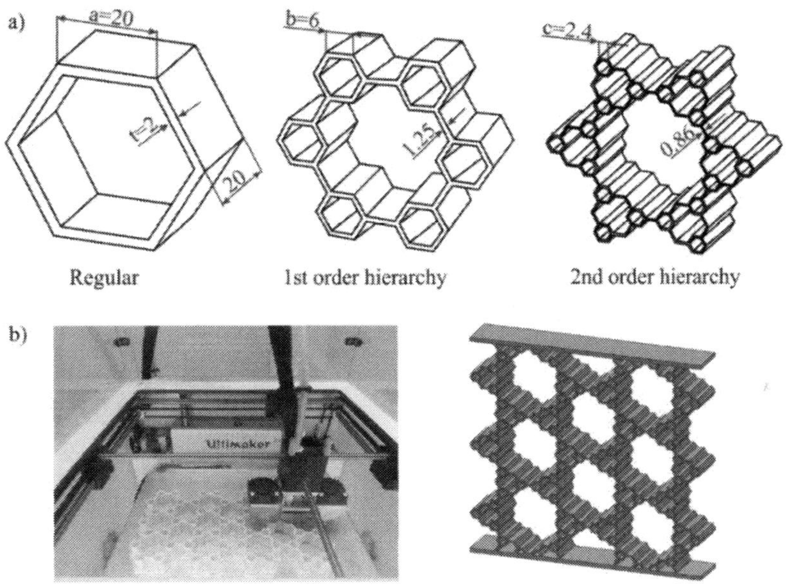

Fig. 1. Hierarchical Honeycombs: a) cell of the regular and hierarchical honeycombs and b) the printing process and the virtual model of 2nd order hierarchy.

3.2 Compression experiments

The in-plane compression tests were conducted for the three different honeycomb structures by using a computer-controlled servo-hydraulic single axial test machine Testometric (UK) equipped with a 50kN load cell. The cross-head speed was set at 5mm/min and at least three specimens of the same level of hierarchy were tested. Upon testing, the load-displacement curves were calculated in order to directly be compared with the predicted ones from the finite element analysis.

4 Results and discussion

The typical load-displacement curves under compressive loading for all materials under study i.e. ABS, ABS/CNTs and ABS/CF and for each hierarchical honeycomb structure under investigation are illustrated in Figure 2. Initially, the results of ABS reveal that the curves have reached a peak compression load of 570.9 N, 586.1 N and 649.92 N for the regular honeycomb, the first order hierarchy and second order hierarchy, respectively. Similarly, such trend was observed from the results of ABS/CNTs with values of 554.16 N, 698.52 N and 783.8 N. At the same way, the results for ABS/CF shown that they have reached a peak compression load of 675.2 N, 778.7 N and 924.63 N as the hierarchy was getting higher. The highest deformation was up to 8 mm. The results portray clearly that the second order hierarchy structure samples for ABS/CF show higher stiffness than the other materials and hierarchies. This phenomenon was expected since the effective elastic modulus is expressed as the ratio of mean stress and mean strain by the equation [5]:

$$\frac{E}{Es} = (^{t}/_{a})^3 f(\gamma_1) \tag{2}$$

where

$$f(\gamma_1) = \sqrt{3}/(0.75 - 3.525\gamma_1 + 3.6{\gamma_1}^2 + 2.9{\gamma_1}^3) \tag{3}$$

According to A. Ajdari et al. [5], in order to find the maximum normalized elastic modulus for 1st order of hierarchy structures with constant relative density, t/a has to be excluded from Eq. (2), using through Eq. (1) the relative density expression. This expression for E/E_s is a ρ^3 times function of γ_1, while setting $(\partial(\frac{E}{Es})/\partial\gamma_1)_\rho = 0$ leads to $E_1/E_s = 2.97\rho^3$, two times the stiffness of the zeroth level of hierarchy honeycomb structure, for which $E_0/E_s = 1.5\rho^3$ (regular honeycomb's effective modulus can be calculated by setting $\gamma_1 = 0$ in Eq. (2), and using Eq. (1) to eliminate t/a). For the 2nd order hierarchical structure, a stiffness almost three point five times that of regular honeycomb was observed. The experimental load-displacement results of the honeycombs manufactured with neat ABS were in agreement with these theoretical values; 1.91 times higher stiffness for the 1st level of hierarchy and 3.33 times for the 2nd level, compared with the stiffness of the regular honeycomb. Considering the honeycombs manufactured using ABS/CNTs the stiffness increase was 1.92 and 3.27 for 1st and 2nd level of hierarchy, respectively. Such values were 2.0 and 3.50 in the case of honeycombs manufactured from ABS/CF [6].

Overall, as the load increases and the honeycomb structure is under compression, the hierarchy cells tend to sustain deformation. This results in smaller yield displacement compared to the 2nd order of hierarchy, 1st order of hierarchy and regular honeycomb. Therefore, hierarchical honeycombs resist higher values of applied loads compared to the regular honeycomb system and increase the stiffness.

The effect of hierarchy on the elastic properties of 3D printed hierarchical honeycombs has been also investigated in this paper. In order to accomplish that, the compression experimental results were simulated utilizing FEA. A computational model was introduced and a static structural analysis was performed using the commercial code ANSYS. Assumptions of initial material values for the tangent moduli E_i of the multilinear stress-strain curves of the honeycombs' were made in the FEA model. On the top plate of each honeycomb structure, a vertical displacement was applied in steps and the reaction force was obtained at the bottom, which was considered with a fixed boundary condition. The values of this vertical displacement were acquired by the experimental results shown in Figure 2. Considering the deformation, a static force was determined and compared to the experimental one. If the calculated force does not converge with the measured then the values E_i are approximated and the finite element model is solved again. This procedure is repetitive until the last pair of values has converged and the loop ends. The final force-displacement values calculated from the FEA are illustrated in Figure 2, where it can be easily seen that these values converged with the experimental results. Thus, the assumptions of the multilinear material model of each hierarchy were considered accurate.

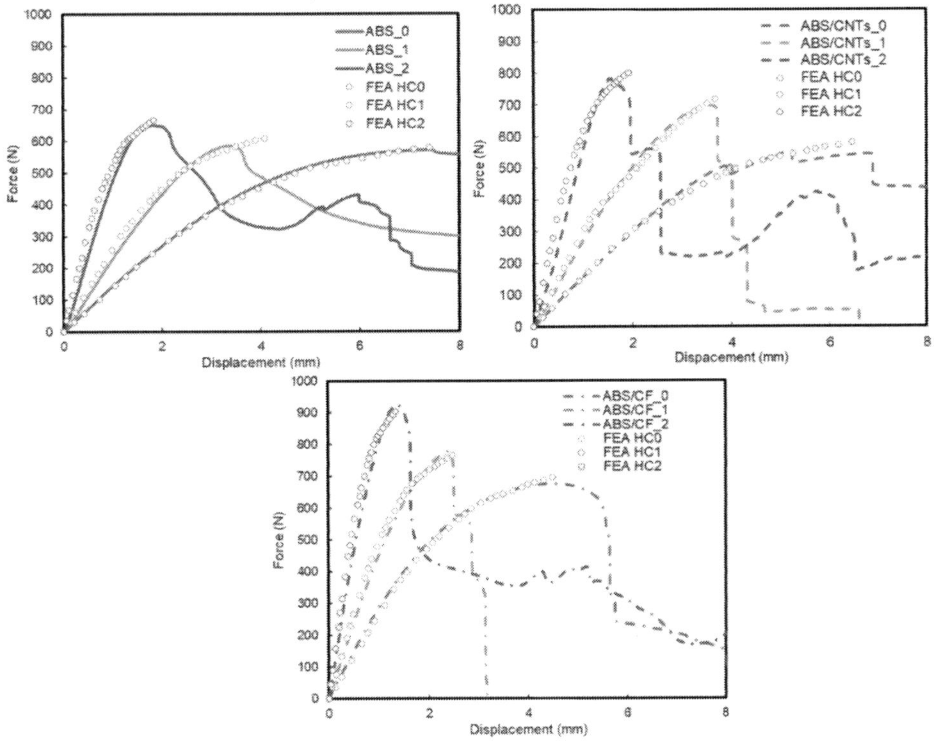

Fig. 2. Typical experimental load displacement curves and the curve-fit utilizing a FEA model of HC0, HC1 and HC2 structures for ABS, ABS/CNTs and ABS/CF materials.

Finite Element Analyses were performed on honeycomb structures over the three levels of hierarchy, as shown in Figure 3, where the stress strain behaviour of ABS was compared with the ABS/CNTs and ABS/CF at different levels of hierarchy. The E_1 value (initial slope) of the elastic modulus was determined to be 2050 MPa, 2350 MPa and 3210 MPa for ABS, ABS/CNTs and ABS/CF, respectively. The level of hierarchy did not have any significant effect on the Young's moduli in the elastic region. First order hierarchy and second order hierarchy structures show higher strength than the regular honeycomb without compromising significantly their ultimate strain, even at higher levels of hierarchy.

5 Conclusions

The mechanical performance of the three different materials and the three levels of hierarchical honeycombs was evaluated through static compression tests. The results revealed increased stiffness and strength for the ABS/CF and the 2nd order hierarchy of honeycombs in comparison with the other materials and the 0th and the 1st hierarchies, respectively. A procedure is introduced in order to extract the hierarchical honeycombs' constitutive laws based on the evaluation of uniaxial compression test results through a developed FEA simulation. The FEA model utilizes the compression test results as input data to the described model and extracts the stress-strain curves of the 3D printed hierarchical honeycombs. The procedure followed shows that the presented compression test result evaluation method is a very efficient procedure to depict the stress-strain behaviour of the 3D printed hierarchical honeycombs.

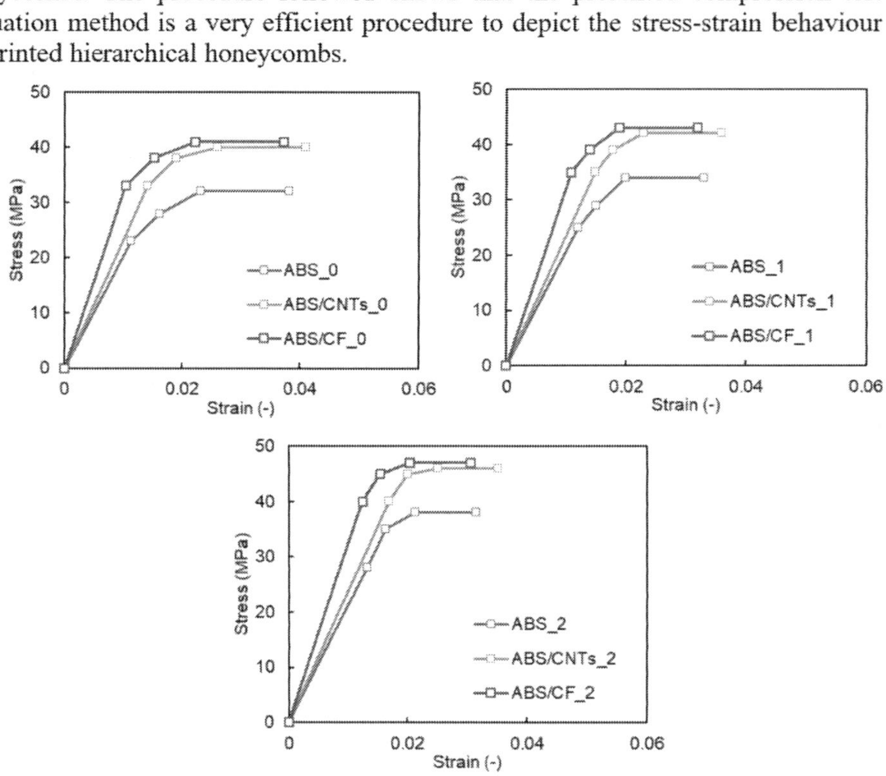

Fig. 3. Comparison among the ABS, ABS/CNTs and ABS/CF materials for different levels of hierarchy.

This research is co-financed by Greece and the European Union (European Social Fund - ESF) through the Operational Programme «Human Resources Development, Education and Lifelong Learning» in the context of the project "Strengthening Human Resources Research Potential via Doctorate Research - 2nd Cycle" (MIS-5000432), implemented by the State Scholarships Foundation (IKY). Also, the authors would like to thank Mrs. V. Papadopoulou for her assistance with the computer generated algorithm in relation to the design of the honeycombs.

References

1. M. Mansour, K. Tsongas, D. Tzetzis, A. Antoniadis, P Polym Plast Technol Eng, **57**, 1715-25 (2018)
2. M. Mansour, K. Tsongas, D. Tzetzis, Polym Plast Technol Eng, **58**, 1234-44 (2019)
3. M. T. Mansour, K. Tsongas, D. Tzetzis, A. Antoniadis, *IOP Conf. Ser.: Mater. Sci. Eng.*, **564** (2019)
4. L. Sang, S. Han, X. Peng, X. Jian, J. Wang, Compos. Part A Appl. Sci. Manuf., **125** (2019)
5. A. Ajdari, B. H. Jahromi, J. Papadopoulos, H. Nayeb-Hashemi, A. Vaziri, Int. J. Solids Struct., **49**, 1413-19 (2012)
6. R. T. L. Ferreira, I. C Amatte, T. A. Dutra, D. Bürger, Compos Part B-Eng, **124**, 88-100 (2017)

Vibration isolation performance of an elevator motor using Nitrile-Butadiene Rubber /Multi-Walled Carbon Nanotube composite machine mounts

Konstantinos Tsongas[1,*], *Gabriel* Mansour[1]

[1]Laboratory for Machine Tools and Manufacturing Engineering, Department of Mechanical Engineering, Aristotle University of Thessaloniki, University Campus, 54124, Thessaloniki, Greece.

Abstract. The objective of this paper is to evaluate the vibration isolation performance of an elevator motor mounted on elastomeric nanocomposite mounts. A series of conventional acrylonitrile-butadiene rubber (NBR) mounts have been reinforced with 20wt% concentration of multi-walled carbon nanotubes (MWCNTs). The vibration isolation capacity of the machine mounts was calculated through the transmissibility of an elevator motor test system. A Finite Element Model (FEM) was introduced and a harmonic analysis based on the ANSYS code has been utilized to investigate the modal behavior of the nanocomposite machine mount/elevator motor system and extract a representative model of the vibrational behavior. The cyclic compression results have revealed that the stiffness and damping capacity of the conventional elastomers can be modified by adjusting the proportion of MWCNTs. Elastomers' vibration isolation performance of the motor was ameliorated with the inclusion of MWCNTs, signifying that the enhancement of the elastomers with MWCNTs was rather effective. The vibration level of the elevator motor was decreased to 90% by incorporating the optimal concentration of MWCNTs in NBR mounts.

1 Introduction

Conventional elevators move vertically by the traction machine, which suspends the car and the counter weight by the rope. Elevators in residential buildings present the potential for intrusive noise and vibration to the dwelling units. The adverse noise and vibration can result from the elevator equipment located in the machine rooms or the shafts. The impacts can be significant issues related to the sound quality, sleeping conditions, and enjoyment within residences. Resilient mounting is widely used in vibration and noise control to isolate a vibratory source from a receiving structure or from a structural component with a significant radiating area. In order to design an effective isolator or to predict the level of isolation, an accurate measurement of the dynamic behavior, i.e. dynamic complex stiffness and damping, is necessary. The concept of complex stiffness with viscous and hysteresis damping has been

* Corresponding author: ktsongas@auth.gr

exquisitely explained for a single degree of freedom (SDOF) system [1]. Numerous test methods have been developed in order to characterize the complex modulus properties of elastomeric machine mounts [2, 3]. In the past, the damping capacity of conventional engineering materials does not always provide sufficient energy dissipation to limit resonant or near-resonant amplitudes of vibration [4]. High damping capacity is an important factor for various industrial applications; therefore, it is necessary to develop structural components with a high level of structural damping. It would therefore be of interest to investigate new materials that simultaneously exhibit high damping capacity with high stiffness and low density which includes rubber matrix nanocomposites and porous metals.

High loss factor polymers have been used in the past for modern damping applications [5-7]. Nowadays there are various other advanced materials that could be utilized in vibration isolation applications. Researchers have concentrated their efforts towards the development of polymers in which nanomaterials are embedded in polymer substrates and much promising results have been established primarily for improvement of the mechanical properties [8-12]. The interest in porous metals arises from the interesting combination of characteristics and the easy adjustment of their properties to meet the desired needs by tailoring porosity characteristics.

In order to assess the vibration isolation performance of such high stiffness and high damping materials, a modal testing method proposed by Mansour et al. [13-18] derives modal parameters from experimental transfer functions by a curve-fitting technique, in order to determine preliminary modal parameters, from which analytical expressions for the TFs are obtained.

The vibration isolation behavior of an elevator motor machine isolated by MWCNTs/NBR elastomers is the subject of the current investigation. In order to obtain the modal properties of the elevator motor machine with nanocomposite isolators, a finite element model was developed. Loading-unloading (cyclic) compression tests were performed in order to determine the stiffness and damping capacity, as input for the finite element model. The vibration isolation performance of the elevator motor was characterized by analyzing the resonant frequencies, the damping and the transmissibility of the nano-modified elastomeric system. This procedure was repeated for the different nanofiller concentrations in the elastomeric matrix. The nanocomposite elastomers were considered optimal for the vibration isolation of the elevator motor machine.

2 EXPERIMENTAL DETAILS

2.1 Cyclic compression tests of NBR/MWCNTs nanocomposite machine mounts

At least five nanocomposite specimens were prepared with 20wt% MWCNTs/NBR, as well as 0wt% MWCNTs/NBR which served as a control sample. Subsequently, MWCNTs/rubber elements were bonded between a steel pedestal and steel cap. The cap provides protection for the rubber element from diesel, oil, ozone, UV rays that are harmful to rubber. The final dimensions of the NBR/MWCNTs mounts were 60mm diameter and 27mm height.

Steady state, strain rate–controlled cyclic compression tests at ambient temperature were performed with constant strain rates in loading and unloading. The measurement was carried out using a material testing system (Testometric, UK equipped with a 50 kN load cell) at a frequency of 0.01 Hz and up to a 10 kN load. The loss of energy in the loading-unloading cycle for each material was calculated from the hysteresis loop. The stress-strain behavior during loading and unloading was obtained and each rubber demonstrated a hysteresis behavior. The loading and unloading speeds were set constantly at 5 mm/min.

3 MECHANICAL AND HYSTERETIC BEHAVIOR OF THE MWCNTS NANOCOMPOSITE ELASSTOMERIC MOUNTS

3.1 Mechanical and damping properties of the NBR/MWCNTs nanocomposite mounts

Under alternating stress, hysteresis occurs when the rate of deformation is less than the rate of stress variation. In this case, since the absorbed and released energies are not balanced in each cycle, the stretching and recoil curve form a closed loop, which is known as a hysteresis loop as shown in Figure 1. The area within the loop represents the energy loss. For elastomeric materials, a larger hysteresis loop means higher damping, which more effectively reduces vibration [19]. The damping constants may be derived from the area surrounded by hysteresis loops. Based on the theory of free vibration, the vibration-isolating capacity of materials can be evaluated from the damping constant and the hysteresis damping characteristics [20].

The specific damping capacity (SDC) is given by:

$$SDC = \frac{\Delta W}{W} \times 100\% = \left(\oint \sigma d\varepsilon \Big/ \int_{\omega t=0}^{\pi/2} \sigma d\varepsilon \right) \times 100\% \\ = (\pi \cdot \sigma_0 \cdot \varepsilon_0 \cdot sin\,\delta)/(\frac{1}{2} \cdot \sigma_0 \cdot \varepsilon_0 \\ \cdot cos\,\delta) = 2\,\pi\,\tan\delta \tag{1}$$

where σ and ε are the stress and strain, δ is the phase angle between them, ΔW is the energy dissipated in any one cycle and W is the maximum energy associated with that cycle [21]. The specific damping capacity can be related with the loss factor [22] by:

$$n = \tan\delta = \frac{\Delta W}{2\pi\, W} \tag{2}$$

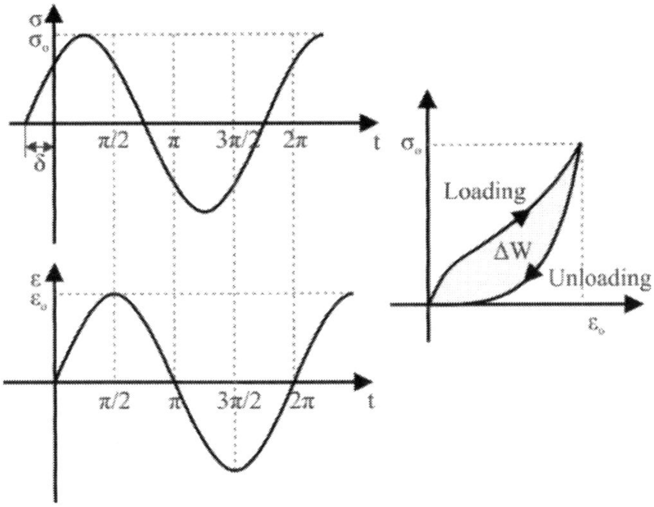

Fig. 1. Stress-strain hysteresis loops under fully reversed strain-controlled tension-compression.

The stress-strain curves of conventional NBR and NBR/MWCNTs mounts under uniaxial compression, as shown in Figure 2. For the same stress, it was found that the corresponding strain of the nanocomposite mount is lower than that of conventional NBR. Evidently, the stiffness of the nanocomposite mount was also improved by the addition of the MWCNTs, as shown in Table 1. This indicates that the NBR/MWCNTs composite system exploits the properties of one component to overcome the weakness of the other in mechanical performance.

Figure 2 presents the hysteresis loops curves of NBR/MWCNTs specimens under compressive vibration at 0.01 Hz with an ultimate force of 10 kN. Considering Eq. 2, the energy loss over a cycle (ΔW), the maximum energy of that cycle (W) and loss factor (n) were used to measure the material damping of the loading–unloading tests. ΔW indicates that the anti-vibration property of conventional NBR mounts is improved as the amount of MWCNTs increases. Therefore, the material is assumed to enhance the ability to transform its kinetics to those of thermal dissipation upon the application of an external load. Nevertheless, higher damping constant n, which is the ratio of ΔW to W, is observed for the NBR/MWCNTs mount and this indicates faster energy dissipation at particular amplitudes, which become stable with less vibration.

Table 1. The stiffness and damping values of NBR and NBR/MWCNTs mounts.

Specimen	Stiffness (N/mm)	Loss factor n
Conventional NBR mounts	585	8.9%
20wt% MWCNTs/NBR	935	18.2%

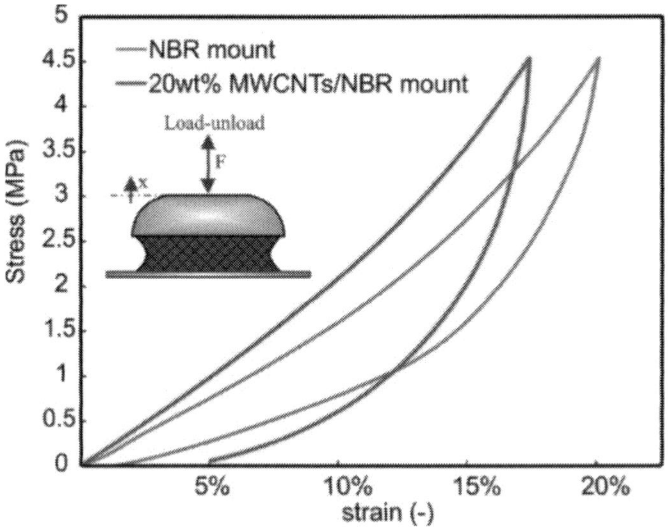

Fig. 2. Loading-unloading stress-strain curves of the mounts for different content of MWCNTs.

3.2 Assessment of the vibration isolation performance of the elevator motor machine using finite element analysis (FEA)

An elevator motor machine was used to apply the under study nanocomposite mounts. Firstly, a finite element model of the machine-isolator-ground system was introduced in order to assess the modal parameters and the vibration isolation performance of the existing mounts and the proposed nanocomposite ones. The computational model implemented in order to study the dynamic response of the elevator motor machine was achieved with the use of the finite element software code ANSYS APDL. Both a modal and a harmonic mode superposition analysis were performed. The total mass of the elevator motor machine was

106.8 kg, which was used as gravitational static load. In order to determine the natural frequencies and participation factors, the modal analysis is carried out using the block Lanczos method with ANSYS for fixed-free boundary conditions. The modal superposition method enables one to express the response of a multi-degree-of-freedom system as a linear combination of its corresponding modal responses; to use free vibrations mode shapes to uncouple equations of motion. For the harmonic analysis, the motor load was applied and the response was acquired under the mounts at points 1, 2, 3 and 4, as shown in Figure 3(a).

The results of the transmissibility curves as estimated by the FEM are shown in Figure 3(b). The vibration isolation behavior of the nanocomposite 20wt% MWCNTs/NBR mount presented moderately higher resonant frequencies at 25Hz, 134Hz and 260Hz, since MWCNTs increase the stiffness of the elastomer in the isolator. More importantly, the transmissibility value at resonance is significantly decreased to 90%. Thus, the 20wt% MWCNTs/NBR mounts are considered as the optimal solution to reduce the vibration level of this elevator motor machine.

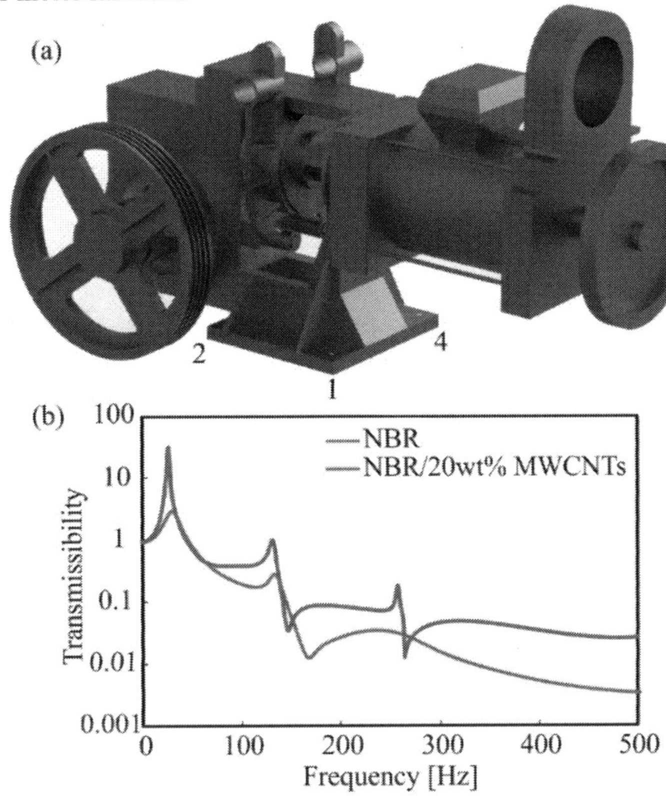

Fig. 3. (a) Characteristic measurement points for the dynamic analysis of the elevator motor machine and (b) evaluation of the transmissibility curve of the NBR and the nanocomposite mounts 20wt% MWCNTs/NBR using FEA.

4 Conclusions

The overall objective of this paper is the estimation of the vibration isolation performance of an elevator engine machine mounted on conventional NBR isolators reinforced with MWCNTs. Cyclic compression test results showed higher stiffness and damping constant for the nanocomposite mount, which indicated faster energy dissipation at particular amplitudes, so they become stable with less vibration. The vibration isolation properties of the elevator motor machine were assessed by the transmissibility curves. The FEA results indicated that the addition of MWCNTs in the rubber-based composites significantly improved the vibration isolation performance of the machine compared with NBR mounts. Overall, the resonant frequencies, the damping and the capacity to isolate vibrations can be adjusted by modifying the content of MWCNTs in NBR mounts. The maximum percentage change of the transmissibility value at resonance reached a decrease of 90%. Therefore, the 20wt% MWCNTs/NBR mounts are considered effective vibration isolators for this elevator motor machine.

References

1. S. Neumark, Aer. Res. Coun. rep. and mem. no. 3269; (1962)
2. S.O. Oyadiji, and G.R. Tomlinson, J. Sound Vib.**186** (4), 623-647 (1995)
3. U.J. Kurze, Act Acustica **2**, 483-490 (1994)
4. R. Adams, J de Phys. Col.; **44** (C9), C9-29-C9-37 (1983).
5. JB Kosmatka, SL Liguore, J Aerosp Eng **6**:268–283 (1993)
6. A Baz, J. Ro, J Sound Vib **28**, 18–21 (1994)
7. WH Liao, KW Wang, J Sound Vib 207:319–334 (1997)
8. M. Mansour, K. Tsongas, D. Tzetzis, A. Antoniadis, Polym Plast Technol Eng, **57**, 1715-25 (2018)
9. M. Mansour, K. Tsongas, D. Tzetzis, Polym Plast Technol Eng, **58**, 1234-44 (2019)
10. M. T. Mansour, K. Tsongas, D. Tzetzis, A. Antoniadis, *IOP Conf. Ser.: Mater. Sci. Eng.*, **564** (2019)
11. D. Tzetzis, K. Tsongas, G. Mansour, Mater Res, **20**(6), 1571-1578 (2017)
12. G. Mansour, K. Tsongas, D. Tzetzis, ICCM20, 20th Intern.l Conf. on Comp. Mat., 19-24 July 2015, Copenhagen, Denmark.
13. K. Tsongas, D. Tzetzis, G. Mansour, Plast Rubber Compos, **46** (10) 458-468, (2017)
14. K. Tsongas, G. Mansour, Trib in Ind, **41** (3), 433-442 (2019)
15. G. Mansour, K. Tsongas, D. Tzetzis, Measurement **91**, 31-38 (2016)
16. G. Mansour, K. Tsongas, D. Tzetzis, Compos Part B-Eng **94**, 152-159 (2016)
17. G. Mansour, K. Tsongas, D. Tzetzis, J Reinf Plast Comp **35**(19), 1401–1410 (2016)
18. G. Mansour, K. Tsongas, D. Tzetzis, K. Tzikas, , Polym Plast Technol Eng, **56**(14), 1505-1515 (2017)
19. Sperling LH. Introduction to physical polymer science. 2nd ed. New York (NY): Wiley; 1992.
20. Rao SS. Mechanical vibrations. 5th ed. New Jersey (NJ): Prentice Hall; 2011.
21. Wevers M. NDT&E Int. 1997;30(2):99–106.
22. G. Mansour, K. Tsongas, D. Tzetzis, J Balk Tribol Assoc, **21**, 575-584, (2015)

Prediction of the main cutting force in longitudinal turning of AISI D6 tool steel bars by applying full and fractional experimental design

Kyriaki-E. Aslani[1,2], *John* D. Kechagias[1], *Nikolaos* A. Fountas[3], *Nektarios* Vidakis[4], *Angelos* Koutsomichalis[5], *Dimitrios* E. Manolakos[6] and *Nikolaos* M. Vaxevanidis[3,*]

[1]General Department, University of Thessaly, Larissa, Greece.
[2]Department of Mechanical Engineering, University of West Attica, Athens, Greece.
[3]Department of Mechanical Engineering Educators, School of Pedagogical and Technological Education (ASPETE), Amarousion, Greece
[4]Department of Mechanical Engineering, Hellenic Mediterranean University, Iraklion Crete, Greece
[5]Department of Aeronautical Studies, Hellenic Air-Force Academy, Dekelia Air Force Base, Greece
[6]School of Mechanical Engineering, National Technical University of Athens (NTUA), Athens, Greece

Abstract. This investigation concerns the cutting force magnitude during turning of AISI D6 tool steel using both Taguchi Experimental Design (TED) and full factorial design (FFD). Three main cutting parameters namely spindle speed, feed rate and depth of cut were considered as the cutting parameters, each one having three levels, while the cutting force (Fc) was selected as the machinability process output. The full factorial design of the 27 (33) experiments was splitted in three sub-arrays, each one having 9 experiments. These three sub-arrays were orthogonal and were treated as Taguchi L9 orthogonal arrays. The performance of the FFD and each TED was analysed using stem-and-leaf plots, box plots, as well as analysis of means (ANOM) and analysis of variance (ANOVA). The results obtained indicate the suitability of all proposed experimental designs for machinability studies.

1 Introduction

Turning is a type of material processing operation where a cutting tool is used to remove unwanted material to produce a desired product and it is generally performed on either conventional or computer numerically controlled (CNC) lathes. In recent decades, considerable improvements were achieved in turning, enhancing machining of difficult-to-cut materials and resulting in improved machinability in terms of surface finish and cutting force [1]. Cutting force estimation and modeling are of major importance for the metal cutting theory. There is a great number of inter-related parameters that affect the cutting

* Corresponding author: vaxev@aspete.gr

forces, such as operational parameters, cutting tool geometrical characteristics and coatings, etc; therefore the development of a proper model is a quite difficult task [2]. Although that an enormous amount of related data is available in machining handbooks, the majority of these data attempt to define the relationship between only a few of the various cutting parameters whilst keeping the other parameters fixed [3]. ANNs are one of the most powerful computer modeling techniques, currently being used in many fields of engineering for modeling complex relationships which are difficult to describe with physical models. ANNs have been extensively applied in modeling many metal-cutting operations either conventional (turning, milling, grinding) or unconventional (EDM, AWJM, etc) [3-7]. This study evaluates the full factorial design (FFD) and the Taguchi experimental design (TED) during longitudinal turning of tool steel material. In this first attempt only cutting force is evaluated.

2 Experimental details

The experimental data were extracted from a previous investigation regarding the evaluation of machinability in turning of AISI D6 tool steel, applying a feed forward back propagation (FFBP) artificial neural network (ANN); see [7]. In this study, only the data concerning the main cutting force of the AISI D6 tool steel are utilized. The cutting parameters are spindle speed (n) in rpm, feed rate (s) in mm/rev and depth of cut (a) in mm, each one having three levels. The original experimental results as reported in the corresponding L_{27} array can be found in [7]. Turning experiments were conducted using a Kern Modell D18L conventional lathe. The cutting tool was a SECO® coated tool insert, coded as TNMG 160404 – MF2 with TP 2000 coated grade. Sverker-3® tool steel was used as a cutting material. It is a high-carbon (2.05%), high-chromium (12.7 %) tool steel alloyed with tungsten (1.1 %) identical to AISI-D6 grade with hardness 240 HB. All test materials were supplied as Ø45 mm bars.

3 Results and Discussion

In this investigation, Taguchi experimental design (TED) and full factorial design (FFD) are compared with the use of analysis of means (ANOM) and analysis of variance (ANOVA). Both methologies are applied in order to determine the factors which affect the main cutting force Fc. Minitab 17 Statistical Software was utilized for the statistical analysis.

3.1 Full Factorial Array

In the present investigation, the original L_{27} array is considered as the full factorial array [7]. Analysis of Variance (ANOVA) is tabulated in Table 1, while Analysis of Means diagram is presented in Figure 1. The parameter values which minimize the cutting force are speed = 850 rpm, feed = 0.06 mm/rev and depth of cut = 0.5 mm. In case of ANOVA, F and P values and contribution rate should be considered for determining the parameters which affect a variable. High F-values and contribution rate imply significance; so, do small P-values ($P < 0.05$). According to the above, depth of cut (a) is the cutting parameter which affects more the cutting force (F = 72.89, P = 0.000, CR = 67.45%), followed by feed rate (s) which is almost as important as feed of rate in terms of influence (F = 24.12, P = 0.000, CR = 22.32%) and by speed (n) which doesn't affect Fc (F = 1.06, P = 0.365, CR = 0.98%).

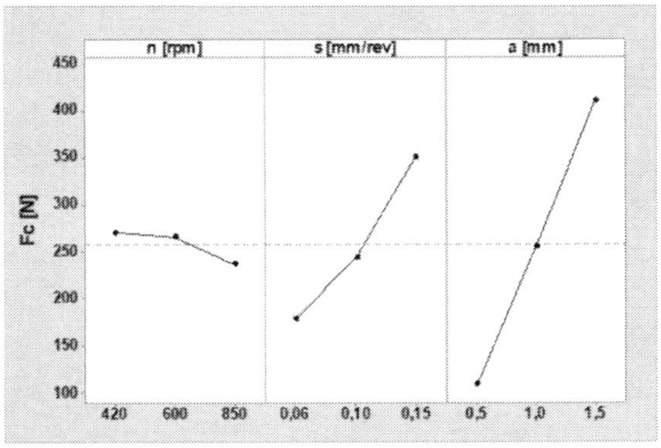

Fig. 1. ANOM for main cutting force Fc.

Table 1. ANOVA for main cutting force Fc.

Source	DF	Adj SS	Adj MS	F-Value	P-Value	Contribution Rate [%]
n [rpm]	2	5915	2957	1.06	0.365	0.98
s [mm/rev]	2	134430	67215	24.12	0.000	22.32
a [mm]	2	406287	203143	72.89	0.000	67.45
Error	20	55743	2787	-	-	9.25
Total	26	602375	-	-	-	-

3.2 L_9 (1) Array

The first L_9 Orthogonal Array for the main cutting force Fc is tabulated in Table 2. As indicated by ANOM diagram (Fig. 2) the parameters that minimize the cutting force are: speed = 600 rpm, feed = 0.06 mm/rev and depth of cut = 0.5 mm. ANOVA (Table 3) revealed that depth of cut (a) is the dominant parameter (F = 6.87, P = 0.127, CR = 70.78%), followed by feed rate (s) (F = 1.72, P = 0.367, CR = 17.76%) and by speed (n) (F = 0.11, P = 0.898, CR = 1.17%).

Table 2. L_9 (1) experimental array.

	n [rpm]	s [mm/rev]	a [mm]	Fc [N]
1	420	0.06	1	188
2	420	0.1	1.5	452
3	420	0.15	0.5	146
4	600	0.06	1.5	284
5	600	0.1	0.5	96
6	600	0.15	1	364
7	850	0.15	1.5	536
8	850	0.06	0.5	138
9	850	0.1	1	184

Table 3. ANOVA for main cutting force Fc.

Source	DF	Adj SS	Adj MS	F-Value	P-Value	Contribution Rate [%]
n [rpm]	2	2216	1108	0.11	0.898	1.17
s [mm/rev]	2	33731	16865	1.72	0.367	17.76
a [mm]	2	134411	67205	6.87	0.127	70.78
Error	20	19555	9777	-	-	10.29
Total	26	189912	-	-	-	-

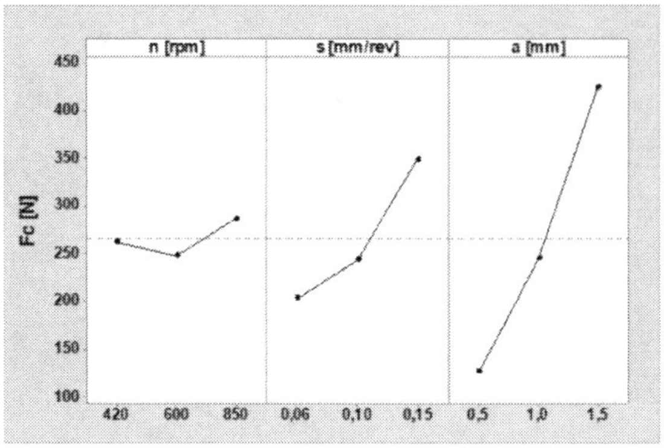

Fig. 2. ANOM for main cutting force Fc.

3.3 L_9 (2) Array

The second L_9 Orthogonal Array for the main cutting force Fc is presented in Table 4. ANOM diagram (Fig. 3) illustrates that the parameters which minimize the cutting force are: speed = 850 rpm, feed = 0.06 mm/rev and depth of cut = 0.5 mm. ANOVA (Table 5) revealed that depth of cut (a) is the dominant parameter (F = 28.97, P = 0.033, CR = 63.41%), followed by feed rate (s) (F = 13.26, P = 0.07, CR = 29.01%) and by speed (n) (F = 2.46, P = 0.289, CR = 5.39%).

Table 4. L_9 (2) experimental array.

	n [rpm]	s [mm/rev]	a [mm]	Fc [N]
1	420	0.06	1.5	260
2	420	0.1	0.5	112
3	420	0.15	1	376
4	600	0.06	0.5	72
5	600	0.1	1	272
6	600	0.15	1.5	536
7	850	0.06	1	160
8	850	0.1	1.5	344
9	850	0.15	0.5	136

Table 5. ANOVA for main cutting force Fc.

Source	DF	Adj SS	Adj MS	F-Value	P-Value	Contribution Rate [%]
n [rpm]	2	9632	4816	2.46	0.289	5.39
s [mm/rev]	2	51915	25957	13.26	0.070	29.01
a [mm]	2	113419	56709	28.97	0.033	63.41
Error	20	3915	1957	-	-	2.19
Total	26	178880	-	-	-	-

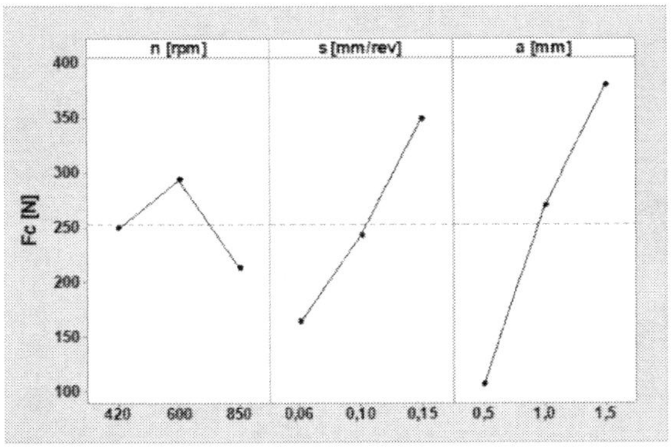

Fig. 3. ANOM for main cutting force Fc.

3.4 L_9 (3) Array

The third L_9 Orthogonal Array for the main cutting force Fc can be found in Table 6. ANOM diagram (Fig. 4) shows that the parameters which minimize the cutting force are: speed = 850 rpm, feed = 0.06 mm/rev and depth of cut = 0.5 mm. ANOVA (Table 7) revealed that depth of cut (a) is the dominant parameter (F = 26.75, P = 0.036, CR = 70.53%), followed by feed rate (s) (F = 8.26, P = 0.108, CR = 21.76%) and by speed (n) (F = 1.92, P = 0.342, CR = 5.07%).

Table 6. L_9 (3) experimental array.

	n [rpm]	s [mm/rev]	a [mm]	Fc [N]
1	420	0.06	0.5	68
2	420	0.1	1	246
3	420	0.15	1.5	584
4	600	0.06	1	184
5	600	0.1	1.5	428
6	600	0.15	0.5	152
7	850	0.06	1.5	256
8	850	0.1	0.5	56
9	850	0.15	1	320

Table 7. ANOVA for main cutting force Fc.

Source	DF	Adj SS	Adj MS	F-Value	P-Value	Contribution Rate [%]
n [rpm]	2	11793	5896	1.92	0.342	5.07
s [mm/rev]	2	50652	25326	8.26	0.108	21.76
a [mm]	2	164118	82059	26.75	0.036	70.53
Error	20	6134	3067	-	-	2.64
Total	26	232697	-	-	-	-

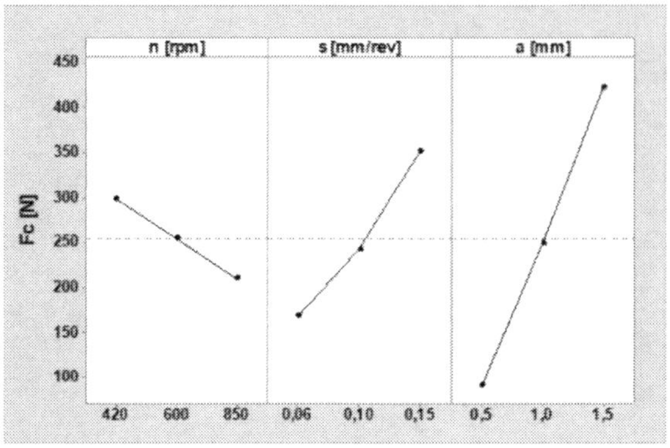

Fig. 4. ANOM for main cutting force Fc.

4 Conclusions

Data analysis applied on the above two experimental design methodologies and the following conclusions are revealed:

- The cutting force ANOM diagrams have the same trend lines as it concerns feed rate (s) and depth of cut (a). These two parameters are the most important for cutting forces during turning and the results are in accordance with the orthogonal cutting theory. When these two parameters are increased, the cutting force increased too.
- The contribution of the cutting parameters in cutting force are: (i) depth of cut: FFD (67.45%); TED (70.53%, 63.41%, 70.53%), (ii) feed rate: FFD (22.32%); TED (17.76%, 29.01%, 21.76%) and (iii) speed: FFD (0.98%); TED (1.17%, 5.39%, 5.07%)

The above analysis approve that the two experimental design methodologies have very similar results and can be safely concluded that TED methodology is as reliable as FFD when used for cutting force predictions and data analysis in turning of tool steel materials.

References

1. N.M. Vaxevanidis, N. Galanis, G.P. Petropoulos, N. Karalis, P. Vasilakakos and J. Sideris, Surface Roughness Analysis in High Speed-Dry Turning of Tool Steel, *Proc. ESDA2010 (paper ESDA2010-24811)*
2. G. Boothroyd and W. Knight, *Fundamentals of Machining and Machine Tools*, CRC Press, (2005)
3. T. Szecsi, J. Mater. Process. Technol., 92-93, (1999)
4. G. Dini, CIRP Ann., **46**, 2 (1997)
5. T.M.A. Maksoud, M.R. Atia and M.M. Koura, Mach. Sci. Technol. **7**, 3 (2003)
6. E.O. Ezugwu, D.A. Fadare, J. Bonney, R.B. Da Silva and W.F. Sales, Int. J. Mach. Tools Manuf., **45** (2005)
7. N.M Vaxevanidis, J.D. Kechagias, N.A. Fountas, D.E. Manolakos "Three component cutting force system modeling and optimization in Turning of AISI D6 tool steel using design of experiments and Neural Networks" *Proc. WCE 2013*, 1 (2013)

MATEC Web of Conferences **318**, 01052 (2020) https://doi.org/10.1051/matecconf/202031801052
ICMMEN 2020

A microcontroller-based software framework for controlling a mechatronic system

Nikolaos Nikolaidis[1], *Nikolaos Evgenidis*[1,*], *Dimitrios* Bechtsis[1], *Fotis* Stergiopoulos[1], *Apostolos* Tsagaris[1], *Dimitrios* Triantafyllidis[1], *Asterios* Papaoikonomou[2] and *Anastasios* Filelis[2]

[1]International Hellenic University (IHU), Department of Industrial Engineering & Management, PO Box 141, Sindos, Thessaloniki, 57400, Greece
[2] Evresis S.A DA 12a Block:39b Industrial Area, 57400, Sindos, Thessaloniki

Abstract. The proposed software framework is presented and an Application Programming Interface (API) is developed based on the Arduino Mega 2560. The API processes external commands that follow the operational logic of a gel electrophoresis device. The API acts as an intermediary layer between the gel electrophoresis mechatronic system's microcontroller and the motors' controllers. The microcontroller enables the basic functionalities of the gel electrophoresis system while the use of 2 axis (X, Z) motor controllers is necessary for controlling the moving parts of the mechatronic system. We control the movement's direction, position, speed, and acceleration. The developed API controls the stepper motors drive axles and the DC motors for opening and closing the drawers and other moving parts of the mechatronic system.

1 Introduction

An Application Programming Interface (API) has been developed in order to control a gel electrophoresis device that is currently under development. Electrophoresis has been known for about a century and involves the movement of proteins when an electric current is applied in a compartment filled with gel [1]. A agarose enhance solution is used as a substrate to facilitate the proteins movement, as a common practice in many Gel Electrophoresis (GE) devices The device (Fig. 1) consists of the following basic mechanical components [2]: (i) tube drawer system, (ii) bar code scanner, (iii) wash tank, (iv) robotic arm/gripper, (v) sample carrier, (vi) electrophoresis chamber, (vii) frame/film manipulator, (viii) staining-unstaining chamber, (ix) drying chamber, (x) camera analysis chamber. In this context, the typical steps of the gel electrophoresis process are: (i) take the sample and place it at a gel substrate, (ii) Apply an electric current, (iii) Dry the sample, iv) Stain and un-stain the sample, (v) Dry the sample and (vi) Analyze the sample and print the results. In order to automate the typical steps of the gel electrophoresis process, a mechatronic system was used, and the proposed API has been developed for controlling the mechanical arm of the mechatronic system. The use of an API for controlling a microcontroller is widely used when a lot of external sources must be coordinated [3].

*Corresponding author: evgenidisnikos@gmail.com

Fig. 1. The overall mechatronic system

2 The microcontroller's connections

Microcontrollers can provide accurate control for the moving parts of a mechatronic system [4]. For controlling the mechatronic system of the electrophoresis device an Arduino microcontroller has been used and the schematic diagram of the controller is presented at Fig. 2.

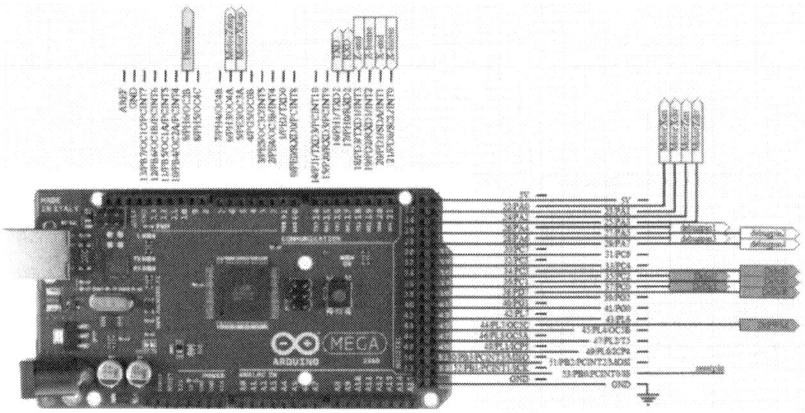

Fig. 2. The connections of the microcontroller

The yellow colored ports stand for the stepper motor connections, the blue for the power connections and the orange ports control the drawer movement. Moreover, the diagram indicates the debugging signals and an additional serial port for future use. Ports 18 to 21 are selected in order to provide an independent external stop signal INT0 to INT3, ports A12 to A14 are used as the drawers limit switches in a Pin Change INTerrupt (PCINT[2:0]) business logic. The drawers' velocity is controlled by port 44 PWM OC5C signal and finally the actuation of the motors MotorXen (X axis) and MotorZen (Z axis) uses ports 22 and 24 respectively. On overall microcontroller operated control circuits that drive motors and gear-bearing arrangements are considered of major importance in custom devices [5].

3 The messaging scheme of the developed API

For every input message the API provides an answering message using the USB port as an interface. Warning messages are also sent at an error handling situation when the software identifies an error. In general, 4 types of answering messages are provided: (i) Information Messages, (ii) Warning Messages, (iii) Error Messages and (iv) Setup Messages.

Table 1. The 4 types of Answering Messages: Information, Warning, Error and SetUp Messages

<table>
<tr>
<td rowspan="2">I1 Info: Motor X initialized
I2 Info: Motor Z initialized
I3 Info: motor X started
I4 Info: motor Z started
I5 Info: motor X finished
I6 Info: motor Z finished
I7 Info: X enable output is high active
I8 Info: X enable output is low active
I9 Info: Z enable output is high active
I10 Info: Z enable output is low active
I11 Info: X drive is always active
I12 Info: X drive is in auto mode
I13 Info: Z drive is always active
I14 Info: Z drive is in auto mode
I15 Info: Motor X drive set to manual mode and enabled (use S,XE,M for permanent change).
I16 Info: X drive set to manual mode and disabled. X,L and X,R commands will be ignored until re-enabled.
I18 Info: Motor Z drive set to manual mode and enabled (use S,ZE,M for permanent change)
I19 Info: Z drive set to manual mode and disabled. Z,L and Z,R commands will be ignored until re-enabled
I30 Info: Drawer 0 is opening
I31 Info: Drawer 0 stopped
I32 Info: Drawer 0 is closing
I33 Info: Drawer 0 closed
I34 Info: Drawer 1 is opening
I35 Info: Drawer 1 stopped
I36 Info: Drawer 1 is closing
I37 Info: Drawer 1 closed
I38 Info: Drawer 2 is opening
I39 Info: Drawer 2 stopped
I40 Info: Drawer 2 is closing
I41 Info: Drawer 2 closed
I42 Drawer 0=status, 1=status, 2=status</td>
<td>W0 Warning: you must connect pin 53 to RESET to activate reset command
W1 Warning: motor X already stopped
W2 Warning: Motor Z already stopped
W3 Warning: speed of X exceeds max and replaced with 800
W4 Warning: speed of Z exceeds max and replaced with 800
W5 Warning: speed of X cannot be 0 - replaced with 1
W6 Warning: speed of Z cannot be 0 - replaced with 1
W7 Warning: acceleration of X lower than min - replaced with 10
W8 Warning: acceleration of Z lower than min - replaced with 10
W30 Warning: Drawer # is already closed
W31 Warning: Drawer # is already closing
W32 Warning: Drawer # is already opening
W33 Warning: drawer # is already stopped
W34 Warning: drawer # is closing now and cannot be opened
W35 Warning: drawer # is neither opening nor in unknown status - stop command is not allowed</td>
<td rowspan="2">S1 Setup: Enter the active level of the enable input for drive X (H for high, L for low) and press
S2 Setup: X enable output set to high active
S3 Setup: X enable output set to low active
S4 Setup: enter the active level of the enable input for drive Z (H for high, L for low) and press
S5 Setup: Z enable output set to high active
S6 Setup: Z enable output set to low active
S7 Setup: enter the enable mode for drive X (M for manual, A for Automatic) and press <Enter>
S8 Setup: X drive set to manual mode
S9 Setup: X drive set to auto mode
S10 Setup: enter the enable mode for drive Z (M for manual, A for Automatic) and press <Enter>
S11 Setup: Z drive set to manual mode
S12 Setup: Z drive set to auto mode
S13 Setup: X enable set to high active
S14 Setup: X enable set to low active
S15 Setup: X drive set to auto mode
S16 Setup: X drive set to manual mode
S17 Setup: Z enable set to high active
S18 Setup: Z enable set to low active
S19 Setup: Z drive set to auto mode
S20 Setup: Z drive set to manual mode</td>
</tr>
<tr>
<td>E0 Error: unknown command received
E1 Error: motor X already running
E2 Error: motor Z already running
E3 Error: direction X must be L (left), R (right), 0 (stop) or E (enable)
E4 Error: direction Z must be L (left), R (right), 0 (stop) or E (enable)
E5 Error: minimum number of steps in X is 1 - received 0
E6 Error: minimum number of steps in Z is 1 - received 0
E7 Error: Valid S,ZE parameters are S,ZE,H S,ZE,L S,ZE,A and S,ZE,M
E8 Error: Valid setup commands are S,aE,H S,aE,L S,aE,A and S,aE,M where a = X or Z
E9 Error: Valid S,XE parameters are S,XE,H S,XE,L S,XE,A and S,XE,M
EA Internal error: OCR and prescaler arrays are not of the same size
E10 Error: Motor X enable must be ON or OFF
E11 Error: Motor Z enable must be ON or OFF
E30 Error: Wrong drawer number. It must be 0, 1 or 2
E31 Error: Wrong drawer command. Available: H=Home, O=Open, S=Stop or U=statUs
E32 Error: you must wait for drawer # to finish moving
E33 Error: Max time of 3000 ms exceeded in move of drawer # and move cancelled</td>
</tr>
</table>

4 Software flowcharts for controlling the motors

Arduino supports the development of the software program in two distinct phases. The first phase is the program setup which is executed once either at the initialization step or after every reset signal. The second phase includes a continuously running loop, the main program, that continuously monitors all the ports and reacts at every signal. Our main goal was to use a serial programming architecture and as a result the routines are controlled with interrupt routines. The main program monitors the ports for an input message string (a comma separated character string) from the USB port and makes a call to the parser in order to select and execute the proper interrupt routines. The main program also controls the motors (Stepper and DC motors) for the mechatronic system (arm, drawer's and other moving parts). The flag stringcomplete is updated from Arduino's serialEvent() that is called after the end of a single loop and indicates the routines that must be executed. The main loop checks if stringcomplete is activated and calls the serialEvent() routine (Fig. 3a) in order to get the characters from the input message string. For every distinct character at the string, the system safely stores the character and returns stringcomplete =1 for indicating that a input was successfully inserted. The next step is to execute the parser routine (Fig. 3b) for indicating the procedure and the commands that should be followed. The procedure includes a maximum of 5 substrings that are separated with a comma (the message format is: scmd, s1, s2, s3 and s4). Depending on the first substring (scmd values are R,X,Z,D) the main loop executes the corresponding routines.

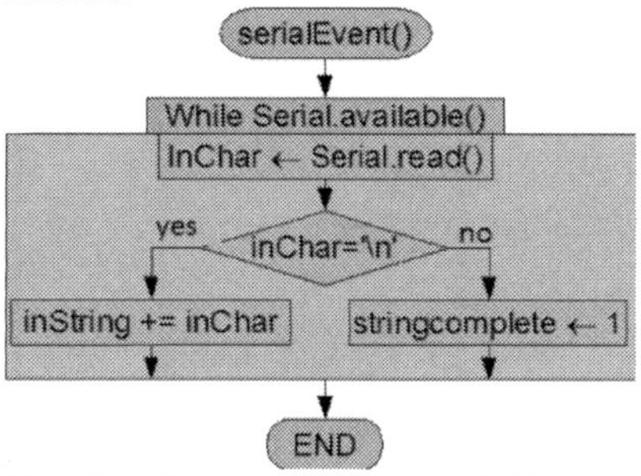

Fig. 3a. Flowchart of the serialEvent routine

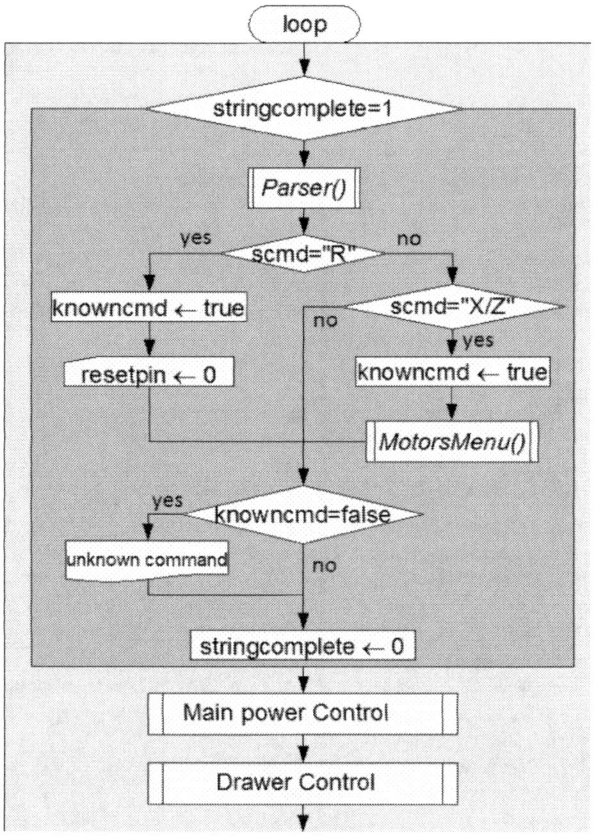

Fig. 3b. Flowchart of the main loop

The controller is based on Arduino Mega 2560 and uses three (3) digital outputs for each stepper motor: (i) enable, (ii) direction and (iii) moving step. The outputs are used both for the X and the Z axis. The serial port of the controller is used for the communication with the mechatronic system in order to control the motors. The control message setup includes information for (i) the movement at the X or Z axis (ii) the left (L), right (R) or (0) stop signals (iii) the maximum or min speed (vi) the acceleration at a certain level (10 to 65.535) and (v) the number of steps from 1 to 65.535 (Fig. 4a). For example, a message that states (X,R,100,200,800) would inform the system that the motor that controls the X axis should turn right at a speed value of 100 and an acceleration value 200 for 800 steps. Furthermore, we could handle the motors by using a setup command in order to enable either the manual or the automatic mode. For example, a message setup message that states (S,XE,M) would inform the system for controlling the axis X in manual mode. At the automatic mode the output signal becomes active just before the start movement and inactive just after the end of the movement. This reduces the energy concussion of the system, the motor's temperature and the restraint torque at our prototype. At the manual mode, the output signal is activated until the user sends a deactivation command.

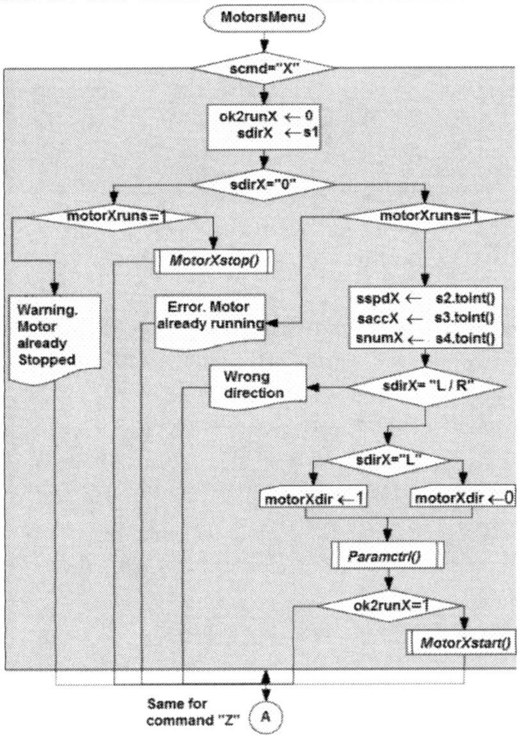

Fig. 4a. The motors basic menu

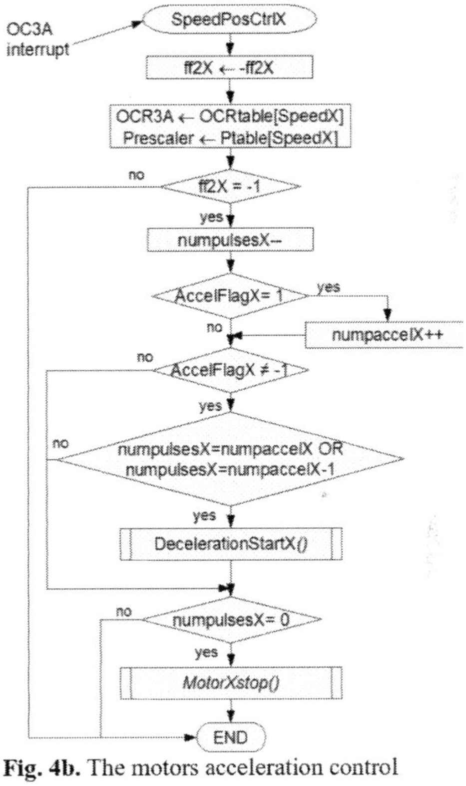

Fig. 4b. The motors acceleration control

At Fig. 4b we present the control flowchart for the movement of the motor. To set the motor movement in a specific position we need timers to control the speed, the position and the acceleration. For speed control and position identification at the X axis we use a timer3 in clear time (CTC) mode and for acceleration we use timer1 in compare register (OCR) mode. For moving at a specific position, we first identify the total number of pulses needed and the exact steps for the motor. In order to reach the total number of pulses the motor gradually accelerates and waits for the timer 3 who controls axis X to stop the motor using the routine MotorXStop. While AccelFlagX equals to one, the motor's acceleration gradually increases the pulses at numpaccel. As a next step in order to smoothly stop the motor the DecelerationStartX routine is used. If AccelFlagX is not equal to minus one (-1) and the number of pulses X (numpulsesX) equals the number of acceleration pulses (numpaccelX) the DecelerationStartX routine is activated in order to gradually stop the motor. For the Z axis the routine uses timer 4 and repeats the logical steps.

5 Conclusion

In the previous sections, a software framework based on the Arduino microcontroller for controlling the mechatronic system of an innovative fully automatic electrophoresis system has been presented. The proposed API takes into consideration all the functionalities of the mechatronic system and responds well in real world conditions. The total size of the software is 11.162 bytes that is about 4% of the total available storage (253.952 bytes). The global variables are using 1208 bytes that is about 14% of the total RAM memory (8.000 bytes). The results showed high accuracy in operation and efficiency in the control of the mechatronic system's moving parts, according to the requirements.

This research has been co-financed by the European Regional Development Fund of the European Union and Greek national funds through the Operational Program Competitiveness, Entrepreneurship and Innovation, under the call RESEARCH – CREATE – INNOVATE (project code: T1EDK-02403)

References

1. B. Aslam, M. Basit, M. A. Nisar, M. Khurshid, & M.H. Rasool, J. Chromatogr. Sci., **55(2)**, 182-196 (2017)
2. K. Theodoridis, F. Stergiopoulos, D. Bechtsis, N. Nikolaidis, D. Triantafillides, A. Tsagaris, A. Filelis, A. Papaikonomou, An innovative and fully automated system for gel electrophoresis, *ESCAPE 2020 Conference, Italy (*Accepted*)* (2020)
3. L. Sungchul, J. Juyeon , K. Yoohwan , St. Haroon, A Framework for Environmental Monitoring with Arduino-Based Sensors Using Restful Web Service, *Proceedings of the 2014 IEEE International Conference on Services Computing* (2014)
4. R. Md. Kamruzzaman, H.B. Muhibul, Microcontroller Based DC Motor Speed Control Using PWM Technique, *International Conference on Electrical, Computer and Telecommunication Engineering* (2012)
5. V. K. Singh, A. Sahu, A. Beg, B. Khan and S. Kumar, "Speed & Direction Control of DC Motor through Bluetooth HC-05 Using Arduino", *International Conference on Advanced Computation and Telecommunication (ICACAT)*. (2018)

MATEC Web of Conferences **318**, 01053 (2020) https://doi.org/10.1051/matecconf/202031801053
ICMMEN 2020

Growth of hydrophobic membrane on copper surface and study of its corrosive behavior

Aikaterini Baxevani[1], *Fani* Stergioudi[1,*], *Azarias* Mavropoulos[1], *Nikolaos* Michailidis[1]

[1]Physical Metallurgy Lab, School of Mechanical Engineering, Aristotle University of Thessaloniki, GR541 24 Thessaloniki, Greece

Abstract. Hydrophobic films were produced on a copper surface by chemical methods. The whole process consisted of three stages. Initially the hierarchical structure of the surface was achieved by in situ reduction of Ag^+ ions to Ag using silver nitrate solution, thereby creating a nanostructured surface. The copper surface was then coated with polydopamine by a simple immersion procedure. Finally, it was decorated with thiols, to achieve a hydrophobic surface. The wetting angle was used to evaluate the hydrophobicity of the produced surfaces. The anti-corrosion behavior of the produced hydrophobic films, were evaluated on the specimens that showed higher hydrophobicity by electrochemical methods. Electrochemical techniques (potentiodynamic polarization and resistance spectroscopy tests) were performed in 3.5% NaCl solution. Morphology was investigated by scanning electron microscopy (SEM). All experimental results showed a clear improvement in the corrosion behavior of the hydrophobic copper surfaces while the hydrophobicity was maintained after the corrosion tests.

1 Introduction

In recent years a plethora of metallic surfaces with special wetting ability have been manufactured through a combination of surface micro- and nanostructures and specific chemical composition, inspired by structures that exhibit materials of nature. Some of the mostly known water-repellent natural surfaces, are the lotus leaf, the shark skin and the skin of some insects. These metal surfaces with tailored wettability, are used in significant applications in corrosion protection systems, such as organic-water separation, fluid transport, self cleaning and anti-icing surfaces.

The wettability could be characterized by the contact angle (CA), between the surface and the droplet. When the CA of a surface is lower than 90◦, the surface is described as a hydrophilic surface. When the CA is between 90◦ and 150◦, the surface is described as a hydrophobic surface and for contact angles more than 150◦, the surface is super-hydrophobic. For rough surfaces, the wetting state can be described by the Wenzel and Cassie–Baxter models.

[*] Corresponding author: fstergio@auth.gr

There are various coating methods to produce super-hydrophobic surfaces that can be categorized as: i) rough surface fabrication with low surface energy material and ii) rough surface modification with low surface energy material [1,2]. The most well-known methods for developing hydrophobic surfaces are sol-gel, electrospining, layer-by-layer, plasma spray coating. These methods, although widely used, have many disadvantages. Typical is the sol-gel method, which involves multiple and complex steps, the electrospinning method, which requires specialized equipment, consumes high amounts of energy and can be produced only one specific nanostructure (fibers). Special equipment is also required in the plasma spray method while the spin coating method is unfriendly due to the use of environmentally harmful solvents [3,4]. Another method which is widely used, in order to create hydrophobic surfaces, is the immersion method. In the literature, this method either involve multiple steps, or require long residence times in solutions, or even if they meet the criteria of the simple and short method, the coatings produced do not have satisfactory mechanical stability [5-8].

In this research work, hydrophobic films were developed on copper sheets used as substrate due to its excellent properties (low weight, electrical conductivity) by a simple immersion method and with short immersion durations in solutions. The hydrophobic surfaces produced showed very good resistance to aqueous corrosive sodium chloride (NaCl).

2 Experimental methodology

Copper sheets (20x20 mm) were used as substrate. Before coating, the substrates were mechanically polished down to 1200 grit size using abrasive SiC papers. The copper substrates were then ultrasonically cleaned in acetone solution for 30 min for degreasing, followed by oxide removal by immersion in 0.1 M HCl for 5 min and subsequently rinsing in water and ethanol. The cleaned substrates were immersed in the silver nitrate ($AgNO_3$) solution (4 mM) for durations of 5, 10, 15, 20, 30, 40 and 50 minutes, in a temperature of 30°C while stirring at 300 rpm.

To achieve a coherent and uniform silver (Ag) coating, a polydopamine aqueous solution (PDA) was used. The production of polydopamine was as follows: dopamine (DOPA) (45Mm), was added to a Tris-buffer solution which was prepared by dissolving 10 mM tris-hydroxymethyl-aminomethane ($C_4H_{11}NO_3$) in distilled water. The obtained pH of the solution was measured to pH=10. Samples were immersed for 20 min in the polydopamine aqueous solution, at a temperature of 40°C while the solution was constantly stirred at 300 rpm. Finally, for surface modification and reduction of surface energy, all specimens were immersed in an ethanolic solution of dodecanethiol (DDT) (0.1 M) for 1 hour.

The wetting characteristics of the sample surfaces were carried out by measuring static water contact angles. A water droplet of about 100 μl was used for each measurement.

The microstructure of the films developed, was examined by scanning electron microscopy (SEM). The corrosion behavior of the coated copper specimens that showed better hydrophobicity (at the initial characterization via contact angle measurement) was tested in 3.5% w/v NaCl aqueous solution by electrochemical techniques (potentiodynamic polarization and resistance spectroscopy (EIS) tests). Uncoated copper samples were also studied for reasons of comparison. All electrochemical experiments were performed in a three-electrode cell containing 3.5% w/v aqueous NaCl, according to ASTM D1193 IV. A platinum wire was used as an auxiliary electrode. The experimental installation of the electrochemical corrosion apparatus was in accordance with ASTM G 69-97 and ASTM

G71-81. All electrochemical measurements were performed at room temperature (25°C). In the case of polarization, the potential scan rate was 0.1 mV/sec and the scan range was -800 to 600 mV relative to the reference electrode. Prior to the measurements of the polarization curves and the EIS, the working electrode was immersed in the test solution in open circuit (OCP) for 30 min to achieve steady state. The EIS was performed in the 0.01 Hz-100 kHz frequency range using peak-to-peak voltage of 10 mV.

3 Results and Discussion

3.1 Hydrophobicity and structural characterization

The contact angle between a water droplet of about 100 μl and the surface of the coated copper, with a thin layer of silver, for different immersion times in the silver nitrate solution ($AgNO_3$) is shown in Figure1. Each value is an average of 10 measurements. The contact angle exhibits superhydrophobic behavior and maximum contact angle of 153° for a 15 min immersion time, in the silver nitrate solution. Figure 1b shows typical images of water droplets of about 100 μl on the surface of silver-coated copper corresponding to different times of growth of the silver films.

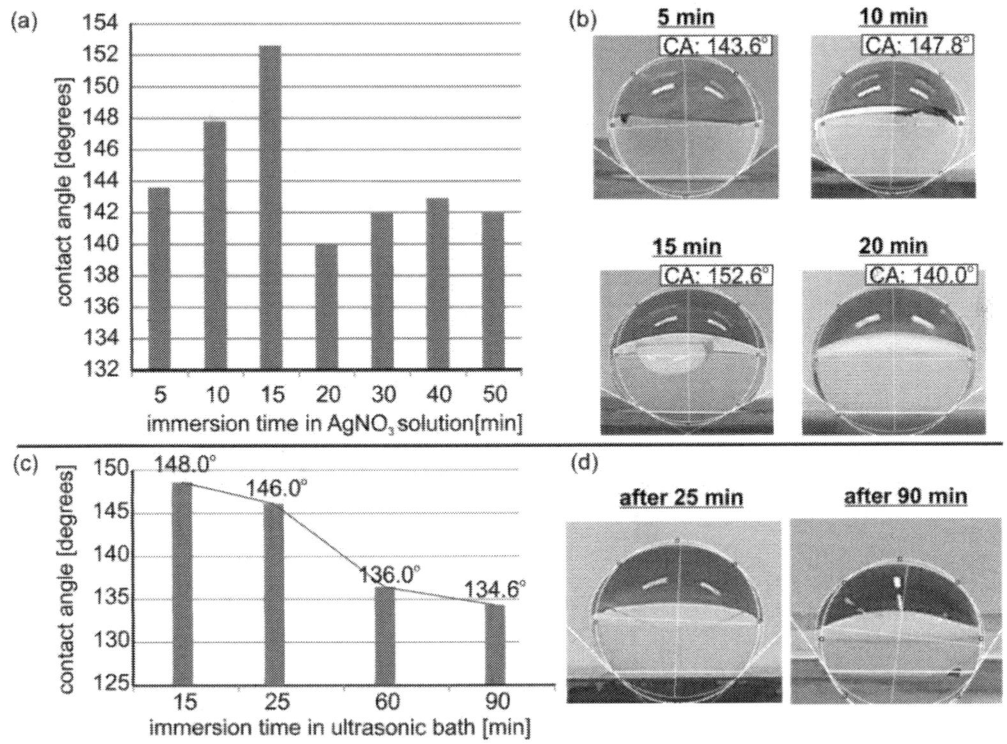

Fig. 1. (a) Contact angle relative to the immersion time in silver nitrate solution, (b) contact angle after various times in an ultrasonic bath for the optimal coated superhydrophpobic sample (15 min), (c) images with water droplets of about 60 μl on the surface of silver-coated copper for various times of hydrophobic silver films growth, (d) images with water droplets of about 60 μl on the surface of silver-coated copper after 25 and 90 min in an ultrasonic bath.

For the initial testing of silver coating stability, the optimally coated samples (15 min) were immersed in an ultrasonic bath for various times and the probability of their hydrophobicity loss was investigated by measuring the contact angle as well as the possible abruption of the hydrophobic film. Figure 1c. shows that the contact angle decreases with increasing the immersion time in an ultrasonic bath. However, the hydrophobicity remains after 60 minutes and the contact angle remains almost constant at about 122°. It is noteworthy that no coating abruption was observed at any point on the coated specimens after 90 minutes in an ultrasonic bath. The results showed that there was a satisfactory stability and affinity of the developed silver films to the copper substrate.

The microstructure of the developed hydrophobic films is shown in Figure 2. A copper substrate in which the hydrophobic silver coatings were developed is also presented, for comparison. When the immersion time in the silver nitrate solution is 5 min, the coating shows discontinuities, as shown in the SEM image and the copper substrate. By increasing the immersion time to 10 min, the self-organization of silver ions and the development of needle-like nanostructures are observed. There are several clusters (of various sizes) so that the surface exhibits morphologically sufficient heterogeneity.

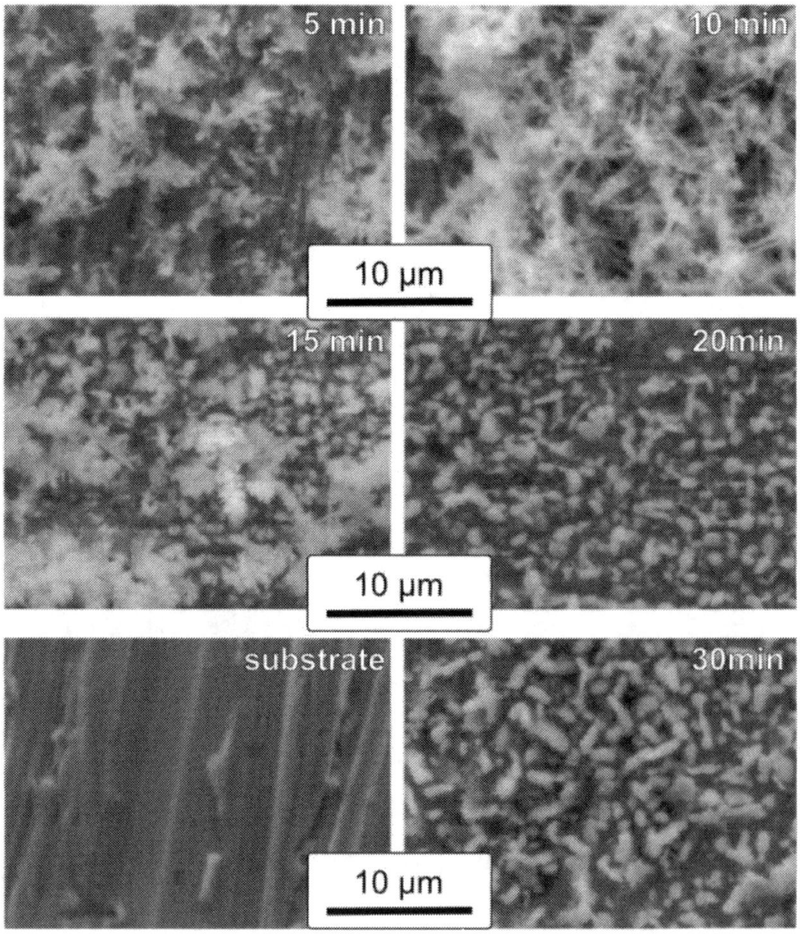

Fig. 2. (a) SEM images of hydrophobic films for different immersion times in silver nitrate solution

Increasing the immersion time to 15 min, has as a result the breakdown of agglomerates and the creation of new structural morphologies. Needle-like morphologies are now scarce and the sample exhibits dendritic or leaf-like morphologies. In addition, a second morphology of spherical agglomerates of approximately 1 μm is shown. As the immersion time in the silver nitrate solution increases to 20 min, the leaf-like structures decay and larger clusterls of silver with polyhedral morphology are appeared. This morphology (polyhedral aggregates) remains for 30 min in silver nitrate solution. The homogeneity and uniformity of the coating are excellent for immersion times greater than 20 min. However, due to the polyhedral morphology and relatively large (but uniform) aggregates the hydrophobicity decreases with the contact angle ranging from 140-143 °.

3.2 Corrosion behavior

Figure 3 indicates that the character of the anodic and cathodic polarization curve for the silver-coated copper, is similar to that of the copper substrate, which indicates that the reactions, which occur in the two samples are the same. The same figure shows the values of corrosion potential and current. The hydrophobic silver layer, which was developed on the copper substrate, shifted the anodic and cathodic polarization curve to lower current densities, therefore increasing the corrosion resistance. It is worth noting that the hydrophobicity of the samples was maintained at a contact angle of 142 °.

Figure 4 shows the Nyquist and Bode diagrams from EIS measurements. As observed, the copper substrate had a value of /Z/ = 6 ohm cm^2, at a high frequency of 10^4 Hz. However, the coated copper showed a value of /Z/ = 21 ohm cm^2 which is almost 4 times higher than that of the substrate, on the same frequency. Similarly, at the low frequency of 0.01 Hz, the /Z/ value of the copper substrate is 1.4 kohm cm^2, while that of the coated copper is 7 kohm cm2. The high value of the composite impedance /Z/ of the hydrophobic coating, compared to that of the copper substrate, indicates that the developed hydrophobic films are more corrosion resistant. The Nyquist diagrams show higher ohmic and capacitive resistance (real and imaginary part respectively) for the hydrophobic silver film.

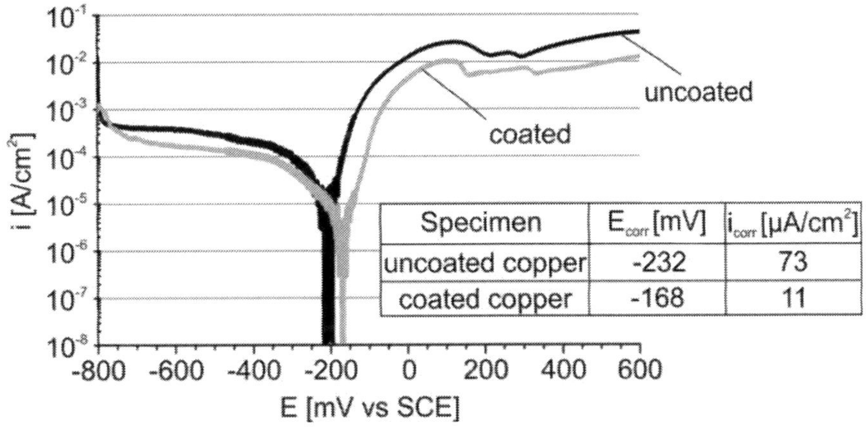

Fig. 3. Polarization curves for the silver-coated copper and uncoated copper.

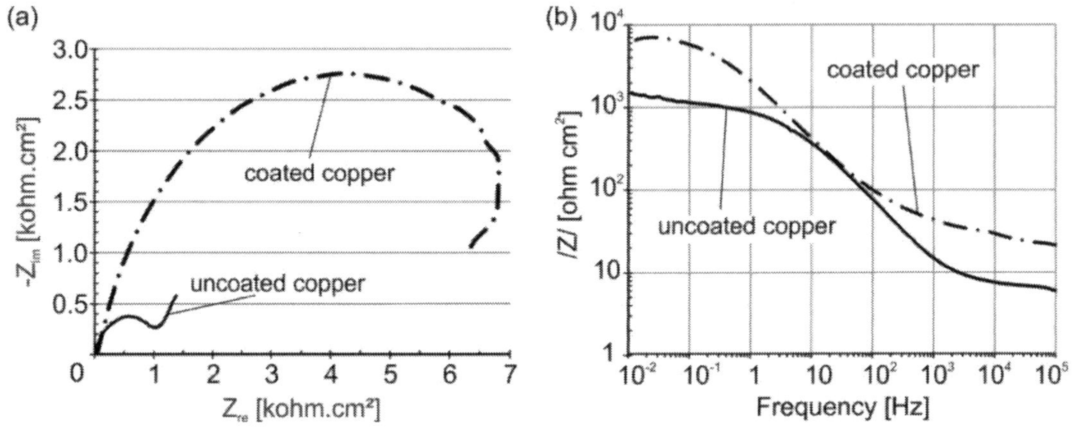

Fig. 4. (a) Nyquist and (b) Bode diagrams for silver coated copper and for copper substrate in 3.5 wt.% aqueous solution NaCl at temperature of 25°C.

4. Conclusions

Hydrophobic coatings were grown on a copper surface by chemical methods. The whole methodology is fast and simple, non-time-consuming and low-cost. By adjusting the immersion time in the silver nitrate solution, silver films of varying morphologies were studied which were studied for their hydrophobicity. Optimum contact angle at 152.6 ° was achieved for a dip time of 15min. In this case the uniformity of the coating topology was achieved while its morphology reveals feather-like nanostructures. Developed hydrophobic silver film improves the corrosion resistance of copper thus demonstrating the multifunctionality of the developed nanostructured films.

References

1. A.M.A. A.M. Mohamed, Abdullah, N.A. Younan, Arab. J. Chem. **8**, 749 (2015).
2. Gh.B. Darband, M. Aliofkhazraei, S. Khorsand, S. Sokhanvar, A. Kaboli., Arab. J. Chem., DOI: 10.1016/j.arabjc.2018.01.013.
3. Y. Chen, Sh. Yang, Q. Zhang, D. Zhang, Ch. Yang, Z. Wang, R. Wang, R. Song, W. Wang, Y. Zhao, Coatings **9**, 69 (2019).
4. P. Varshney, S.S. Mohapatra, A. Kumar, Intern. J. Smart Nano Mater. **7**, 248 (2016).
5. H.M. Ali, M.A. Qasim, S. Malik, G. Murtaza, *Techniques for the Fabrication of Super-Hydrophobic Surfaces and Their Heat Transfer Applications. Heat Transfer - Models, Methods and Applications* (Kingston University London, London, England, 2018)
6. P. Che, W. Liu, X. Chang, A. Wang, Y. Han, Nano Research, DOI 10.1007/s12274-015-0925-5 (2015).
7. D.K. Sarkar, N. Saleema, Surf. Coat. Techn. **15**, 2483 (2010).
8. K. Liu, L. Jiang, Nanoscale **3**, 825 (2011).

AUTHOR INDEX

AUTHOR INDEX